FOR REFERENCE

Do Not Take From This Room

National Safety Council

SUPERVISORS'
Safety Manual

7th EDITION

Project editor: Patricia M. Laing
Technical advisor: Carlton D. Piepho
Cover and interior design: Frank Waszak
Composition: Impressions, Inc.
Cover photograph courtesy of Gatorade Industrial,
© 1989 The Quaker Oats Company.

Library of Congress Cataloging in Publication Data
National Safety Council
Supervisors' Safety Manual
International Standard Book Number: 0-87912-151-3
Library of Congress Catalog Card Number: 90-062580
10M1090   Product Number: 15141

Contents

Preface to the 7th Edition

The roles and responsibilities of the first-line supervisor have changed. You are still the direct link between management and the workforce. You must still produce quality goods and/or services. You are still responsible for quality job training, employee motivation, development of good safety attitudes, and detection of hazardous conditions and unsafe work practices. And you must do all these things in an environment that is in the midst of the "information explosion."

You need to understand the techniques and psychology of human relations. You should know the fundamentals of loss control, the applicable industry standards and federal and local safety and health regulations. You must know all the potential hazards your workers can encounter, how to prevent them, and/or what safeguards and personal protective equipment are needed, how to use them, and how to enforce use of them. You must be able to assist in accident investigations that occur in your area of responsibility. And you should be able to recognize when your workers need to seek help for stress, problems in adjusting to workshift changes or job changes, personal problems, or drug or alcohol abuse—and how to encourage them to get the help they need.

The seventh edition of the *Supervisors' Safety Manual* provides the basic information, references, and resources you will need to meet these responsibilities and duties.

This edition was completely revised by experienced and technically knowledgeable safety and health professionals. New material on employee assistance programs (EAP), stress management, the effects of shiftwork and shift changes, ergonomics, and industrial hygiene augment the completely revised text. The text is written in easy-to-understand language with many bulleted lists for quick reference. New study aids highlight features of each chapter. Closing summaries of key ideas, accident case studies, and a new safety and health glossary support the ideas presented. New charts, tables, and photographs present information in easy-to-understand formats. Finally, the chapter order now reflects an arrangement from general supervisory and management information to specific technical information.

Even though much detailed information has been included, the purpose of the book is not to serve as a complete handbook. Rather, it is designed to emphasize important issues to be considered by the supervisors who are responsible for safety and health in their organizations.

A completely rewritten and updated Supervisors' Development Program is also available to support your training for your responsibilities and duties. It consists of printed instructional manuals, workbooks, and 13 videos with leaders' guides.

The National Safety Council appreciates the contributions of many supervisors and safety and health professionals who have helped in the preparation of this edition. A special thank you to Edgar Mendenhall, Mendenhall Technical Services, Bloomingdale, IL, for his contributions to Chapter 14 and James G. Gallup, PE, CSP, Professional Loss Control, Inc., Vernon Hills, IL, for his contributions to Chapter 15.

Staff contributors include: Thomas Danko, Barbra Jean Dembski, John Fogel, Gary Fisher, James Kaletta, Joseph Kelbus, Ronald Koziol, Joseph Lasek, Randy Logsdon, Robert LoMastro, Jill Niland, Donald Ostrander, Michael Peltier, Carlton Piepho, Douglas Poncelow, and Philip Schmidt.

Introduction

As modern production processes have become more highly sophisticated, so has our understanding of how to create a safe, healthful work environment. In the past, "accident prevention" simply meant freedom from serious injuries and property loss. Today, however, "loss control" covers not only injury but occupational disease and environmental concerns along with fire and property damage control. This new understanding of safety is reflected at all levels of management from the CEO to the line supervisor. Loss control is considered so essential a function that in most companies safety performance is included as part of a supervisor's overall job evaluation.

As a result, today's supervisor must become skilled in three aspects of loss control: (1) learn to recognize hazards, (2) learn the acceptable level of risk for department operations, and (3) learn how to control these hazards to prevent injuries, illness, and property damage. The National Safety Council's *Supervisors' Safety Manual* presents the skills a supervisor will need to perform these responsibilities successfully.

SUPERVISOR: MANAGER OF SAFETY PROBLEMS

The supervisor, as a member of the company's management team, shares responsibility for maintaining a safe, productive workplace. He or she must communicate and enforce rules and procedures, train workers, and represent the interests of both the organization and employees.

The supervisor must constantly watch over and inspect both the workplace and work procedures, keeping in mind the three E's of safety: engineering, education, and enforcement. It is the supervisor's job to work with safety and health professionals, designers, engineers, maintenance, and personnel staff to engineer as many hazards out of the workplace as possible, to educate employees in safe work practices and procedures, and to enforce all safety rules and policies. In this role, the supervisor acts as investigator, safety researcher, and advocate.

For example, if an employee complains of headaches while performing a degreasing procedure, it is up to the supervisor to investigate the situation. It may be that a degreasing chemical is causing the employee's headaches. If an investigation by the safety and health professional shows this is the case, there are three options: *eliminate* the hazard by substituting a different chemical, *isolate* the hazard by installing surface or point-source ventilation, or *compensate* for the hazard by issuing personal protective equipment.

The supervisor may be involved in determining which of the three options is the best solution to the problem. For example, in this case, personal protective equipment is desirable only when all other options have failed. The equipment does not remove the hazard— it only decreases the risk for the worker.

Once the best solution has been chosen, the supervisor may be involved in presenting the plan to upper management for approval of the required action. At this point, the supervisor becomes an advocate, presenting the benefits of the plan on behalf of the workers as well as the organization. *Supervisors' Safety Manual* is designed to take you step by step through the process of learning how to manage and control safety problems.

HOW THIS BOOK IS ORGANIZED

The supervisor manages safety problems by managing the people, equipment, and environment in the workplace. To assist you in these tasks, this book has been organized as follows:

- Chapters 1 through 7 focus on the basic safety knowledge and skills that every supervisor must acquire: an understanding of loss control and the fundamentals of communication, human relations, employee safety training, inspection, and accident investigation. These chapters orient the supervisor to his or her management responsibilities on the job.
- Chapters 8 through 10 cover what a supervisor must know to meet OSHA and other regulatory standards to maintain a safe, healthful workplace: industrial hygiene, personal protective equipment, and ergonomics. Understanding the worker-machine-environment relationship can help to reduce hazards, accidents, and job-related illnesses considerably.
- Chapters 11 through 15 focus on the general principles for safeguarding, maintaining, and using tools and equipment safely to prevent accidents and property damage and loss. In these chapters, you will learn how to make the day-to-day production process safer for all concerned and how to work with various safety experts from within and outside the company to solve safety-related problems.

No safety program can eliminate all the risks and hazards from a job, but *Supervisors' Safety Manual* can help you come as close to this ideal as possible. The more you know about the hazards in your workplace and how to control them, the more everyone in the company will benefit.

Safety Management

After reading this chapter, you will be able to:

- Understand the terms *accident, hazard, hazard control,* and *loss control*
- Understand the safety responsibilities of front-line supervisors and upper management
- Know the procedures for accident prevention and the steps to take should any accident, serious or minor, occur
- Realize the direct and indirect costs of accidents
- Know the best approach to safety performance on the job

The bottom line of all safety programs is accident prevention, more often called "loss control." Primary responsibility for safety and accident prevention rests with top management. They, in turn, share this responsibility with middle management and, through them, with first-line supervisors. Your manager should hold you accountable for accident prevention, because he or she is also held accountable, and so on up the line.

Unfortunately, there are still some supervisors and managers who do not consider accident prevention an important part of their jobs until after an accident causes a serious injury or illness. The concerned manager or supervisor then investigates to determine how and why the accident occurred.

This approach is not accident *prevention*—it is accident *reaction*. Certainly, if an accident occurs, we must identify the causes and eliminate them to prevent a recurrence. However, as supervisors, our job is to *prevent* accidents—and the toll they take in human and cash losses—by controlling the hazards that produce them.

This chapter will help you to understand your safety responsibilities, take positive actions to prevent accidents, and give you a way of measuring how well you perform your safety duties. The chapter emphasizes preventing all types of accidents, not just those that result in serious injuries. By taking this approach, you will perform the safety portion of your supervisory job in the most efficient manner.

And by doing it well, you'll have more time for the other important parts of your job.

DEFINITIONS OF TERMS

Let's define our terms before we go any further.

> **Accident:** An unplanned, undesired event, not necessarily injurious or damaging, that disrupts the completion of an activity.

Let's examine this definition. Accidents are clearly unplanned events. When they occur, they not only upset your schedule but demand all of your attention. You must stop what you are doing and handle the many problems caused by the accident. Notice also that what some people still call incidents or "near misses" *are* accidents (see Figure 7-1, in Chapter 7).

A "near" accident or "near miss" is an example of an incident resulting in neither an injury nor property damage. However, a near accident has the potential to inflict injury or property damage if its cause is not corrected. About 75 percent of industrial injuries are forecast by near accidents or near misses. It's in a supervisor's best interest to find and eliminate their causes to keep near misses from recurring or becoming serious accidents.

For example, an employee feels the tingle of a slight electric shock while using a defective portable drill. This is a near accident because no injury or property damage results. If the cause—the defective drill—is removed from service, a potential injury or fatality can be prevented. Therefore, supervisors must establish a system that encourages the reporting of near accidents so the cause can be determined and appropriate corrective action taken.

Such a reporting system may take time to develop and implement, but it will pay great dividends in the long run. Employees learn that they help their supervisor and themselves by reporting all near misses. They act as additional sets of eyes to help the supervisor identify hazards so he or she can take the necessary action to eliminate them. Obviously, employees must have confidence in the program and learn they will not be criticized or be considered troublemakers by the supervisor or upper management for reporting problems. Supervisors should make it clear that employees who point out hazards will be valued and praised for their cooperation. Also, all levels of management must support the program and ensure that any hazard reported is quickly remedied.

The reporting system can be initiated by using a simple report form, as shown in Figure 1-1. The program should be announced by top management and actively supported by all management levels. Reports could be handed in to the employee's supervisor or, if an employee wished to remain anonymous, could be placed in a suggestion box posted in a central area.

Hazard: A hazard is any existing or potential condition in the workplace that, by itself or by interacting with other variables, can result in death, injuries, property damage, and other losses.

Keep two factors of this definition in mind. First, potentially hazardous conditions, as well as those that exist at the moment, must be considered. Second, hazards may result not from independent failure of workplace components but from one workplace component acting upon or influencing another.

Hazard control: Hazard control involves developing a program to recognize, evaluate, and eliminate (or at least reduce) the destructive efforts of hazards arising from human errors and from conditions in the workplace.

As a necessary part of the management process, hazard control is made up of safety audits and eval-

---

Figure 1-1. "Near Miss" Report Form

I. What happened:

II. What could be done to prevent it from happening again?

Signature (optional) _____

---

uations; sound operating and design procedures; operator training; inspection and testing programs; and effective communication regarding hazards and their control. A hazard control program coordinates shared responsibility among departments and underscores the interrelationships among workers, their equipment, and the work environment.

Loss control: Loss control is accident prevention, achieved through a complete safety and health hazard control program. Loss control involves preventing employee injuries, occupational illnesses, and accidental damage to the company's property. It also includes preventing injuries, illnesses, and property damage that may involve visitors and the public.

Many supervisors mistakenly believe that accidents are only those incidents that result in serious injuries. If a minor injury or property damage results, some supervisors shrug off the incident and return to their routine. They let the results of an accident determine their level of interest in investigating its causes and preventing a recurrence. But we know that the results of an accident (the degree of loss resulting from it) are a matter of chance. It would be better to try to control the hazards that lead to accidents than try to minimize the damage done once an accident occurs. That is why many safety professionals prefer the term "hazard control" to "loss control." If you ignore the warnings of minor and near accidents, you are neglecting an important part of your supervisory responsibilities.

You must look for the cause(s), regardless of the results. The hazardous condition or action that causes a near accident one time may bring about a serious injury or fatality the next. Likewise, the hazard that produces only minor property damage at one time may result in serious property damage at another time.

AREAS OF RESPONSIBILITY

Let's look at the responsibilities of supervisory jobs. There are four major areas that supervisors must control:

1. Production
2. Quality
3. Cost
4. Accident/illness (loss)

Many supervisors willingly accept responsibility for the first three areas, but ignore or procrastinate about the fourth. This is because you may have assumed that the responsibility for accident loss control belongs to a safety director or someone in the human resources department. This assumption is incorrect. Loss control is *your* job. If line managers and supervisors do not assume responsibility for safety, no company program, no matter how good, will work.

Loss control through accident prevention must be accomplished at all times, not merely on some convenient Friday afternoon. For instance, ask yourself this question: "When should supervisors perform safety inspections?" If your answer is "once a week" or "once a month," you do not understand your responsibility. You should conduct an informal safety inspection every time you walk through your department, even if your primary purpose is only to check attendance or to determine whether supplies are adequate (Figure 1-2).

During your inspection, be alert for anything that may cause an accident, such as tripping hazards, fire hazards, poorly stacked materials, poor housekeeping, safeguards missing from machines, and/or unsafe worker practices. Safety responsibilities cannot be separated from the other parts of your work. In fact, the best way to describe your job is to say that you are responsible for *safe production*. With this idea in mind, you will soon handle your accident prevention responsibilities almost automatically.

Safety Responsibilities as Performance Measure

Many progressive, successful firms include the supervisor's performance of safety responsibilities as part of the performance evaluation. Production, quality, cost, and loss control are of equal importance in measuring job performance and cannot really be separated. When you accept a supervisory job, you also assume responsibility for the safety of your people, whether you supervise a group of machinists, assemblers, a construction crew, or an office staff. Every supervisor in any company is responsible for the safety of his or her workers.

Yet the concept of safety responsibility is far broader than just the supervisor's duties. Who is responsible for safety and accident prevention? Although accident prevention is a line function (Figure

Figure 1-2. A supervisor should make an informal safety inspection during every department walk-through.

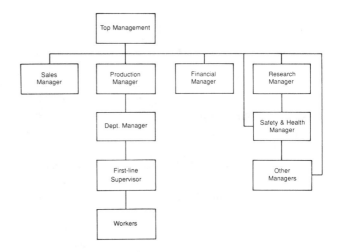

Figure 1-3. The responsibility for safety flows from top management to every level of the organization.

1-3), responsibility begins with the top person in the organization, who is held accountable for loss control throughout the company. The top manager cannot handle all of the details of every job, so he or she delegates accident prevention responsibilities, along with commensurate authority, to various middle managers.

In turn, the middle manager, such as your boss, will delegate responsibility for safety and accident prevention in each department to someone like you, the first-line supervisor. Thus, you are accountable to your manager for accident prevention, just as he or she is accountable, in turn, to the top manager.

Let's define the principal terms we are using:

- *Responsibility* is having to answer to higher management for activities and results.
- *Authority* is the right to correct, command, and determine the courses of action.
- *Delegation* is sharing authority and responsibility with others. Even though we delegate responsibility, we cannot be completely relieved of it.
- *Accountability* is an active measurement taken by management to ensure compliance with standards.

So, the answer to our original question, "Who is responsible for safety?" is that we *all* are. People at every level of supervision and management in an organization are responsible for controlling losses due to accidents in their areas. You should expect to have your safety performance measured by your boss, because your boss is held accountable for you. In turn, your employees are accountable to you for performing their jobs safely.

The Cost of Accidents

Another area of major concern to supervisors is the cost of accidents. Many people fail to realize how much accidents *really* cost. In fact, many expenses are not always obvious. Therefore, attention to loss control can also improve your department's financial performance and enhance your company's overall success.

Let's consider the question of "where does the money come from to pay for the results of accidents?" Some people believe that organizations have money set aside to pay for accident costs. However, supervisors know that the money must come from profits.

Thus, a series of costly accidents can reduce profits radically. Accidents have obvious, direct costs, such as medical, hospital, and rehabilitation expenses; Workers' Compensation payments; and higher insurance premiums or even loss of insurability. But there are less obvious, indirect costs that are usually uninsured. These include the various disruptions of normal work procedures, as when employees stop to help the injured, or even a drop in

production that can mean late delivery of products. If profits are not sufficient to cover costs, the company may be forced to defer the procurement of new equipment and facilities.

People often try to minimize the costs of accidents by saying that they are covered by insurance. But insurance covers only a portion of the total accident cost. Moreover, as accident losses increase, so will a company's insurance premiums. It is clear that directly and indirectly, accidents reduce profitability.

Total accident costs can be compared to an iceberg (Figure 1-4). The part of the iceberg that we can see above the surface is like the smaller (direct) portion of the total accident costs. An examination of a serious accident can give you a better understanding of what makes up total accident costs.

Suppose one of your people receives an electric shock from faulty equipment and is seriously injured. Many people in the department stop working; some rush to give first aid to the injured person, while others call you to the scene. When help arrives for the injured worker, do other employees immediately return to work? Not always. They may continue to help or remain merely to watch. In this situation, all idle work time is included as part of the accident costs.

As soon as the injured person receives proper medical treatment, your next job is to investigate the accident. You must isolate the area, interview witnesses (see Chapter 7), and determine what caused the accident. All time spent on accident investigation and reporting, as well as wages paid to witnesses, are included in the total costs.

If the injured person misses work for only a short time, you may be able to make up for the production loss by having the rest of the department work overtime. Overtime premiums paid are included as accident costs. On the other hand, if the injured person is gone for months, you may have to hire and train a replacement worker. Not only can the cost of replacing your skilled operator be prohibitive, but the new operator may not be as efficient as the previous one. If a machine is damaged in the accident and the job must be performed with less effective equipment, your output will drop further. All this reduced efficiency represents another indirect cost. (A more complete list of factors adding to indirect costs is provided in Figure 1-4.)

As with the example of an iceberg, we find that the indirect costs of accidents are usually greater than the direct costs. A conservative estimate is that for every $1 of direct accident costs, there are $3 of indirect costs. Robert E. Sheriff, in his article in *Professional Safety* (1980), showed that indirect costs may be four to ten times the insured costs.

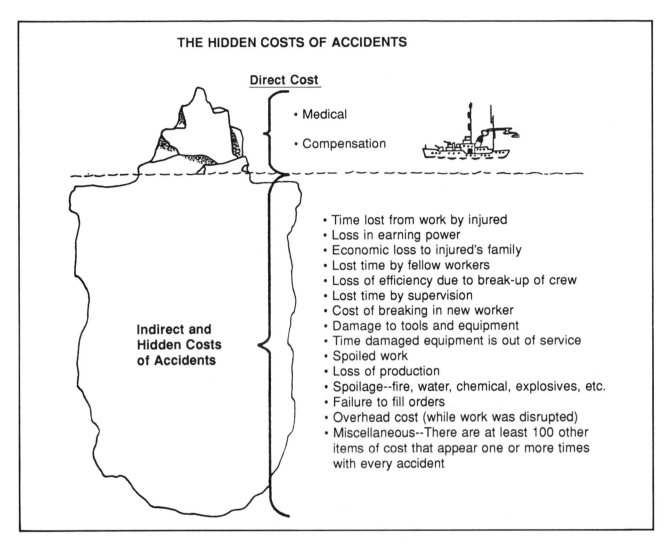

Figure 1-4. Like an iceberg, the hidden costs of accidents are not visible on the surface but are there just the same.

Frank Bird (1976) calculates that for every $1 of direct costs, there may be $5 to $50 of uninsured ledger costs and $1 to $3 of uninsured miscellaneous costs.

Just as there are many hidden *costs* due to accidents, there are hidden *savings* in accident prevention. For this reason, the phrase "loss control" is often used in safety management. Every accident you prevent saves direct and indirect accident costs—money that remains in profits. Other benefits of accident prevention efforts include:

- Workers will not be injured or killed.
- Property and materials will not be destroyed.
- Production will flow more smoothly.
- You will have more time for the other management duties of your job.

THE "OLD" APPROACH TO SAFETY PERFORMANCE

Let's take a closer look at the way many supervisors have measured their safety performances in the past. The measurement was made according to the number of lost-time accidents. As long as no one was injured seriously, supervisors felt they were doing a good job. Many times minor injuries, property damage, or near-miss accidents were brushed aside or ignored.

In 1931, H. W. Heinrich (1980) conducted a now-famous accident study. He showed that for every accident resulting in a serious injury, there are approximately 29 resulting in only minor injuries and 300 producing no injuries (Figure 1-5). If you react only to major-injury accidents, you are ignoring 99.7 percent of the accidents that occur in your operation! Heinrich stressed the fact that the same

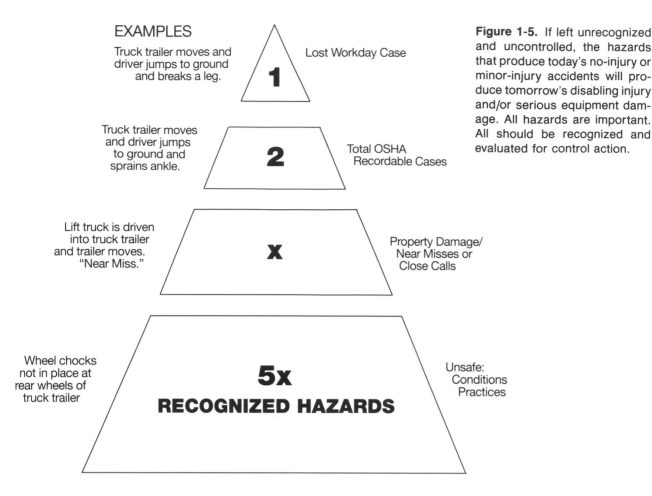

EXAMPLES

Truck trailer moves and driver jumps to ground and breaks a leg. — **1** — Lost Workday Case

Truck trailer moves and driver jumps to ground and sprains ankle. — **2** — Total OSHA Recordable Cases

Lift truck is driven into truck trailer and trailer moves. "Near Miss." — **x** — Property Damage/ Near Misses or Close Calls

Wheel chocks not in place at rear wheels of truck trailer — **5x RECOGNIZED HAZARDS** — Unsafe: Conditions Practices

Figure 1-5. If left unrecognized and uncontrolled, the hazards that produce today's no-injury or minor-injury accidents will produce tomorrow's disabling injury and/or serious equipment damage. All hazards are important. All should be recognized and evaluated for control action.

Safety is the **Control** of recognized **Hazards** to attain an **Acceptable** level of risk.

factors causing a near-accident at one time can cause a major injury the next time.

If you look only at the major injuries in your department, you will miss many opportunities to find and eliminate the causes of near-accidents and property-damage accidents. Effective hazard or loss control requires being aware of the possibilities for all types of accidents and knowing how to prevent them from occurring.

A BETTER APPROACH TO SAFETY PERFORMANCE

Include loss control as a regular part of your job and expect to have this part of your performance measured. For example, you are expected to perform periodic safety inspections of the areas for which you are responsible (see Chapter 6). Your manager can verify that you conduct these inspections, check on

their quality, and determine how well you follow up on the items needing attention.

Similarly, other areas of your safety activities should be measured, such as housekeeping (discussed in Chapter 6). Pay attention to the issues and details involved with housekeeping, such as (a) are housekeeping inspections being performed on schedule? and (b) is the supervisor taking positive action to improve housekeeping?

Although safety and housekeeping inspections and the problems you discover are important, what you *do* about them is more important. If a problem can be corrected by your people, assign the appropriate tasks as soon as possible. If, on the other hand, service or maintenance personnel must be involved, issue a work-order request immediately. Be sure to follow up, as needed, to see that the job is done. It may be necessary to have your manager help expedite the work by getting assistance from other departments.

Job Safety Instruction

You are responsible for training the workers in your area, and it's up to you to monitor their work habits (Figure 1-6). One of the most effective ways to avoid accidents is to make sure that employees are following the safe work procedures in which they have been trained (see Chapter 4).

Some points to consider are:

- What is the quality level of the job safety instruction training (JIT)?
- How many people in the department are responsible for the training?
- Are all new employees trained?
- Are transferred employees trained?

Job Safety Analysis

You are also the one who looks for ways to improve operating procedures in your area while maintaining safe and healthy working conditions (see Chapter 4). Occasionally you will conduct job safety analyses (JSAs) or assign them to others.

Points to consider include:

- Are the assigned number of job safety analyses being performed?
- Is the quality improving?
- Are JSAs reviewed as operations are revised?
- How well are they being used?

Other Measures

Another positive measure of safety management is to have your people use the appropriate personal protective equipment (PPE). When new employees

Figure 1-6. The supervisor trains workers in good work procedures and monitors for safe work habits.

are trained, they should be informed about the need for personal protective equipment. Some PPE, like those for eye protection, should be fitted; others, like some respiratory protective equipment, should be demonstrated so that the workers know how to use them properly. You should emphasize that people will be expected to wear and use safety equipment on the job.

It is easy to determine whether your people are following these rules. Whenever you enter your area, make a quick check to see if the required equipment is being used. Any violations of these rules must be dealt with at once. Your people must understand that following these safety rules is a condition of their employment.

Likewise, anyone visiting your area must also comply with the requirements for wearing the proper equipment. Make sure all visitors put on the appropriate gear before they enter the work area.

We have discussed only a few of the positive approaches to loss control that you can take. You can probably add other items to the list. The more you incorporate these actions into your normal routine, the more easily you can accomplish loss control work.

As the number of accidents in your department declines, department operations will run more smoothly. You can devote more time to other parts of your job, such as production planning, quality improvements, and other cost controls.

Help with Your Loss Control Work

Where can first-line supervisors turn for help and guidance with their safety activities? First, look to your manager for help. He or she should be most concerned with your control of losses. This is part of your manager's job-performance measurement, as well as your own. Your efforts in this area can play an important part in helping both of you successfully carry out employment safety performance on the job.

The safety director or manager in your company can be another source of help and can serve as a catalyst for your program. His or her job is to work with company management to plan the overall loss (hazard) control program and to assist supervisors in carrying it out. It is wise for you to cooperate completely with the safety director's program for your area of responsibility.

You, as a supervisor, can also seek to prevent hazards from entering your area in the first place. Work closely with those who design machinery, equipment controls, and safeguards that will be used in your area to be sure the best designs and controls are used (Figure 1-7).

Figure 1-7. The supervisor should work closely with the design department to assure all possible safeguards and controls are incorporated into designs.

SUMMARY OF KEY POINTS

To summarize what we've covered in this chapter:

- Many supervisors measure their safety performance by the number of lost-time accidents that occur in their departments. This is accident *reaction*, not accident *prevention*. To do your job properly, you should work to prevent accidents from occurring.
- Treat all near misses and incidents, not just more serious mishaps, as accidents and investigate their causes. Minor accidents or incidents provide an "early warning system" that can help you to prevent more serious accidents later on.
- The responsibility to prevent accidents through a hazard (loss) control program is a line function. The company's top manager delegates the responsibility to your boss, who in turn delegates responsibility to you.
- Hazard or loss control is as important a part of your job as your production, cost, and qual-

ity control responsibilities. Your objective is *safe production*. As a result, you must establish effective job safety training programs for your workers and see that safety rules and policies are observed.
- Indirect and direct accident costs are higher than most people realize. Direct costs represent only a small portion of the total. Indirect costs, such as worker downtime, accident investigation and reporting, and equipment replacement or repair add significantly to total accident costs. It is estimated that for every $1 of direct costs, there are $3 to $10 of indirect costs—all of which must be taken out of company profits.
- An effective approach to safety includes several features:

  1. Conduct safety inspections as a regular part of your day-to-day routine and take immediate steps to correct any problems.
  2. Train your employees thoroughly in safe work procedures.

3. Conduct job safety analyses or assign them to others.
4. Make sure employees and visitors wear safety gear.

5. Ask your boss or the company director for help in addressing your safety problems.

2

Communications

After reading this chapter, you will be able to:

- Explain the elements that comprise good communication and understand the need for developing your communication skills
- Understand the filters that interfere with good communication
- Know the various methods of communicating messages—oral, written, and nonverbal
- Understand the importance of listening skills and how to overcome the barriers to effective listening

The success of your accident prevention efforts depends, to a great extent, on how well you communicate with your people. Your plans for preventing accidents, your ideas for creating a safer workplace, and the feedback you get about how to improve operations depend on how well you communicate. Good communications are vital to your success as a supervisor—they are essential to your accident prevention program (Figure 2-1).

Remember that communication involves not only what you send or say but how well you receive or listen. Research reveals that most supervisors and managers spend 50 percent more time on the job listening then they do speaking.

A good definition of communication is: "Sharing information and/or ideas with others AND BEING UNDERSTOOD." The last three words are especially important. If there is no understanding, then we have failed to communicate. Remember that the receiver may not agree with what is being said, but he or she must understand it for communication to occur. Once understanding takes place, the receiver can begin to accept the message.

Make sure that the communications you send can be understood, and develop ways to improve your understanding of the communications you receive. Ask questions, clarify wording, or request further details of any oral communications or memos you receive to enhance your understanding of the message. As you improve your techniques for comprehending the messages you receive, you will develop skills in helping others to understand your communications.

Work diligently to improve your communication skills. They will be an asset in your accident prevention work. Good communication will pay dividends in all aspects of your supervisory job.

ELEMENTS OF COMMUNICATION

Communication consists of three basic elements: the sender, the message, and the receiver. However, making sure the message is received and understood may not be a simple matter. Problems in communicating can result from the method chosen by the sender, the form of the message, or the filters or barriers of the receiver.

Sending the Message

Communication involves sending messages in a variety of ways. The simplest form is "one-way communication": one person sends a message and another receives it. This can be illustrated as follows:

SENDER → MESSAGE → RECEIVER

This type of communication has several problems:

1. Information flows in only one direction.

Figure 2-1. Safety depends upon good communication between the supervisor and workers. Safety communication on the job site is reinforced by safety talks in the classroom or conference room.

2. The lack of feedback means the sender will not know if the message has been received and/or understood.
3. The receiver may not understand the message (e.g., the sender and receiver interpret words differently).

Feedback

To be effective, a communication system must provide for feedback; it must be two-way. The diagram can be expanded to include this:

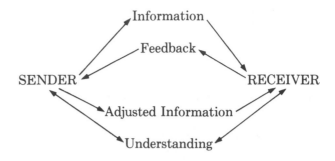

When you are communicating, whether orally or in writing, always provide for feedback. This is the only way to know if your message got through and was understood by the receiver.

Communication improves if the receiver has an opportunity to ask questions and to express how he or she understands the message. The sender will then know when a misunderstanding has occurred and can adjust the communication accordingly. In this way true communication is more readily achieved. One-on-one, face-to-face, two-way communication is best to use in the majority of situations that will arise on the job.

After giving oral instructions, for example, it is

Figure 2-2. This supervisor asks for feedback when he explains a new procedure.

good to say something like, "Just to be sure I've covered the material completely, can you repeat what I just told you?" The response you get will tell you how well you've communicated. It is also wise to provide for feedback in your written communications; for example, include a closing sentence such as, "Please call me by Friday to give me your view of these proposals."

Getting good feedback can enhance understanding. The feedback and questions you get will also help you to improve the future communications you send (Figure 2-2).

Communication Filters

An important factor to consider when "sending" is always to communicate through the receiver's filters. Filters can be barriers to good communications, as shown here:

SENDER → **Message** → **RECEIVER**

Some of the most common filters include the following:

Knowledge. To communicate effectively, we must know something about the receiver's level of knowledge. We can then adapt our communication to the appropriate level—not above or below it.

Bias. People are biased by what has happened to them in the past. Biases have a definite bearing on people's attitudes and on the way in which they hear and understand communications. In fact, biases may result in a person's listening to only part of what is being said. In extreme cases, people may "tune out" entirely and fail to hear any of the message.

Mood. A person's mood can be one of the most serious filters to consider in communications. If listeners have something else on their minds at the time your communication is transmitted, they may not get the message at all. This is another reason for obtaining feedback from the receiver to determine if your communication has been received and understood. For example, if you are instructing one of your people about the hazards of a task he is about to start, and you suspect the individual has something troubling him, ask the person to repeat the hazards back to you so you know he understands. A serious accident could result if you neglect this step. Don't simply ask, "Do you understand?" because the answer is likely to be, "Yes." Rather, ask people to repeat what you have told them. Similarly, you must not let your problems distract you so that you fail to listen to what people tell you about their jobs.

Your Audience

Take a moment before communicating a message to consider the needs of your audience. What are their concerns, problems, levels of knowledge, age ranges, or other important qualities? Whether it's oral instructions to one of your people, a safety meeting you plan to lead, or a memo to your boss, thinking of your audience (your receiver) can help to improve your communication. By mentally putting yourself in the place of your receiver, you will have a better understanding of how he or she is likely to react and respond to your message. You can then tailor your message to your audience's needs. In turn, your audience is more likely to understand and accept what you have to say. Achieving understanding is the key to solid communication.

Positive Reinforcement

Another effective use of communication lies in motivating workers. Many supervisors make the mistake of failing to "communicate the positive." If workers do something wrong, most supervisors are quick to point it out. Although this is proper, many supervisors fail to point out when a job is done correctly. Why not take a minute to praise someone for a job well done? You will find it pays considerable dividends in improved human relations. People know when they have done good work. Your praise tells them you know as well. Giving positive feedback to your people can improve their morale. By offering reinforcement, you can make your supervisory work more effective. Remember the old cliche, "Praise in public, reprimand in private."

METHODS OF COMMUNICATION

Your choice of a communication method is important: oral, written, or nonverbal. In some situations, oral communications are appropriate, while in others, you may prefer to write your message. When giving job instructions, you can combine both methods. You may discuss the job procedures face-to-face, pointing out the hazards and showing the operator the safe job steps. You may also give the worker a Job Safety Analysis Form (see Chapter 4) as a reference and reminder of points covered orally. Thus, oral and written communications reinforce each other.

Oral

Oral communications may take the form of a "tailgate" meeting, in which you discuss with several people a job they are about to start. During and after the discussion they can ask questions about the job. The questions asked show you how well you explained the operations to them. If there are no questions, don't assume you've explained everything perfectly. You should then ask questions to determine whether they understood what you said.

Written

In some circumstances, you may prefer to use written communications. Generally, this method is appropriate when you are dealing with complicated or technical subjects. Written communications can be used as references in the future. However, getting feedback is more difficult using written communication than it is using face-to-face communication. You may want to follow up with verbal questions to be sure your receiver has a clear understanding of your written communications.

A good example of written communication is a Work Order Request sent to maintenance asking the department to eliminate a problem you found on a safety inspection. You will probably want to follow up such a request orally to be sure that your message was understood and to determine when the job can be performed.

Nonverbal

A great deal can be communicated without spoken or written words—through actions. People watch your actions very carefully. Because of this, the example you set as a supervisor is as important as the words you speak. For instance, if people hear you say that wearing personal protective equipment in your department is essential but notice that you don't wear it, they receive a contradictory message. Your words say, "Protective gear is important." Your actions tell them, "It's not important to wear personal protective equipment." People will usually take your actions more seriously than your words.

Likewise, when operating under pressure, be careful not to pass on those pressures to the people reporting to you. If they see that under pressure you take shortcuts and work in an unsafe manner, they may do the same. Serious accidents may result. As a supervisor, you must realize that your nonverbal messages are "read" and understood by workers just as readily as the verbal communications you send.

EFFECTIVE LISTENING

Thus far we have concentrated only on the sending portion of communication. We have emphasized the importance of making messages clear so they will be understood. But no oral communication takes place unless someone listens with understanding. To be an effective leader, you must be a skillful listener. It is difficult to think of a profession that doesn't require good listening skills.

Certainly the first-line supervisor must be a skilled listener. In addition, each of us plays a number of listening roles in our family and social life. We interact with parents, children, spouses, friends, and neighbors. In each capacity, it is important to listen with understanding. How well do we carry out this essential task? If a supervisor says, "My people just don't listen to me!" those employees may complain, "My boss says she has an open-door policy. But what good does it do when I go into her office, and she doesn't listen to me at all!" Fathers say, "My kids won't listen to me." But those children may be saying to their teachers, "Will you please listen to me? My Dad won't." Within the many roles we play in life, most of us can improve our listening skills.

Listening can be classified in three ways:

- *On-the-job listening.* Listening to the boss, people who report to us, other supervisors, and all others at work.
- *Social listening.* Listening we do off-the-job, outside the family circle—listening to friends, neighbors, social acquaintances.
- *Family listening.* Listening to our spouses, children, parents, and other family members.

Improving your listening skills in all three areas can help you become a better communicator.

The Importance of Listening

How important is listening in your supervisory job? A manager spends about 70 percent of the workday communicating. Studies have classified this communication as follows:

Writing	9%
Reading	16%
Speaking	30%
Listening	45%

How well did our educational system prepare you for a job involving communication? Some parts of a 12-year education were designed to teach reading and writing. These two combined represent only 25 percent of your communication job in business. You may have had one course in speech. Most schools have no courses in listening, but listening is involved in almost half of our communication time. It's not surprising, therefore, that we don't listen as well as we should; we haven't had much listening training.

How good a listener are you? When asked that question, many of us rate ourselves as "average" or "below average" listeners. How much does the "average" listener retain of what he or she hears? Tests conducted at the University of Minnesota showed that immediately after hearing a 10-minute reading, the average listener retained 50 percent of the material read. Several days later the retention rate dropped to 25 to 30 percent. If we are, in fact, "average" listeners, we are retaining only 25 to 30 percent of the oral communications directed our way. We can do a lot to improve our listening, but we must work hard to do so.

Some of us are inclined to say, "I may not listen well all the time, but when something really important is being said, I make an effort to listen better." This is a mistake. Few people realize that it is when the average listener tries harder to listen that he or she functions at the 25 to 30 percent level. Clearly, we all have considerable room for improvement.

The Cost of Not Listening

It would be difficult to estimate accurately how many accidents have been caused by poor listening. Perhaps the worker was distracted when the supervisor discussed the hazards of a job and how to avoid them. Or perhaps the supervisor did not carefully listen when the worker mentioned a problem the last time the work was done. The fact is that a better understanding of a problem by both supervisors and employees can reduce accident potential.

In addition to accidents, consider the costs of rework or scrap caused by poor listening: retyped letters, poor customer relations, and perhaps even the loss of business. When we consider the costs that result when people do not listen, we can begin to appreciate the importance of improving listening skills.

Authorities say that we can double our listening ability, but that it will not occur overnight. If a non-runner begins preparing for a 100-mile race, it will take considerable practice to reach even the first steps toward the goal. Similarly, we can find ways to improve our listening skills, but it will take time and diligent effort.

Steps in the Listening Process

There are four distinct steps in the listening process. By studying each of them, we can start improving our listening skills.

- *Sensing.* The first step is purely mechanical: Did the listener hear the words that were spo-

ken? If he or she can repeat the sense of the words, this step has taken place.
- *Interpreting.* The next step begins to complicate the process. How did the listener interpret or understand the words spoken? Do the words mean the same to both speaker and listener?
- *Evaluating.* At this stage the listener determines whether he or she agrees with what has been said. Remember that before evaluation occurs, understanding must take place.
- *Responding.* In the final step the listener responds to the message. Response may be a simple nod or shake of the head or a verbal "I see." Before going to a lengthier response, the listener should be certain that the speaker has finished a particular point.

Barriers to Effective Listening

There are many factors that keep us from listening as well as we should. By looking at several of them, we may be able to determine some ways to improve in this important area. The barriers fall into two areas—words and emotions.

Word barriers. Hearing certain words can shut off our ability to concentrate on what else is being said. People have different "turn-off" words or phrases. When one of these is spoken, the listener stops paying attention to what is being said and focuses on what the "turn-off" word brings to mind. For example, the word "death" is a turn-off word for some people because it may remind them of a disturbing family tragedy. When that turn-off word is heard, the listener may stop listening for 15 seconds to 15 minutes. For some people, turn-off words include lay-off, panic, grievance, abortion, and taxes (Figure 2-3).

What words turn off your listening? To improve your listening ability, you must discover your turn-off words, realize why these words affect you, and work to overcome the problem. When you hear one of these words, concentrate on what else the speaker is saying to keep your mind from being distracted by thoughts that the turn-off term evokes. You are the only one who can reduce the effects of turn-off words on your listening ability.

Emotional barriers. The listener's emotions can also block listening. For example, when a person becomes angry, he or she concentrates on the source of the anger, rather than on what is being said. People think of things to say that support their argument and often prepare questions to "trip up" the person at whom the anger is directed. They may try to embarrass that person in front of others. During this time they are not listening to what is being said.

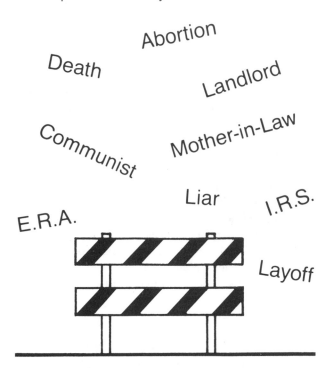

Figure 2-3. Turn-off words that recall emotional memories can interrupt your listener's concentration for 15 seconds or even 15 minutes. Avoid using negative words.

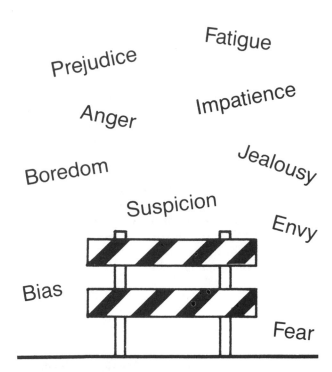

Figure 2-4. The listener needs to control emotional barriers in order to hear and understand the entire communication.

Anger is just one emotion that can be a barrier to listening. Others include hate, fear, suspicion, jealousy, overenthusiasm, and distrust (Figure 2-4). Any one of these responses can keep us from listening to what is being said. To improve our listening, we must control our emotional reactions and concentrate on what is being said, even though we may be angry at the speaker or suspicious of his or her motives. By keeping emotions under control, we can do a lot to improve listening skills. Consider how emotions affect your listening during conversations with your people, your spouse, your union steward, and others.

Distractions. Another reason for failure to listen is that we allow distractions to disrupt our thoughts. External noises, such as machines or equipment operating, can affect our listening and derail our train of thought. Many times a personal problem can be a distraction and interfere with effective listening. We must overcome the many distractions from within and without so that we can listen attentively. We must keep these facts in mind when setting up communications with our people and try to keep distractions to a minimum.

Heated Discussion Rule

A good tool to use in situations in which anger may cloud your listening is the "Heated Discussion Rule." Of course, participants must agree to the pro-

visions in advance. If the speaker in the meeting makes a statement with which you disagree, signal that you wish to speak. When the other person has completed her statements, she turns the discussion over to you. You may then make your statements, provided that you can first state the speaker's position and tell why she feels as she does about the matter. If the speaker agrees with your statement of her position, you may continue. This process forces you to listen in order to have a turn to speak.

When a heated discussion occurs, the points of disagreement often are not as serious as the combatants believe them to be. If both people would listen carefully to each other, they may find surprisingly little difference in their positions. When responding in a sensitive situation, it is wise to begin by stating the other person's position. A good way to start is, "As I understand what you have said, you feel this way about the matter." Frequently we will find that our own and the other person's viewpoints are not so far apart. By listening to someone's supporting comments, rather than thinking of a rebuttal, we can do a lot to improve understanding.

Five Keys to Improved Listening

A substantial difference exists between the rate at which people speak and the rate at which they listen. Most people speak at a rate of approximately 125 to 150 words per minute. Listening occurs at rates in excess of 600 to 700 words per minute. For some people that time differential interferes with good listening. Time is wasted because people think of other topics to bring up or they let their minds wander, occasionally returning to the speaker. Consequently, only 25 to 30 percent of what was said is heard and remembered.

On the other hand, the good listener thinks along with the speaker, mentally outlining the points and evaluating his or her credibility. The good listener will use the time differential to analyze the nonverbal messages sent by the speaker to determine if the verbal and nonverbal messages agree.

You can do a lot to improve your listening skills by making the maximum use of this time differential. It can help you to improve your understanding, which is the key to good communications. Here are five general rules to improve your listening:

1. *Stop talking.* You can't listen while you are talking. In two-way communications, when you are the listener, stop talking so that you can listen to all that is being said.
2. *Empathize.* When you put yourself in the other person's place, you can get a better understanding of why he or she feels a certain way. Remember, understanding is the secret to successful communications.
3. *Maintain eye contact.* This serves a dual purpose. First, it helps you to concentrate on what is being said; second, it shows the speaker that you are listening.
4. *Share responsibility for communication.* The "receiver" is just as responsible as the "sender" for good communication.
5. *Clarify.* When listening, if you do not understand any part of the message, be sure to ask questions until the meaning is clear.

SUMMARY OF KEY POINTS

To summarize what we've covered in this chapter:

- Good communication is essential for accident prevention. How well you send messages and how well you receive them, by listening and reading, will determine your success.
- Communication involves a sender, a receiver, and a message *that is understood by both.* Two-way, face-to-face communication is the best way to convey messages on the job. This method allows for feedback to let the sender know that the message has been received and understood and allows the receiver to ask questions about the message.
- Because messages can be filtered by the receiver's level of knowledge, biases, or moods, it is important to tailor your message to meet the needs and characteristics of your audience. Use a combination of oral and written communications for technical or complex messages.
- In communicating with others, make sure your verbal and nonverbal messages are consistent. Also, use positive reinforcement as well as criticism and correction.
- In supervisory work, listening is a key component of good communication. Supervisors spend about 45 percent of their time listening, yet the average person retains no more than 30 percent of what he or she hears.
- Listening can be classified as on-the-job, social, and family listening. Steps in the listening process involve sensing, interpreting, evaluating, and responding to messages.
- To improve listening skills, supervisors must overcome word and emotional barriers and eliminate distractions. Five general rules to better listening include not talking when others speak, practicing empathy, maintaining eye contact, sharing the responsibility for communicating effectively, and clarifying the message.

3

Human Relations

After reading this chapter, you will be able to:

- Understand the importance of human relations on the job
- Explain Maslow's hierarchy of needs and how supervisors can help employees meet these needs within the organization
- Know various leadership styles and the qualities of an effective leader
- Know how to detect employees' social problems, such as substance abuse, that can pose a threat to job safety and productivity
- Explain the objectives and scope of employee assistance programs

Some industry experts believe that the general shift to high technology has created a high-tech/high-touch era. They mean that the more sophisticated the technology, the more we need the "personal touch" among people. What is abundantly clear is that on balance high technology has had an adverse impact on human relationships.

HUMAN RELATIONS CONCEPTS

Because human relations concepts have not changed much over the years, they are just as applicable to today's high-tech society as they were back in the 1930s. What has changed drastically is the general lack of human interaction. Thus, the supervisor in the workplace is in an ideal position to promote and implement contacts among the workers in an area or department. The problem is knowing how to promote such "high-touch" principles in a way that increases worker satisfaction and productivity.

The Hawthorne Effect

Between 1924 and 1932, Elton Mayo, a clinical psychologist working at Harvard University Business School, conducted a series of experiments, carried out in the wiring rooms of Western Electric's Haw-

thorne Plant, located just west of Chicago. The research constitutes a landmark study. To determine whether there was any correlation between the amount of illumination in a workplace and productivity, Mayo turned up the lights. Production went up. When the lights were turned down, production went up again. Mayo conducted a second series of experiments varying the temperature, the length of rest periods, humidity, and other factors. No matter what was changed, productivity invariably went up. His conclusion: Attention to employees, not work conditions, was the dominant influence on productivity. The result of these experiments is referred to as the *Hawthorn effect* and has established a base for a number of studies in industrial social psychology.

The Hierarchy of Needs

One of these studies, conducted by Abraham Maslow, suggested that there is a hierarchy of needs among people (Figure 3-1). At the base are physiological needs, including food, clothing, and shelter, often referred to as the tissue needs. In this society, adequate wages satisfy most basic needs, leaving social and personal needs to be fulfilled. This is why so much emphasis is placed on motivating employees through human relations efforts in organizations. There are a number of well-documented practices and procedures widely used to motivate

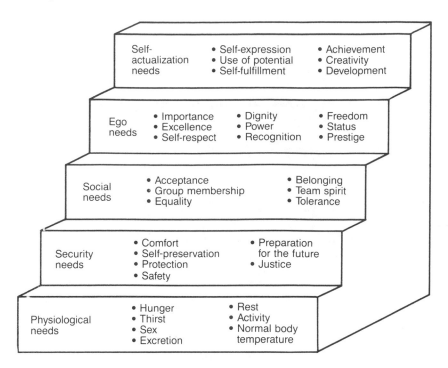

Figure 3-1. Maslow's hierarchy of human needs begins with the basic physiological needs and progresses to the self-actualization needs.

employees to increase production, enjoy greater job satisfaction, and reduce frustration. One does not have to be a psychologist to implement these practices, just have a sincere desire to help.

In studies done to determine what basic practices tend to make a company successful, it was found that the best companies implemented a network of informal, open communications. The first step for supervisors in establishing informal communications is to get out of the office. Institute an open-door policy. In some organizations, doors are not even permitted, since they are considered barriers to informal communications.

The second step is learning to be a good listener. Spend some time each day just listening to employees. Find out about their interests, goals, and ambitions. Getting out of the office and learning to listen are important first steps in establishing a good base for human relations in any organization. Moving about lets your people know that you are approachable and interested in them as individuals (Figure 3-2).

The following sections describe how these principles can be applied to help workers meet their own hierarchy of needs on the job.

Safety and Security

In Maslow's hierarchy, safety and security needs come first. On the job, this means people want to know what is expected of them and how to do their jobs safely and efficiently.

Supervisors are responsible, therefore, for seeing that their workers are thoroughly trained. This is essential to safe, efficient work and contributes to the individual's feeling of security. Whether you do the training yourself or assign another to do it, the ultimate responsibility is yours (Figure 3-3).

Injuries, work spoilage, and misused or broken equipment are almost certain to result when workers have been inadequately trained. Such casualties and production delays reflect on the individual, supervisor, and department and reduce the security of all. Workers who have had poor training and feel as though they have never mastered the job will continually be uncertain and hesitant. They may give up trying to do good work and lose interest in their jobs. This defeatist attitude leads to general job dissatisfaction and deteriorating performance.

Training should also include accident prevention. Workers should know that the company wants to protect them against injury and what they can do themselves for their own protection. Special emphasis on safety creates an employee interest in safe practices for their own sake as well as for satisfying workers' need for security.

Group Identity

Once safety and security needs are met, workers seek to satisfy social needs. People want to belong to their work group. That's why successful supervisors will

Figure 3-2. A supervisor needs to get out of the office, talk with, and listen to the employees.

do their best to create and maintain a feeling of group solidarity and friendliness. Watch for opportunities to talk with your group and use these times to build a sense of team identity. Words such as "we," "our department," and "our safety record" encourage group interest and team spirit.

The supervisor, when seeing that people are left out or excluded from the group, can look for ways to include them. A small clique that excludes other members of the work group is harmful. The best way to combat this tendency is to give all members the feeling that they belong to the larger group. This group feeling promotes good performance and safe practices.

Once the group accepts them, members discipline one another. Often, group standards are higher than the standards of individuals by themselves. The supervisor can appeal to people's pride in their safety record and in consistently following safe practices. However, supervisors who want to preserve a healthy group feeling should not tell people, "You have got to . . ." or "Now listen, I want you to. . . ." Instead, the supervisor must appeal to the self-interest of the group. When you appeal to each member to follow a line of action or a pattern of conduct, you must make sure that such action or conduct is in the member's and the group's best interest.

Because the need to belong is so strong, most people do not want to stand out as exceptions. They want to be liked by others, so they try to think and act as others do. Individuals usually conform because of peer pressure. This does not mean that individuals don't ever want to be considered as exceptional—to be singled out for their unusual abilities. A man may be proud of his strength or a woman of her ability to produce more than anyone else in the department; he or she may want others to recognize this. But even these workers will generally exercise their special talents only within limits allowed by their group.

As a result, people who consistently try to prove their abilities may do so because they feel unaccepted by the group. Some may consider lack of acceptance to be a real handicap. Others view this as less important. By and large, though, you will find that people need to be part of the group. This desire can affect their actions for better or worse.

As supervisor, you can use this desire for peer

Figure 3-3. Training includes accident prevention. The supervisor on the right shows a worker the safe procedure; the supervisor on the left reminds a worker to remove a dangling necklace.

approval to your own and to the group's benefit when training new workers. If the group works safely and efficiently, the new member will be likely to work this way as well. The fact that the new person wants to be a member of the group will be a powerful incentive to follow safety procedures and to become an efficient worker.

Recognition

Along with social needs, people want to know that their efforts are appreciated. To belittle a person's work is to belittle the person. Praise good work and give credit when it is due—but with caution. Giving a worker a slap on the back and saying, "Well done," only means something if it is done sincerely, not mechanically. Likewise, praise for work that the employee knows is poor not only makes the supervisor look foolish but reduces the value of future praise, even when it is justified.

A beginner's work might not be up to the quality standards of an old timer, but beginners should be encouraged. Tell them that they are making progress (providing they are), and give them further instruction to help them attain acceptable work.

Although experienced workers generally know when their work is good, they are pleased when the boss knows it, too. Perhaps the best general rule is for supervisors to maintain a positive interest in the

workers in their department. If you take such an interest, you will know what job each worker does, praise those who do well, and encourage those who are still learning or do not do quite as well as they should.

By the same token, praising workers for following safety practices will help to impress upon them how important you consider safety to be, as well as to encourage them to continue to do the job safely.

If people know their supervisor is not just looking for faults but also gives them credit for good work, they, in turn, will have a better attitude toward their jobs. On the other hand, workers who do not feel appreciated or recognized are likely to do little more than the absolute minimum to satisfy their job requirements.

Given employees' need for recognition and self-esteem, how you correct workers can make a big difference in their performance. As a supervisor, you know your own reactions when you are roughly criticized by your boss. Often it makes little difference whether it is a decision or your general competence that is questioned. Justified or not, criticism is a threat to one's self-respect. A worker feels the same way when you are doing the criticizing.

The supervisor who attempts to maintain standards by using a "meet-them-or-else" attitude is more than likely to find people busily looking for ways to get around those standards. If, however, you explain goals or standards to workers and encourage compliance, this approach is likely to be more effective than laying down the law and demanding conformity. Workers usually meet hard-line tactics with passive resistance, work slowdowns, equipment failure, and the like. No one wins. An approach that recognizes workers' efforts and encourages high productivity benefits everyone.

Fulfilling Work

Maslow found that people want jobs that provide opportunities for personal satisfaction and growth. Employees are more likely to have this feeling if they know how their job fits into the total pattern of the company. Their work is often so small a part of the whole operation that it is hard for them to see what contribution they are making.

Supervisors can help by explaining the company's overall objectives and by showing people why their jobs are valuable to the company. The vital role that safety plays in the entire production picture must also be emphasized.

A supervisor might think that employees ought to realize they would not be doing a job unless it has value. But this is not always true. When people do the same thing over and over again—especially when a task does not produce a complete product—they may begin to feel their jobs lack meaning. Although an individual usually feels a sense of accomplishment on learning a job, after a certain amount of time, this feeling needs to be reinforced. People may come to believe that their efforts are swallowed up in the vast quantities of similar work turned out by others.

As a supervisor, you can help workers maintain job satisfaction by showing them how their work is a vital part of the final product. You can also regularly discuss working conditions, growth possibilities, or the work itself, to provide opportunities for workers to develop their abilities or to expand their jobs.

LEADERSHIP

Since the Hawthorn studies were made, a number of management leadership studies have been conducted. One of these, by Douglas McGregor (see Bibliography), created the Theory X and Theory Y concepts. Theory X assumes the average human has an inherent dislike of work and will avoid it if at all possible. Therefore, people must be controlled, directed, and/or threatened with punishment to get them to meet the goals of the organization.

In McGregor's Theory Y concept, the assumptions are that people do not inherently dislike work and that threats of punishment are among the least effective means of getting people to work toward organizational goals. According to Theory Y, people not only accept responsibility but actively seek it. Satisfaction results from reaching one's goals.

Regardless of which theory one believes, leadership is a complex function. It is impossible to recommend a particular style. What works for a supervisor in one organization may be disastrous if applied in another. People are diverse; they seek different goals and behave in different ways to achieve those goals. Consequently, good leadership depends, to a large degree, upon knowing yourself, your group, and your particular situation.

Leadership Styles

Following are some leadership styles that are in vogue today. Some of these may be useful, depending upon your perception of your leadership problems. If you don't have a problem, count your blessings. There is no need to change styles.

Autocratic. A leader following McGregor's Theory X concept, in which people are closely controlled and directed, is autocratic. Such leaders identify the problems and the objectives, then determine the course of action. Subordinates have few opportunities to participate in the decision-making process. In some cases, such a leader may even use threats to obtain selected goals.

Benevolent/Autocratic. Such a leader will identify problems, set objectives, and make plans for subordinates. In some cases, this leader may ask for input that may or may not be followed. The flow of information is mostly downward, and subordinates have rather limited freedom.

Consultive. A consultive leader presents ideas, invites questions from subordinates, and considers their input. This style enables subordinates to participate to some degree by identifying problems and offering suggestions for solutions. A consultive leader considers alternative courses of action.

Participative. A leader following this style allows subordinates a great deal of freedom in identifying problems and solutions. The leader and the subordinates jointly set objectives, develop plans, and evaluate alternative courses of action. This style allows for a free flow of information.

Democratic. A supervisor following this leadership style would also be following McGregor's Theory Y concept. A democratic leader uses less authority and gives more freedom than the participative leader. This type of leader solicits the group's consensus and may allow the group to override the leader's decision.

Permissive. A leader following this style may make few decisions, giving wide latitude to subordinates. Such a leader makes no systematic effort to set objectives. Subordinates may do as they wish. A permissive leader exercises little control and does not hold subordinates accountable for results.

Leadership Qualities

Existing literature on leadership reveals that there are certain qualities expected of good leaders in a work environment. An effective leader develops most, if not all, of the following qualities:

1. Recognize and acknowledge that many of the ideas you use come from the people in your group. In this connection, many studies show that leaders do not develop all of the plans they use; in fact, many good leaders know that the majority of the ideas originate from the group. It is a common failure of new leaders to feel that they lose face if they accept suggestions from subordinates. On the other hand, experienced leaders know the value of having group members feel they are participating and that goals reflect their ideas and contributions.
2. Show respect for employees, both as individuals and as a group. You should be sensitive to, and understand, workers' needs.
3. Support, guide, train, and counsel group members.
4. Have confidence in your own abilities and judgment. You should not feel that you have to prove others wrong.
5. Act as a buffer between employees and higher management by not passing on the pressures you feel to those who work for you. This includes taking personal responsibility for the directions and orders you are required to give, even when they are not what you, personally, would like to do or would have chosen. You should never say, "The front office says we have to do this," because it gives others the feeling that you are shirking your responsibility.
6. Leave your personal problems and feelings out of your relationships with the group and its members by trying to maintain a calm, understanding approach at all times. A lack of emotional balance stirs up tension and stress in the group, particularly when you are grumpy, irritable, and irascible one day and pleasant, sensitive, and good humored the next.
7. Try to be fair at all times and play no favorites—always in keeping with good leadership.
8. Live up to what you say by setting the right example. The old cliche, "Don't do what I do; do what I say," does not work in on-the-job human relations.
9. Be available and understanding when people come in for help. Avoid the impression that you don't have time to help or to listen to problems. (When a supervisor is really under pressure and just can't take time at the moment, it is appropriate to say that and tell people you will be with them as soon as you can. Make sure that you live up to this promise.)
10. Be accepted as one of the group, but not so much that your status as a leader is destroyed. Unfortunately, some leaders try so hard to belong, particularly in off-the-job get-togethers, that they cross the indefinable but very real line that separates leaders from those they lead. People know that this line exists and expect it to be maintained; when it is not, your status as a leader is damaged or destroyed.
11. Create a job climate that is generally harmonious and that minimizes conflict and confusion. A hostile or chaotic environment generates fear and anxiety in group members and reduces efficiency in operations.
12. Become a good one-on-one, face-to-face communicator. Good leadership and sound communication skills are strongly related. As a good leader you should ensure that workers' basic needs are met.

Measuring Leadership Effectiveness

You can measure your effectiveness as a leader by how well you are meeting the goals and objectives of the organization. If you are satisfied, and the organization is satisfied with your performance and that of your subordinates, there is no problem. However, if you are not achieving what you or the organization wants, there is a problem. Review the following areas to determine if your performance level meets organizational norms and standards:

- Results of accident investigations—are underlying causes identified and corrected?
- Total cost of accidents—increasing or decreasing?
- Cost of production—increasing, decreasing, remaining the same?
- Quality of production—improving or declining?

- Safety inspection follow-up—are problems corrected or ignored?
- Safety training for all employees—is training thorough or haphazard?
- Job safety analysis—are analyses conducted?
- Quantity of production—increasing or decreasing?
- Absenteeism—increasing or decreasing?
- Employee turnover—increasing or decreasing?

If you decide that you have a problem in any of the above areas, keep in mind that the issue will not go away and cannot be ignored. Supervisors have to decide how to solve their problems. One solution might be to change your leadership style. Perhaps you need a more autocratic style to get results, or perhaps you have been too autocratic. In any event, supervisors have to decide how to solve the problems since they are in the best position to know the work environment, the conditions involved, and workers' attitudes.

COPING WITH DIFFICULT PROBLEMS

At times you will need to use your human relations skills to deal with employees' special problems that affect their job performance and relationships with other workers. As a supervisor you should be alert to when these problems are temporary and can be handled on the job and when they are more serious, requiring stronger measures. Your concern is not to diagnose employees' conditions nor to try straightening out their private affairs or personality problems. Any attempt to do so on your part could cause more harm than good.

Your primary concern is the safety and productivity of your workers. If an employee exhibits antisocial behavior and interrupts the work group, he or she should be referred to the human resources department for help. Some of the problems employees experience could be caused by their resistance to change, personal circumstances, a change in work conditions, or job-related or personal stress. The following sections discuss measures you can take to address these problems.

Resistance to Change

Leadership skills are particularly valuable in the face of today's fast-paced, high-tech world. With so many technological advances, many jobs change in character, in some cases, drastically. Old jobs are abolished and new jobs are created. As jobs change,

so do employees' work habits and feelings about their jobs. Change threatens peoples' feelings of security. Workers should always be notified in advance about job changes, as well as policy or organizational changes. If people are not informed until after decisions are made, they have no time to adjust. Supervisors should explain the reasons for changes and how decisions were reached.

People, in general, tend to resist change because it usually means that they must learn something new. Old skills may no longer be needed. This can lead to frustration; it may also make people feel unimportant and unwanted. When new skills are required, workers should be given an ample opportunity to learn them, providing that they have the capabilities.

As a good leader, you can provide advance knowledge of job changes and help to lessen resistance. Let workers participate, whenever possible, in the decision leading up to the change. Participation makes people feel more important and increases their sense of security.

Temporary Problems

At times, people experience money or marital problems, personal tragedies, or other crises that make them temporarily inattentive. Supervisors who know their people well can detect when a person is not acting normally. If supervisors are friendly and approachable, workers will, in many cases, say that they are not able to perform certain exacting or hazardous work for that day. On the other hand, nonhazardous physical work is often just what a person needs to escape temporarily some personal problem. Such a reprieve will often allow the person to find a solution to the problem. Fortunately, much undesirable behavior is temporary. Because workers' behavior varies from day to day, be flexible in your approach to their problems so that employees have some leeway within the limits of doing safe, efficient work.

Strong Emotion

Although no one expects supervisors to become personnel counselors, they can help to reduce strong emotions in group members. Supervisors are in the best position to evaluate the emotional level of an individual worker and to do something about it.

Everyone is familiar with the emotional outbursts of people who have had a trying and upsetting experience. Such expressions of emotion may take many forms, from an outburst of invective or a torrent of tears to physical assault on inanimate objects. The end result is the same—relief from the intolerable tension and emotional relaxation.

To reduce the intensity of such emotional responses, supervisors can engage in permissive listening, that is, giving employees an opportunity to talk about their problems. The freedom to say what the individual wants can reduce tension that otherwise might lead to an accident. It is important that supervisors do not attempt to evaluate or judge an employee's remarks. An emotionally upset employee needs an opportunity to "blow off steam" without retaliation or evaluation.

The stresses that cause emotional upsets can occur within the work environment or away from it, but each kind of stress has an effect on the other. A simple question such as "How are you today?" may be enough to set off an outburst of pent-up emotion. After the discussion, employees often can evaluate more clearly and realistically the conditions that upset them in the first place. This sensitive relationship between the supervisor and the worker is one of the supervisor's everyday functions. Diligent observation and effective, sensitive listening are the tools required.

SHIFTWORK AND SHIFT CHANGES

In our round-the-clock society many industries such as chemical plants; health, safety, and transportation services; the armed services; and nuclear power plants must be in constant operation. They require that workers be on a night shift (a fixed night schedule) or rotating shifts (a combination of day, afternoon, and/or night shifts). Such shiftwork and shift changes create stress and pose serious health and safety problems.

The human body has a biological clock, known as the circadian rhythms. This internal mechanism is capable of measuring time and regulating involuntary responses, such as reaction time, respiration rate, digestion, body temperature, and so on. When workers are on night shifts or rotating shifts, their schedules violate both the body's natural rhythms and the dominant social rhythms. Serious negative effects of this disruption may include:

- Higher risk of accidents and injuries because of slower reaction time or lower attention and motivation levels
- Chronic fatigue and poor health
- Lower quantity and quality of sleep
- Digestive problems, poor nutrition, poor eating habits
- Potentially slower action of some prescription medications

- Negative effects on family and safety and thus on mental health
- Poorer working conditions, such as lighting, ventilation, food services

Studies of shiftwork have shown some methods of minimizing these negative effects. Companies can adjust shiftwork schedules, improve workplace conditions, and educate employees in ways to cope with the effects of shiftwork.

Some common periods of time on a given shift schedule are two or three days, one week, two weeks, or one month. Some companies use a compressed workweek of four 10-hour workdays and three days off, 12-hour shifts, or other compressed schedules. There is little agreement regarding the best shiftwork schedule, but researchers emphasize two important points:

- For workers on rotating shifts, the shifts should rotate forward, that is, from day to afternoon, or from afternoon to night. This is because circadian rhythms adjust more easily to moving forward.
- Shiftwork schedules should be as flexible as possible to allow workers to meet their personal needs.

Improving workplace conditions can also minimize the effects of shiftwork. Companies can provide adequate supervision for night-shift workers or ensure that workers know the extent of their authority. Work should be scheduled so that more difficult tasks are done earlier in the shift and that adequate time is allowed for jobs. Remember, body functions and reaction times slow down at night. Make it policy that workers think through tasks before acting and avoid rushing through routine work. Plan breaks and rest periods when workers are most tired and encourage workers to take their breaks. (The incidence of errors and accidents is highest between 4:00 a.m. and 6:00 a.m.) Keep the work area well ventilated and well lit. Reduce noise and isolation of workers. Make available cafeteria services that provide healthy, nutritionally balanced meals.

Finally, employers should educate their workers about shiftwork effects and stress management. Both workers and their families can be informed through employee newsletters, management/union bulletins, discussion groups, and safety talks. (For more details, see sections on stress management and employee assistance programs in this chapter.)

STRESS MANAGEMENT

In recent years companies have come to recognize that stress management is an important element in increasing productivity and avoiding accidents. One definition of stress is "factors which accelerate the rate of aging through the wear and tear of daily living." The combination of job stress and personal stress can increase the probability of accidents occurring and decrease productivity.

Stress Management Principles

Supervisors need to keep in mind some basic principles of stress management.

1. Stress is always present and can be positive or negative. Examples of positive stress include getting promoted, buying a new home, getting married. Negative sources of stress include being the subject of disciplinary action, having unrealistic deadlines or abnormally large work loads, death of a spouse, and divorce. Other sources of negative stress can be as minor as losing a favorite pen or as aggravating as an ongoing conflict with a fellow employee.
2. Anger as a habitual response to stress creates more stress. More effective types of response focus on problem solving and effective communication to reduce the level of stress felt by employees.
3. How employees choose to react to stress on or off the job is more critical than the precise level of stress in their lives. Employees can be taught effective mechanisms to cope with stress even if they cannot eliminate the sources of stress in their lives.

Supervisors should be aware of the following symptoms that could indicate an employee is having difficulty adjusting to the level of stress in his or her life. These include:

- Nervous tics
- Rashes
- Teeth grinding
- Depression
- Anger
- Irritability
- Low self-esteem
- Apathy
- Impatience
- Accidents on and off the job
- Forgetfulness
- Drug use
- Negativity

Poor adaptation to the demands of the workplace is usually a combination of stresses imposed by the demands of job design plus factors that are specific to the individual. Stress reduction requires change by employees and the organization.

Stress Management Skills

From an employee stress-management perspective, employees should be encouraged to develop the following skills. These skills can reduce the level of stress workers are feeling, even if they cannot eliminate the sources of stress. These skills include:

1. **Specific relaxation techniques.** Examples of specific relaxation techniques include:
   - *Autogenic relaxation*—This technique incorporates full, comfortable breathing and concentration techniques to develop a form of self-hypnosis to relax the mind and body.
   - *Positive affirmations*—Developing and practicing positive self-talk to reduce the amount and effect of negative self-talk. This is usually incorporated into autogenic relaxation techniques.
   - *Deep muscle relaxation*—One technique that can be taught very early and can be used at any time by an employee is a breathing technique to reduce stress. Basic breathing can be done in the following manner:
     - Empty your lungs fully.
     - Take a deep, slow inhalation.
     - Make a long, slow exhalation.

     A series of these breaths can immediately produce a more relaxed state.

2. **Values clarification techniques.** These techniques help employees focus on what goals they want to achieve on the job and in their personal life. Values clarification techniques help individuals identify their basic values and determine ways to express these values more frequently in their lives. Skills relating to prioritizing and planning are enhanced along with increasing the likelihood of employees taking concrete steps to reduce the level of stress in their lives.

3. **Exercise.** A sound exercise program will incorporate strength-building, stamina, and suppleness. A minimum of 30 minutes of cardiovascular stamina-building exercise should be incorporated into this program.

These 30-minute segments should be scheduled at least three times a week. Other exercises can be used to build strength and suppleness. It should be noted that values clarification techniques and relaxation techniques are more important with regard to stress reduction than is exercise for most people.

4. **Nutrition.** Good nutrition is also an effective stress-management tool. The basic principles of sound nutrition are simple. They can be achieved by following the guidelines noted below:
   - Eat a variety of unprocessed and little-processed foods.
   - Drink six glasses of fluid daily, two of them water.
   - Take broad-based vitamin-mineral supplements if needed.
   - Eat high-nutrition foods in moderate amounts.
   - Eat three meals a day, particularly breakfast.

Company Stress-Reduction Programs

Equally as important as how the individual deals with stress is how the organization provides programming to reduce employee stress levels. The leading causes of stress in organizations are unrealistic work loads and deadline dates. Listed below are some methods that organizations can use to reduce stress levels in their organization.

- Survey and consider carefully manpower requirements of various departments. Cutting back on staff size can be a mistake when it comes to long-term morale and productivity issues.
- Survey departments to determine typical deadlines employees must meet. Try to determine if deadlines are realistic. If deadlines cannot be altered, develop time-management techniques for employees.
- Provide stress-management classes for employees. These classes should focus on organizational factors contributing to stress and on methods employees can use to reduce levels of stress.
- Provide classes to workers that focus on communication and interaction skills. Much stress results from poor relations and communication between employees.
- Provide classes to employees that focus on conflict resolution.
- Survey the work environment to determine if

there are safety-related issues or occupational health-related issues that have not been addressed. Unresolved concerns in these areas can lead to a high level of stress.

In summary, supervisors should remember that:

1. Stress has both physical and mental effects upon employees.
2. A certain level of stress can be beneficial.
3. Too much stress, negative or positive, can be harmful.
4. Stress is energy and must be used properly.
5. Unchecked stress in the workplace can lead to accidents, poor morale, and reduced productivity.

ALCOHOL AND DRUG PROBLEMS

Employee alcoholism and drug abuse cost industry millions of dollars each year. These problems are often the causes behind lowered productivity, increased absenteeism, inefficiency, increased employee turnover, increased injury rates, and incidents arising from behavioral problems. Employees with an alcohol or drug problem usually have a high absentee rate and frequently engage in unsafe work practices. Their conduct can be disruptive and demoralizing to others and can lead to excessive turnover. As a supervisor, your job is to recognize these problems and take steps to deal with them.

Symptoms of Substance Abuse

Many companies believe that employees with alcohol or drug problems require professional assistance. Early recognition of these conditions and other emotional or behavioral problems by the employee's supervisor is essential to an effective program of evaluation, treatment, and rehabilitation. Concern with alcoholism and drug abuse is a basic management responsibility.

The employee's immediate supervisor is the first member of management most likely to observe unusual employee behavior, especially the possibility of alcoholism or drug abuse. Supervisors are expected, as part of their daily contacts with employees, to be familiar with their appearance, behavior, and work patterns, and to be alert for any changes. Of course, these could result from causes other than alcohol or drug abuse—illness, side effects of prescription drugs, or fatigue.

Because behavioral problems can take many forms and have many causes, supervisors should avoid drawing hasty conclusions about an employee. His or her behavior may or may not reflect a physical or emotional dependency on alcohol or drugs. As a supervisor, you are not expected to make a diagnosis; that is the responsibility of the medical department. Your job is to recognize noticeable changes in an employee's behavior that might indicate a problem for which help is needed. Sudden extremes of behavior in a new employee, for example, may suggest to you that the individual has a problem that should receive medical attention or be brought to the attention of security personnel. For employees who have been on the job for some time, however, marked changes in their behavior and job performance are likely to stand out more clearly. Perhaps the employee's appearance has become more sloppy, or the person's speech is slurred. Symptoms may arise suddenly or gradually, may be episodic or continuous. Regardless, they require immediate attention.

Alcohol and Drugs on the Job

In our society, some people have grown accustomed to the practice of having a cocktail, a bottle of beer, or some other alcoholic beverage at lunch. These individuals believe that a drink or two with a meal will not affect them. There may be no actual drinking on the job, but the effect of alcohol does not end with the meal.

Some individuals simply enjoy drinking as a social activity, just as others enjoy eating. No matter how often you speak about the hazardous effects of alcohol, the person who enjoys drinking will continue to do so. If you are a supervisor, however, make sure that workers reserve their drinking for off-duty hours. Also, see to it that the effects of off-duty drinking do not affect your employees on the job in the form of a hangover or impaired judgment and reaction time.

Treatment for alcohol and drug abuse is not uniformly successful. One medical executive said that, unlike alcoholism, drug addiction is not amenable to outpatient treatment. Although some companies report 60 to 75 percent success rate with alcoholic employees, workers treated for drug addiction experience a high rate of relapse. As a result, companies tend to view drug abuse as a more costly problem.

Interaction of Alcohol and Drugs

Studies indicate that at any one time, from 10 to 20 percent of the population is taking prescription medication. Add to this the percentage of people using drugs illegally, and it becomes obvious that a substantial portion of our working population is exposed to the effects of drugs.

The combination of alcohol and drugs can produce a variety of effects that severely hinder a worker's judgment. Concentrations of alcohol and drugs remain in the bloodstream much longer than most users realize, and the effects may rise unexpectedly. Little scientific study has been conducted on the interaction of alcohol and drugs. However, there is sufficient evidence to conclude that such a combination can lead to increased impairment of judgment and skill and constitute a greater risk of serious accidents on the job.

Countermeasures

Alcoholism and the sale and use of drugs by employees may not be problems in your company at this time, but are you prepared to cope with these problems if and when they arise? Time spent developing countermeasures can pay valuable dividends in the long run.

Alcohol. Many companies have a small group of heavy or problem drinkers who are tempted to drink during working hours. Even if you are fortunate enough to have no workers who drink on the job, remember that the problem drinker can still bring a hangover to work. Some studies also indicate that employees who are alcoholics have more accidents than nondrinkers.

Countermeasures for dealing with alcoholism begin with education. Information on alcoholism and the effects of alcohol is readily available. Write for detailed information on alcoholism programs:

Alcohol and Drug Problems Association of
    North America (ADPA)
444 N. Capital St., N.W.
Washington, D.C. 20001
(202) 737-4340

Supervisors should read as much information as possible in order to be familiar with alcoholism and its effects. One good way to inform your people about alcoholism and the problems it can create is through the five-minute safety talk. Drinking and driving can be discussed during the next safety meeting.

The next step in an alcohol program is the identification of problem drinkers under your supervision. Everyone is familiar with the drunk and can recognize the obvious signs of intoxication. Individuals who have been drinking on the job will probably give themselves away through unsteady actions or by noticeable odors. However, you should also watch for the apparently sober person with the continuous hangover, a signal that may indicate the person has a serious drinking problem.

After identifying a problem drinker, supervisors should examine the ways in which the person can be helped. One way is counseling by people trained to deal with alcoholism. Check your community organizations, churches, and Alcoholics Anonymous (AA) for information on programs to assist alcoholics. (Telephone directories of many cities and towns list a number to call for information about meetings and about the AA program.)

Above all, supervisors must show the alcoholic that they care and have a true desire to help. Simply telling the alcoholic not to drink is like telling someone who enjoys eating not to eat. An admonition not to drink is not the solution. Alcoholism is an illness and must be handled by those trained to deal with it. The alcoholic needs help, not advice.

Drugs. The current increase in the use of drugs is not confined to the younger segment of the population. The individual who often takes an overdose of aspirin or sedatives could, at times, be as much a problem as a heroin or cocaine addict.

One large company tackled the problem of employee drug abuse and designed a policy to cope with it. The company had a high rate of success in dealing with employee alcoholism, so they modified and updated their safety policy to include what they called "drug dependence." Their drug-abuse policy is as follows:

In accordance with our general personnel policies, whose underlying concept is regard for the employee as an individual as well as a worker, we believe that:

- Drug dependency is an illness and should be treated as such.
- The majority of employees who develop a dependency on drugs can be helped to recover, and the company should offer appropriate assistance.
- The decision to seek diagnosis and accept treatment for any suspected illness is the responsibility of the employee. However, continued refusal of an employee to seek treatment when it appears that substandard performance may be caused by an illness is not tolerated. Drug dependency should not be made an exception to this commonly accepted principle.
- It is in the best interests of employees that, when drug dependency is present, it should be diagnosed as such and treated at the earliest possible stage.
- Confidential handling of the diagnosis and treatment of drug dependence is essential.

The objective of this policy is to retain employees who may have developed drug dependence by helping them to arrest its further advance before it renders them unemployable.

Identifying employees with drug problems is much easier if a company takes a positive approach. Since drug dependencies cause marked changes in work behavior patterns, personal relations, and emotional moods, supervisors should be alert for these changes in any employee. When an employee is suspected of drug usage or dependency, the supervisor should seek the assistance of the company staff or a community agency that deals with treating drug dependency.

Your company must first establish a written policy, as did the company described above, when planning countermeasures to deal with drug abuse. This policy must be carefully prepared and made known to all employees in the same manner as company safety policy.

The next step is to locate community or area organizations that offer counseling to persons with drug dependency. Familiarize yourself with the types of drugs that can lead to dependencies and the effects these drugs may produce. There is a vast quantity of material available.

However, supervisors should not become drug detectives, constantly looking for pills and drug substances. Instead, they should watch for changes in employees' work behavior patterns, personal relationships, and emotional moods. Any of these changes in an employee may indicate an alcohol or drug problem.

For more detailed information on alcohol and drug abuse programs, contact:

Alcoholics Anonymous World Services
Grand Central Station, P.O. Box 459
New York, N.Y. 10163
(212) 686-1100

National Institute on Drug Abuse
5600 Fishers Lane
Rockville, MD 20857
(301) 443-6487

NOTE: For further information about Drug and Alcohol Programs and Associations, please contact your local librarian and/or refer to: *Encyclopedia of Associations*, Volume 1, Parts 2 and 3, Karin E. Koek, Susan B. Martin, and Annette Novallo, eds., Gale Research, Inc. Publisher, Detroit, Michigan, 48226.

EMPLOYEE ASSISTANCE PROGRAMS

Employee assistance programs (EAP) are being implemented by enlightened management to help employees with alcohol, drug, and other behavioral or stress-related problems that interfere with job performance. Programs of this type enable a company to retain valued employees. Such programs can assist the supervisor who feels an employee has a serious problem that needs immediate attention.

EAP Objectives and Scope

The function of an EAP is to help employees resolve their problems. In the process, both the employee and the employer benefit. When problems are solved, they no longer adversely affect an employee's job performance. Effectiveness—in personal and professional life—is restored.

When EAPs were first offered, they were composed largely of recovering alcoholic staff members who helped other employees deal with the same problem. Some programs still operate that way, but many have expanded the number and types of problems addressed, and the number of people eligible for the program.

Problems addressed. An EAP can cover any problem that affects a person's lifestyle or job performance. The areas of service may include mental health, financial assistance, substance abuse, legal assistance, family and children concerns, and marital relationships. Some programs provide retirement, relocation, and outplacement services. Health promotion and fitness programs are sometimes offered by companies with the idea that a healthier body makes a healthier, more productive employee. Health promotion services may include exercise, smoking cessation, weight control, nutrition, and other health classes.

Eligibility. Because the problem of an employee's family members can affect the employee and his or her job performance, most programs offer services to dependents as well as to employees. Of course, this will increase the potential number of clients, so staffing and funding must be greater.

Basic EAP Processes

The elements of the day-to-day operation of an EAP depend on the scope of the program. In general, all EAPs will have at least the following operations:

- Assessment and referral of clients
- Confidential records management
- Coordination of benefits

- Evaluating and contracting outside service providers and monitoring the quality of services
- Monitoring follow-up
- Training supervisors and labor representatives about the EAP and how to recognize troubled employees
- Promoting the EAP to employees and others eligible for services
- Interacting with related departments, labor unions, management, and entities inside and outside the organization
- Writing reports on program use and finances. In extensive programs, in-house counseling or teaching health classes would be added to that list, along with other administrative duties. Reports on cost/benefit analysis would probably be more elaborate. Many EAPs publish a newsletter with tips on healthful living and leisure activities and notices of EAP services.

EAPs operate in a number of ways, but their basic processes are the same. After an orientation program introducing the EAP, an employee may enter the program either voluntarily or by referral. In large companies with a full-time staff, an EAP coordinator interviews the employee to assess the problem. Depending on the scope of the particular EAP, the coordinator will begin counseling or treatment, or direct the employee to an outside service, such as a substance abuse program. In a smaller company, the EAP coordinator may refer the employee to an outside, contracted professional for diagnosis and treatment. If this occurs, be sure to submit bid proposals to professional contractors to assure you get what your company needs.

Many of the problems employees encounter require only short-term counseling, three to seven sessions, handled through the EAP system. If treatment must be extensive, however, the employee may receive long-term outpatient treatment or enter an inpatient treatment facility. At this point, health care benefits usually cover the costs.

A company with an EAP does not intrude into an employee's personal life or try to diagnose the person's problems. These areas are not its concern. The company focuses on job performance, which is within its area of concern.

Entry into the EAP Program

When a supervisor or labor representative identifies a decline in an employee's job performance, he or she follows the usual procedure of documentation or remedial action. An employee may enter an EAP either voluntarily or through referral.

Voluntarily. If job performance continues to decline, the supervisor or labor representative may suggest that the EAP could be helpful. At this point, the suggestion is just that—a suggestion. The employee can act on it without telling the supervisor or labor representative.

If the employee's job performance improves, it may or may not be the result of the EAP services. The company will never know. EAP services are completely confidential when initiated voluntarily by the employee.

An employee can also enter the EAP without a recommendation from a supervisor or labor representative. This type of voluntary entry is also completely confidential.

Referred. An employee's job performance may continue to decline, despite documentation and remedial procedures. Because of this, some companies have developed a system for formally documenting referrals to the EAP in the cases of continuing performance decline. Employees are referred through the EAP to counseling or treatment, and the referral is noted in their files. Depending on its policy, the company may or may not be informed of treatment compliance and follow-up.

Types of EAPs

There are various types of EAPs. The type best suited to a particular company depends on the organization's size, financial resources, employee characteristics, work-site locations, and design preference.

Full-time internal programs. This program is seen most often in large organizations (3,000 or more employees) with a central location. The coordinators of these programs have the EAP as their sole responsibility. The coordinators assess problems, refer to appropriate professionals, and monitor treatment and follow-up. The coordinators often have a counseling caseload or class teaching schedule as well.

Part-time internal programs. This type of program is often provided by smaller organizations with 750 employees or less. The coordinator has both the EAP and other responsibilities. Often this person is in the medical or human resources department or has had personal experience with a problem, plus additional training. The EAP coordinator may have other duties in addition to the main EAP duties: (1) assessment, (2) referral, and (3) management of the program.

Consortium. A consortium EAP provides services to a single organization with a number of work sites or to several organizations in the same area. The staff often consists of an independent service provider offering a range of services, including: (1) assessment and referral, (2) treatment, (3) follow-up, (4) aftercare, and (5) management of the program. Serving a number of locations, consortia are usually offsite.

Independent service provider. Smaller organizations often use an independent service provider. Some contract for assessment and referral only. Some purchase the complete range of services: assessment, referral, treatment, aftercare, and management of the program. Independent service providers are usually offsite.

Regional. This type of program is seen in an organization with work sites spread over a region. The coordinator is usually full-time and makes scheduled visits to the sites for assessment and referral, counseling, and coordination of local contracted services.

SUMMARY OF KEY POINTS

The key points covered in this chapter include:

- In our high-technology society, human relations concepts are just as applicable today as they were 50 years ago. These concepts can be used to understand and fulfill many of workers' needs on the job, raise worker morale, and increase safe production.
- The Hawthorne studies showed that attention to employees, not work conditions, is the dominant influence on productivity. This is why many companies emphasize human relations efforts in their organizations. Important first steps include establishing informal communications and developing good listening skills on the part of supervisors.
- Maslow's hierarchy of needs suggests that people seek to satisfy their safety and security needs first, then their need to belong and to be recognized, and finally their need for fulfilling work. These needs should be met on the job, as far as possible, if employees are expected to work safely and efficiently.
- The supervisor, as leader, has a key role to play in creating effective human relations with and among workers. You should learn something about the theories and styles of leadership in order to choose the best one for your situation. By developing the qualities of a good leader, you can help employees manage the work-related stresses and changes common in our fast-paced, high-tech society.
- Coping with employees' difficult problems is an important part of increasing productivity

and avoiding accidents. Supervisors must be trained to recognize the symptoms of stress and other problems that can disrupt the work group and to apply the basic principles and techniques for reducing stress in themselves and their employees.

- Drug and alcohol problems can be particularly costly to a firm in terms of accidents, lost productivity, and personnel conflicts. Supervisors should know how to spot symptoms of drug or alcohol abuse in workers and to take immediate action to address the problem. Above all, supervisors should develop a sincere interest in and desire to help their employees.

- Many companies are creating employee assistance programs to handle a wide variety of problems that workers experience that may be interfering with their job performance. EAPs are voluntary programs and protect the privacy of those who enter treatment. Their purpose is to restore the employee to full functioning in work and personal life.

4

Employee Safety Training

After reading this chapter, you will be able to:

- **Understand the necessity for providing safety training for your workers**
- **Develop a sound orientation program for new and transferred employees in your department**
- **Use job instruction training to provide more detailed training for particular jobs**
- **Understand how to use job safety analysis as an accident prevention tool**

One of the more positive actions you can take as a supervisor in accident prevention is to provide safety training for your workers. The effects of these efforts are observable and measurable, which can serve as a positive evaluation of your accident prevention work.

Employee safety training is now mandated by OSHA and state occupational safety plans. Under the law, employers must provide the following:

- Employee training and education programs
- Pertinent information about the job:
  - Proper working conditions and precautions
  - All hazards employees are exposed to on the job
  - Symptoms of toxic exposure to chemicals or other substances used in the workplace
  - Emergency treatment procedures

The test of a good safety program is whether the employees' new awareness regarding safety results in observable, measurable improvements in job safety performance (for example, fewer accidents, lower absentee rates due to illnesses or injuries). The test for management is higher employee productivity and an improved safety record companywide.

Four steps to changing employee behavior regarding safety are:

1. Providing orientation and training
2. Promoting safety skills
3. Developing "safety awareness" attitudes
4. Modeling good safety practices (supervisor and upper management)

Management must convey to their employees that safety is an important value in the company.

Many studies have been made to determine why people fail to follow safety procedures or to take reasonable precautions on the job. Some of the reasons are that workers have:

- Not been given specific instructions in the operation
- Misunderstood the instructions
- Not listened to the instructions
- Considered the instructions either unimportant or unnecessary
- Disregarded instructions

Any of the above lapses can result in an accident. To prevent such an occurrence, it is essential that safety training work be conducted efficiently. Although as a supervisor you will generally provide the training yourself, you may, in some instances, choose to delegate some of the training to other skilled people who report to you. On technical subjects—for example, fire prevention or first aid—other qualified people may lend their expertise to the training effort. In any event, make a follow-up check to determine that the training achieved its purpose. You have the final responsibility for the effectiveness of training efforts in your department.

Every person who conducts safety training should have the following qualities:

- Thorough knowledge of the subject
- Desire to instruct
- Friendly and cooperative attitude
- Leadership abilities
- Professional attitude and approach
- Exemplary behavior to set an example for others

This chapter covers several areas of safety training of special interest: orientation training, job instruction training, other methods of instruction, and job safety analysis.

Other subjects, such as the care and use of personal protective equipment and proper lifting techniques, are covered in Chapters 9 and 13, respectively.

ORIENTATION TRAINING

The First Day

One of the best ways to involve people in an accident prevention program is to provide a thorough orientation for new employees. One frequently hears of people having accidents the first day on a new job. In many instances, these employees were not given proper instructions.

A good orientation program for new employees should emphasize:

- General company rules and employee benefits. These topics are usually covered by the industrial or human relations department.
- Overall safety rules and accident prevention programs and policies. The safety director usually covers these topics.
- Explanation of the specific hazards in the new employees' department and the applicable safety rules and practices to offset those hazards. This part of the orientation is the most important and should be conducted by you— the first-line supervisor (Figure 4-1).

While on the job, you can show new employees how certain hazards have been eliminated and what precautions have been taken to guard against hazards that cannot be eliminated. Such explanations can yield a number of benefits:

Figure 4-1. Orientation should include showing new employees where to find the relevant hazard information, such as Material Safety Data Sheets.

- New employees will realize the company's commitment to safety and worker protection.
- Supervisors can emphasize the need for safety rules and give reasons why they must be followed.
- New employees, from the first day on the job, will understand how important safety is everywhere in the plant.

Safety Operating Procedures

Before new employees start their first assignment, you should discuss the following procedures:

- All accidents, including near accidents, must be reported immediately.
- Any unsafe conditions also must be reported to you or your staff at once.
- Equipment and tools should be checked before being used, even if they appear to be in good condition.
- No one should be allowed to operate any equipment without specific authorization and instruction from the supervisor.
- Job instruction must always include applicable safety instructions as a matter of course.

Personal Protective Equipment

Safety training sessions or job instruction for new employees is a golden opportunity to demonstrate personal protective equipment and to explain why such equipment is necessary. Knowing and telling new people why they must wear eye protection, for example, will help ensure their compliance with rules.

This is also an ideal time to show new employees how to care for and clean their personal protective equipment and what procedures to follow if any piece needs repair or replacement. From the first day, new employees learn the importance of personal protective equipment and understand that departmental rules regarding its use will be strictly enforced. They will appreciate your no-nonsense approach to accident prevention.

Transferred Employees

You should also conduct orientation sessions for employees transferred into your department. You can cover specific problems that they may not have encountered on their previous job within the company. It's better to alert transferred employees a second time to hazards than to take the chance that they may not be aware of them.

Follow-Up

A follow-up session with new or transferred employees should be scheduled several days after the initial orientation. This session will help to determine how good a training job you have done. It also gives the new people another opportunity to ask questions about their work. Encourage workers to bring up any questions they may have regarding accident prevention. This process helps employees to think about safety from the start.

Many supervisors use a checklist of items to be covered in their orientation for new employees. A sample list is shown in Figure 4-2. It can be used as a basis for making your own checklist. In this way, you can be sure you have covered all the relevant items. Some firms ask new employees to sign the checklist and return it to the human relations department to verify that the points were covered.

JOB INSTRUCTION TRAINING

Once you have completed orientation sessions you can move on to more detailed instruction. Job instruction training (JIT) is a technique for providing on-the-job training for particular tasks. Teaching new and/or transferred employees to do jobs safely and efficiently can improve operations immensely.

Either conduct the job training yourself or delegate it to an experienced operator who relates well to people. Because the responsibility for the quality

Figure 4-2. A checklist helps the supervisor make sure all important information is given to the new employee.

The following information should be covered on the new employee's first day:

JOB	PAY
Specific tasks	Pay rates
Responsibilities	Pay days
Performance standards	Deductions

RULES	ABSENCES
Company policies	Sick policy
Department policies	Tardiness policy
Work rules	Reporting sick days
Safety regulations	Vacations
Probationary policy	Holidays
	Breaks
HOURS	Meals
Hours of work	
Days off	**CHAIN OF COMMAND**
Overtime	Protocol
Compensatory time	Who's who
Flex time	Titles

BENEFITS	EQUIPMENT
Medical	Dress code
Life	Uniforms
Disability	Personal protective
Pension	equipment
Savings	

of the training is vested in you, make sure JIT instructors have a detailed knowledge of the job, a strong demonstrated safety record, a strong desire to teach, ability to communicate, and a friendly, cooperative attitude.

Getting Ready to Teach

Before beginning any job training, you will need to determine several issues:

- *What kind of training is needed.* Find out what trainees already know. It is wasteful to provide training that isn't needed.
- *Set a time table.* On the basis of training needs, determine how much time the instruction will take and plan accordingly.
- *Make sure all equipment and supplies are ready.* To provide uninterrupted training, have on hand all materials, supplies, fixtures, and other necessary items.
- *Arrange the workplace properly.* To develop good housekeeping practices right from the start, make sure that everything is in place. Good housekeeping habits impressed on the trainee on the first day can set the pattern for good housekeeping practices on other jobs.

- *Have the key points firmly in mind.* Key points are items that will enable the operator to do the job better, safer, and faster. Instructors can share their on-the-job experience by demonstrating the safe, efficient way to do a job.

Four-Step Instruction

Four-step job instruction was developed during World War II to help people do training who had not been schooled in instructional techniques. This method worked so well that many people who were trained under this approach applied it to their own instructional techniques. Today it is being used in many countries of the world. It is outlined in Figure 4-3 and examined below.

Step 1. Prepare the worker. New employees are especially nervous on the first day of a new job. You or your trainers must put them at ease. Define the job in detail and show the quality standards that must be met. Make sure employees understand that how well the work is done will affect the quality of the finished product. Knowing the importance of their particular job to the finished product can do a lot for new workers' morale and help to reduce first-day tension. It is recommended that the trainer work alongside the trainees rather than across the work bench or machine. This enables the trainees to see the work done exactly as it will be done on the job.

Step 2. Present the operation. The trainer should demonstrate the job one step at a time. Whenever possible, tell the trainees why a step is done in sequence. This is helpful because it is easier to remember steps when you know why they are performed in a particular way. Trainees should be encouraged to ask any questions regarding any phase of the operation. Failure to ask questions could result in accidents. Be sure to stress key points as they come up in the job steps.

Step 3. Try out performance. At this point in the training cycle, the instructor becomes a coach and watches the trainees perform. Have a new employee describe each step as it is being completed. This will enable the instructor to determine whether the training was effective. In this step, patience and empathy are the most needed qualities. Patience is required when trainees learn slowly. Empathy is the ability to put yourself in the other person's shoes. Remember how you were on your first day on the job? Do you remember all the questions you had regarding the operation? The new individual probably has many of the same feelings and same questions. Try to anticipate those questions and be ready to answer them. There's an old saying in job training, "If the trainee hasn't learned, the instructor hasn't taught!"

Step 4. Follow up. When you have observed enough job cycles to be certain that new employees have mastered the operation, it's time to put the final step into operation: let the workers demonstrate what they have learned. Again, express confidence that quality levels will be met (in time, they will be met).

Let trainees know where they can get help if you're not available. Continue to check back with them as often as is necessary. Encourage the trainees to ask questions at any time.

Which of the steps in job instruction training is the most important? Actually, all four are essential to proper training of new operators. This simple, four-step procedure for training new employees helps to:

- Shorten the learning time
- Reduce scrap and rework
- Reduce injuries among new employees

Many supervisors make the mistake of assuming that they don't need to use training techniques with transferred employees. Just because these people have worked in another area of the organization does not mean they are skilled in your operations. Many accidents have occurred because supervisors assumed that transferred workers knew more than they did.

Take advantage of accident prevention techniques. Use the job safety analysis (JSA) to discover and minimize hazards, then follow up with job instruction training (JIT) for both new and transferred people. The safe way is always the right way.

OTHER METHODS OF INSTRUCTION

The methods discussed so far have been primarily those used to teach job skills in on-the-job situations. However, at times you may need to know how to make a presentation or teach in a classroom situation. You might be asked to teach safety proce-

---

Figure 4-3. The four-step job instruction technique has been very successful.

1. Prepare the worker
2. Present the operation
3. Try out performance
4. Follow up

dures as part of an overall course, such as one designed to train employees to become welders or pipe fitters. You may be requested to conduct a periodic safety meeting. On other occasions, you may need to call your group together to explain a new procedure or method. At such times, the lesson plan format will be useful.

The Lesson Plan

You should become familiar with lesson plans. These plans serve as a blueprint for presenting material contained either in a single unit of instruction or in a course outline. In addition to standardizing training, lesson plans help the instructor to:

1. Present material in proper order
2. Emphasize material according to its importance
3. Avoid omitting essential material
4. Run his or her classes on schedule
5. Provide for trainee (student) participation
6. Increase his or her own confidence, especially if the instructor is new at the job

Names for the part of a lesson plan may vary; even the order may not always be the same. The following, however, is a good example of a lesson plan arrangement:

1. Title: Indicates clearly and concisely the subject matter to be taught
2. Objectives:
   a. State what the trainee must know or be able to do at the end of the training period
   b. Limit the subject matter
   c. Are specific
   d. May be divided into a major and several minor objectives for each session
3. Training aids: Include such items as actual equipment or tools to be used, and charts, slides, films, television, etc.
4. Introduction:
   a. Gives the scope of the subject
   b. Tells the value of the subject
   c. Stimulates thinking on the subject
5. Presentation:
   a. Gives the plan of action
   b. Indicates the method of teaching to be used (lecture, demonstration, class discussion, or a combination of these)
   c. Contains suggested directions for instructor activity ("Show chart, "Write key words on chalkboard")
6. Application:
   a. Indicates, by example, how trainees will apply this material immediately (problems may be worked)
   b. Have employees perform a job
   c. Question trainees on their understanding of job procedures
7. Summary:
   a. Restates main points
   b. Ties up loose ends
   c. Strengthens weak spots in instruction
8. Tests: Help determine if objectives have been reached; should be announced to the class at the beginning of the session
9. Assignment: Gives references to be checked or indicates materials to be prepared for future lessons

Programmed Instruction

Programmed instruction, which is simply professionally prepared training materials, may be used as a supplement to classroom and on-the-job training. Through the use of self-contained teaching materials, programmed instruction permits the trainee to learn at his or her own pace and to absorb knowledge in easy-to-take bits. The learning process is reinforced by requiring the trainee to answer questions and to correct errors before progressing with the course.

A number of programmed instruction courses are available in such areas as safety training, vocational training, and communications. Many of these courses use multimedia materials, such as tapes, slides, films, computers, video disks, and TV monitors (interactive video).

A complete list of courses and devices can be obtained from the National Society for Performance and Instruction, Suite 102, 1126 16th Street NW, Washington, D.C. 20036, (202) 861-0777.

Independent Study

Courses offered through correspondence schools are called home study or independent study courses. Independent study courses combine the fundamentals of good training, guidance, and counseling of a qualified instructor with the convenience of studying at home or in supervised study sessions arranged by the company (often on company time).

A list of subjects taught by accredited private home study schools may be obtained from the National Home Study Council, 1601 18th Street, NW, Washington, D.C. 20009, (202) 234-5100. Extension programs of most major universities offer programs of all types through independent study and will furnish complete information upon written request.

Closed Circuit TV

Closed circuit TV (CCTV) training uses television's "instant replay" techniques to teach employees. The basic technique is to record a work procedure and safety directions on videotape and then play them back later on a monitor. Once the process or manual skills, along with the instructions, are recorded, they can be replayed many times. Videotape can be used to record the steps in a job safety analysis, which can then be used to train employees in safe job procedures.

CCTV is extremely flexible; portions of the tape may be shown for review purposes if the job is not performed regularly. This technique makes uniform training possible, because the job is presented in exactly the same manner to each trainee.

JOB SAFETY ANALYSIS

One of the best accident prevention tools supervisors possess is called a job safety analysis, or JSA. You can use this form to break down any job into steps and check each step for associated hazards. Once these hazards are identified, you can develop solutions to eliminate hazards or to guard against accidents (Figure 4-4). Because any positive action you take can be measured using a JSA, this tool can help you to improve your safety performance record.

JSA can be regarded as a type of "hazard hunt" conducted by the two people best qualified to make operations safer: you—the first-line supervisor—and one of your skilled operators. In most instances, the two of you know more about a particular job than anyone else does and can do more to spot hazards.

Benefits of Job Safety Analysis

Fewer accidents. You should have fewer accidents as a result of JSA efforts. Even if you performed only two JSAs a month, you would be able to cover 24 of the most hazardous jobs in your department during the first year in operation.

Employee involvement. By helping to prepare the JSA, skilled employees will realize that you

National Safety Council JOB SAFETY ANALYSIS *INSTRUCTIONS ON REVERSE SIDE*	JOB TITLE (and number if applicable):		DATE:	☐ NEW ☐ REVISED
	PAGE___ OF___ JSA NO.___			
	TITLE OF PERSON WHO DOES JOB:	SUPERVISOR:	ANALYSIS BY:	
COMPANY/ORGANIZATION:	PLANT/LOCATION:	DEPARTMENT:	REVIEWED BY:	
REQUIRED AND/OR RECOMMENDED PERSONAL PROTECTIVE EQUIPMENT:			APPROVED BY:	

SEQUENCE OF BASIC JOB STEPS	POTENTIAL HAZARDS	RECOMMENDED ACTION OR PROCEDURE

Printed in U.S.A.

Figure 4-4. A job safety analysis.

want to hear their ideas and that you care about what they think. Workers will know you recognize the value of their experience on the job. When these individuals are given an opportunity to make working conditions safer for everyone in the department, they will become more involved in safety training and accident prevention. New employees will look upon the completed JSA as training for their jobs. From the first day new employees join the department, they will be aware of your strong concern for accident prevention. In addition, new workers will generally be more willing to offer their ideas to improve operations. As a result, you'll have them thinking safety right away.

Priority of Job Selection

To conduct a job safety analysis, you need to establish a priority for selecting jobs to analyze. What is a logical way to rank them?

First, select the jobs in which the most accidents occurred. Remember to consider all accidents—injury, property damage, and near accidents. If you choose the jobs to be analyzed in this way, you'll reap immediate accident prevention benefits.

Second, consider jobs that have a potential for severe accidents even if no accidents have occurred as yet. It may be that everyone exercises great care when jobs of that type are running. Even so, such hazardous jobs are splendid candidates for job safety analysis.

Third, study newly established jobs carefully. It is quite possible that a new job contains hazards not previously encountered. Perhaps additional or new personal protective equipment is needed. It is also possible that an old hazard was eliminated when the new operation was put into effect. In these instances, you could strive for wider application of the new process.

Any job for which you conduct a safety analysis should be reviewed whenever a method or process changes. Again, look for hazards that may have been introduced or removed by the change.

Breaking the Job into Steps

You and the operator can work together to break a job down into its various steps. The description of each step should begin with a verb ("turn on the saw") and be as brief as possible. Remember, this is not a detailed industrial engineering analysis that covers every tiny movement of the job.

For illustration, see how the job in Figure 4-4 is broken into steps. The first step is to obtain the proper tools or equipment. It should be standard practice not only to select the proper tools but to be certain they are in good condition.

Once the proper tools and equipment have been selected, the remaining steps list the actions required to do the job. When making a JSA, list the job steps in sequence without, for the moment, mentioning hazards or recommended procedures. These topics will be covered later.

In listing the job steps, avoid giving either too much detail or too little. Too much detail will create an unnecessarily large number of steps, while too little can mean some basic steps are omitted. Either extreme is undesirable. Strive for sufficient detail to cover the job and then move on. Number the job steps in sequence. Remember to begin each description with a verb and keep them as short as possible.

When the job steps have been listed, return to the first step and ask the questions: "What hazards exist in this step?" "What possible accidents could occur?" Some JSA forms provide a list of accidents. This helps the supervisor and operator as they consider possible hazards that may occur with each step. Any hazards they envision should be listed in the column headed "Potential Hazards" (shown in Figure 4-5 in the second column). The hazards should be numbered to agree with the step number. If a step has no hazard, write "none" in the appropriate place in Column 2.

As hazards are listed, the JSA team will next consider the "Recommended Action or Procedure" to overcome or minimize risks. If the recommended procedure involves the use of personal protective equipment, enter that in the third column and in the box provided above the sequence column. In the completed JSA (Figure 4-4), the supervisor or the operator will be able to tell at a glance the kinds of personal protective equipment needed.

When listing the Recommended Action or Procedure, be specific. A statement such as, "Be careful" is of no value. Rather, use such recommendations as, "Lift with your legs, not with your back." The Recommended Action or Procedure should be numbered to agree with the Steps and Hazards to which they apply.

After you complete the form, test the procedure to make sure you have listed all jobs and accompanying hazards, then have the JSA typed and duplicated. The completed job safety analysis should always be available at or near the work area where the job is to be performed. In this way, the JSA will best serve its purpose as an accident prevention tool. Some firms have been successful in developing job safety analysis programs for their maintenance operations as well as for production, another indication of the JSA's versatility.

Periodic Review of Job Safety Analysis

Whenever a job is changed or the process is modified, you should review the job's JSA. A new hazard may have been introduced by the change, or an old

INSTRUCTIONS FOR COMPLETING THE JOB SAFETY ANALYSIS FORM

Job Safety Analysis (JSA) is an important accident prevention tool that works by finding hazards and eliminating or minimizing them *before* the job is performed, and *before* they have a chance to become accidents. Use JSA for job clarification and hazard awareness, as a guide in new employee training, for periodic contacts and for retraining of senior employees, as a refresher on jobs which run infrequently, as an accident investigation tool, and for informing employees of specific job hazards and protective measures.

Set priorities for doing JSA's: jobs that have a history of many accidents, jobs that have produced disabling injuries, jobs with high potential for disabling injury or death, and new jobs with no accident history.

Select a job to be analyzed. Before filling out this form, consider the following: The purpose of the job—What has to be done? Who has to do it? The activities involved—How is it done? When is it done? Where is it done?

In summary, to complete this form you should consider the purpose of the job, the activities it involves, and the hazards it presents. If you are not familiar with a particular job or operation, interview an employee who is. In addition, observing an employee performing the job, or "walking through" the operation step by step may give additional insight into potential hazards. You may also wish to videotape the job and analyze it.

Here's how to do each of the three parts of a Job Safety Analysis:

SEQUENCE OF BASIC JOB STEPS	POTENTIAL HAZARDS	RECOMMENDED ACTION OR PROCEDURE
Examining a specific job by breaking it down into a series of steps or tasks, will enable you to discover potential hazards employees may encounter. Each job or operation will consist of a set of steps or tasks. For example, the job might be to move a box from a conveyor in the receiving area to a shelf in the storage area. To determine where a step begins or ends, look for a change of activity, change in direction or movement. Picking up the box from the conveyor and placing it on a handtruck is one step. The next step might be to push the loaded handtruck to the storage area (a change in activity). Moving the boxes from the truck and placing them on the shelf is another step. The final step might be returning the handtruck to the receiving area. Be sure to list *all* the steps needed to perform the job. Some steps may not be performed each time; an example could be checking the casters on the handtruck. However, if that step is generally part of the job it should be listed.	A hazard is a potential danger. The purpose of the Job Safety Analysis is to identify ALL hazards—both those produced by the environment or conditions and those connected with the job procedure. To identify hazards, ask yourself these questions about each step: Is there a danger of the employee striking against, being struck by, or otherwise making injurious contact with an object? Can the employee be caught in, by, or between objects? Is there potential for slipping, tripping, or falling? Could the employee suffer strains from pushing, pulling, lifting, bending, or twisting? Is the environment hazardous to safety and/or health (toxic gas, vapor, mist, fumes, dust, heat, or radiation)? Close observation and knowledge of the job is important. Examine each step carefully to find and identify hazards—the actions, conditions, and possibilities that could lead to an accident. Compiling an accurate and complete list of potential hazards will allow you to develop the recommended safe job procedures needed to prevent accidents.	Using the first two columns as a guide, decide what actions or procedures are necessary to eliminate or minimize the hazards that could lead to an accident, injury, or occupational illness. Begin by trying to: 1) engineer the hazard out; 2) provide guards, safety devices, etc.; 3) provide personal protective equipment; 4) provide job instruction training; 5) maintain good housekeeping; 6) insure good ergonomics (positioning the person in relation to the machine or other elements in such a way as to improve safety). List the recommended safe operating procedures. Begin with an action word. Say exactly what needs to be done to correct the hazard, such as, "lift using your leg muscles." Avoid general statements such as, "be careful." List the required or recommended personal protective equipment necessary to perform each step of the job. Give a recommended action or procedure for each hazard. Serious hazards should be corrected immediately. The JSA should then be changed to reflect the new conditions. Finally, review your input on all three columns for accuracy and completeness. Determine if the recommended actions or procedures have been put in place. Re-evaluate the job safety analysis as necessary.

Figure 4-5. Instructions for completing a job safety analysis.

hazard may have been removed. To be of maximum value as accident prevention tools, JSAs must be kept current. In addition, establish a regular review cycle for the JSAs. Depending on the nature of your operations, this interval could be every 6 to 12 months.

Training New Employees

The job safety analysis can also be used to train new operators because it spells out the job steps, the associated hazards, and the recommended safety procedures. In addition, the personal protective equipment needed is listed at the top of the form. Use the job safety analysis for refresher training on jobs run infrequently, because the risk of accidents is high for this kind of job. Help to prevent those accidents by having operators review the JSA before they start a job that has not been done for some time.

Accident Investigations

If an accident occurs on a job for which a JSA has already been prepared, the form can be most useful during the accident investigation. The JSA will show either that you missed a hazard in your original analysis or that the procedure prescribed was not being followed. In either event, you can take the appropriate action to rectify the situation.

In conclusion, think about the value of using job safety analyses. Consider how much safer your operations can be. Job safety analysis is one of the best existing accident prevention tools.

SUMMARY OF KEY POINTS

Key points covered in this chapter include:

- Safety training is one of the most positive actions you can take as a supervisor to prevent accidents in your department. Employee safety training is mandated by OSHA and many state occupational plans.

- Under the law, employers must provide training and education programs, pertinent information about the job, including proper working conditions, all hazards, symptoms of toxic exposure, and emergency treatment procedures.
- The test of a good safety program is whether employees' new awareness regarding safety results in observable, measurable improvements in job safety performance.
- Studies show that employees fail to follow safety procedures and guidelines because they were not given proper instructions or they misunderstood, disregarded, or minimized the instructions.
- A sound orientation program is one of the best ways to involve new and transferred employees in an accident prevention program from their first day on the job. Orientation sessions should emphasize general company rules and benefits, safety rules and accident prevention programs, and specific hazards and precautions on the job. Employees will realize the company's commitment to safety and will be encouraged to offer their own ideas on how to improve safety measures.
- Job instruction training (JIT) provides on-the-job training for particular tasks to teach employees to do jobs safely and efficiently. In planning a JIT program, supervisors should determine what kind of training is needed, set a time table, have all equipment and supplies on hand, properly arrange the workplace, and have key points firmly in mind.

- JIT instruction is based on a four-step model: Prepare the worker, present the operation, try out performance, and follow up on training. This procedure can shorten learning time, reduce scrap and rework, and reduce injuries among workers.
- Other methods of instruction can be used in classroom or more informal settings. They include developing a lesson plan, making use of programmed instruction, providing independent study resources, and using closed circuit TV instruction.
- Job safety analysis (JSA) is one of the best accident prevention tools supervisors possess. By using a JSA, supervisors can break down jobs into steps, identify the hazards associated with each step, and formulate solutions or precautions to eliminate or guard against these hazards. Benefits of job safety analyses include fewer accidents and greater employee involvement in safety procedures and policies.
- To conduct a job safety analysis, supervisors must prioritize the jobs to be analyzed according to those in which the most accidents have occurred, those with the highest potential for accidents, or those in which procedures are new or have been changed recently. Whenever a job is changed or a procedure is modified, it should be analyzed for potential hazards. List all hazards and recommended actions or procedures in sufficient detail on the form.
- Job safety analyses can be used to conduct periodic reviews of various jobs in a department, to train new or transferred employees, or to aid in an accident investigation.

5

Employee Involvement

After reading this chapter, you will be able to:

- **Understand how an employee safety committee can get workers involved in safety issues**
- **Know how to hold a successful departmental safety meeting**
- **Understand the motivational value of safety recognition organizations, contests, awards, and other measures**
- **Explain the importance of off-the-job safety and accident prevention programs**

Loss control is management's responsibility. As discussed in Chapter 1, top management formulates safety policies that the supervisor is responsible for implementing and enforcing in the workplace. To the line worker, the supervisor *is* the company.

As a supervisor, you must constantly strive to foster good attitudes toward safety in your staff. You should train workers in safe and healthful work practices and convince them that management is sincerely interested in safety. Emphasize that you expect workers to observe safety rules and guidelines.

Promoting safety and loss control depends a great deal on how well you communicate with your employees. As covered in Chapter 2, communication means much more than tacking up posters, passing out rules books, and talking about safety and health. Success in implementing safety and health programs depends on the dedication of top management and the attitudes of both supervisor and employees.

PROMOTING SAFETY AMONG WORKERS

The supervisor must learn techniques to motivate employees to take an interest in safe work practices. Some of these techniques are described below.

Employee Safety Committees

One of the best ways to create and maintain interest is to get employees involved in running a safety committee. An employee committee gives workers an opportunity to make a personal contribution to the overall safety program. The committee must be given time to do its work and should receive acknowledgment for successful efforts. The committee's duties should be spelled out in writing. The success of an employee safety committee is based on certain fundamental principles. These include the following:

1. A genuine need exists within the department to which the employee safety committee can make a contribution.
2. Although the supervisor delegates various functions to the committee, he or she remains responsible for departmental safety.
3. To carry out committee assignments (inspecting, observing practices, investigating, and making recommendations), members must be given proper instructions, goals, and target dates.
4. Supervisors must communicate directly with other members of the department. The committee should not be used as a buffer.
5. All committee recommendations should receive careful consideration. Worthwhile recommendations should be implemented.

When the decision to adopt a recommendation is beyond the supervisor's authority, he or she should contact a higher authority. If a suggestion or recommendation is not accepted, an explanation should be made to the committee.

6. The committee agenda should be limited to safety matters. The committee should not become involved in personnel actions, labor relations, or in other matters not related to accident prevention.

7. Committee membership should be rotated so that all workers have an opportunity to serve.

8. Written minutes of all meetings should be kept.

9. Meetings should be held regularly, and attendance should be required.

The conventional order of business for a safety committee meeting is as follows:

1. Record of attendance
2. Approval of previous minutes
3. Consideration of unfinished business
4. Review of recent accidents, including near misses
5. Report on special assignments
6. Reports of inspections
7. Progress report on safety program
8. Special features, such as a film, a talk, or a demonstration by a specialist, slides, or similar item
9. Presentation of new business

A top management representative should occasionally be invited to a meeting. This person can provide the front office point of view or explain new plans or policies. A summary of meeting content should be sent to management. You should do all you can to make serving on a safety committee interesting, productive, and rewarding for the members.

Departmental Safety Meetings

In many cases, there will not be a sufficient need for an employee safety committee. Instead, you can hold periodic departmental meetings on safety issues. These sessions can help create a team spirit regarding safe production.

How to hold a meeting. The following guidelines can help the supervisor or employee leaders hold a successful meeting. Keep in mind that every meeting should have as its goal motivating workers to take an interest in safety.

1. The meeting should emphasize no more than three main ideas. Most people can absorb only a few ideas at a time.

2. Choose a comfortable location for the meeting. People become irritated and unresponsive when they stand too long, are too warm or cold, look into strong light, listen over background noises, or if they feel ill at ease. Always consider the attendees when choosing the meeting location.

3. Plan the meeting carefully. Have a written agenda to use as a guide for running the meeting. Preview visual aids to be used and check all equipment and exhibits to make sure everything is operating properly. Have speakers rehearse and time their talks.

4. Keep the session short and simple. Unless there is a good reason, such as an absorbing film or some other visual aid to provide interest, a safety session should not run longer than 30 minutes. Only the most exceptionally appealing material warrants a one-hour session, even if the time is available. These guidelines, of course, apply only to group meetings, not to instructional courses.

The first question you need to ask before holding a safety meeting in the department is, "On what subject?" The safety director or safety department should be able to provide a multitude of ideas, either based on company experience or secured from the National Safety Council (see Bibliography in the back of the book).

If you cannot answer the question, "What subject?," either your department has no safety problems or you have not looked closely enough at your work environment. Take a moment to think about your own safety problems and those of your staff. Chances are, you will not need to look long for a subject.

Visual aids. Departmental meetings can feature a film, videotape, chart talk, or a subject related to the group's work or to off-the-job safety. Models, exhibits, standard equipment, and/or safety equipment can be displayed.

You can easily produce or obtain visual aids on a wide variety of subjects for a group or 15 to 20 people. For example, either you or one of your staff can create a homemade flip chart with little trouble or expense. Supplies required include approximately a dozen sheets of white paper (or a large pad) ranging in size from 18 in. × 24 in. to 24 in. × 36 in. (46 cm × 61 cm to 61 cm × 91 cm) that can be clamped to the top of a backing board; some markers (felt-tip, crayon, or colored chalk); and an outline of points to be discussed. Details are given in National Safety Council Industrial Data Sheet 564, *Nonprojected Visual Aids.*

A set of 35mm color slides (or film strips) is a useful training aid. They can be purchased or produced. See details in the National Safety Council Data Sheet 574, *Projected Still Pictures.* The Council offers sets of professionally produced 35mm slides covering a wide variety of industrial safety subjects. Each set has a reading script and many have audiocassettes. Motion pictures are excellent for larger groups, provided the projection equipment and viewing area are available.

Keep in mind that all prepared visual aids are just aids. Although they can be effective training and motivating devices, they are not intended to communicate the entire message. Visual aids must serve as a basis or a reinforcement for discussion or they lose a lot of their value. They should be selected because they illustrate your main ideas, not simply because they are available.

A large meeting involving more than one department is valuable when company policies require explanation or when general accident causes are discussed. Some meetings may be purely motivational, intended to create an awareness of hazards and stimulate accident prevention. Meetings often are held to present safety awards.

Production huddle or tailboard conferences. This type of meeting is an instructional session about a specific job. Although not restricted to safety, it focuses on the problem of safe work production. Supervisors who conduct these sessions and who give proper emphasis to safety may feel that they have had the best kind of safety meeting possible. Safety has now been "built in" to workers' job routine and attitudes, which is the desired goal.

These sessions can be indispensable, especially in highly hazardous situations. Public utility crews often use this kind of meeting before starting a job. They call it a "tailboard conference." The crew gather around their truck and discuss the work, laying out the tools and materials required and choosing which part of the job each person should handle.

Safety Recognition Organizations

Some organizations have been established in the United States and Canada to recognize people who have avoided serious injuries or minimized them by using certain articles of personal protective equipment. Such awards provide an excellent opportunity to publicize safety performance (Figure 5-1).

Four of these safety recognition organizations are:

Wise Owl Program. Founded in 1947, this organization honors industrial employees and students who have saved their eyesight by wearing protective eye equipment. Address inquiries to Coordinator, Wise Owl Program, National Society to Prevent Blindness, 500 Remington Road, Schaumburg, IL 60173, (708) 843-2020.

The Golden Shoe Club . Awards are made to employees who have avoided serious injury because they were wearing safety shoes. Address inquiries to Golden Shoe Club, c/o Hy-Test, Inc., 130 South Canal Street, Chicago, IL 60606, (312) 559-7424.

The Turtle Club. Founded in 1946, this organization honors people who escaped serious head injury because they were wearing a hard hat at the time of their accident. Address inquiries to The Turtle Club, 2680 Bridgeway, Sausalito, CA 94965-1476, (415) 322-0410.

The Golden Belt Club. This award is presented to an employee or employee's family members in recognition of their wearing a safety belt at the time of a motor vehicle accident. Address inquiries to the National Safety Council, Sales Department, 444 North Michigan Avenue, Chicago, Illinois 60611, (312) 527-4800.

Safety Contests

Safety contests are popular with workers and employers alike. Each year, about 5 million workers in the United States participate in the industrial safety contests sponsored by National Safety Council members. More than 9 billion workhours are accounted for annually in these contests. Hundreds of awards are presented.

Safety contests can be one of the most effective ways to create and maintain employee interest in accident prevention. The supervisor who actively supports the company's contests or who starts a department contest is using one of the best-known safety motivations.

Company or industry contests. Competition can be held between departments, plants, divisions, or among a number of companies within an industry. In addition to the nationwide contests sponsored by the National Safety Council, trade associations and local safety councils often have their own contests. However, it must be emphasized that *safety contests are not a substitute for a good safety program.* Safety contests have no other purpose than to create interest and participation in an organized safety program. If they fail to achieve this goal, they are meaningless.

Some contests are based on accident experience and are conducted over a six-month to one-year period. Winners are determined by relative standards, by performance improvement, or by other factors agreed upon in advance. Many effective local con-

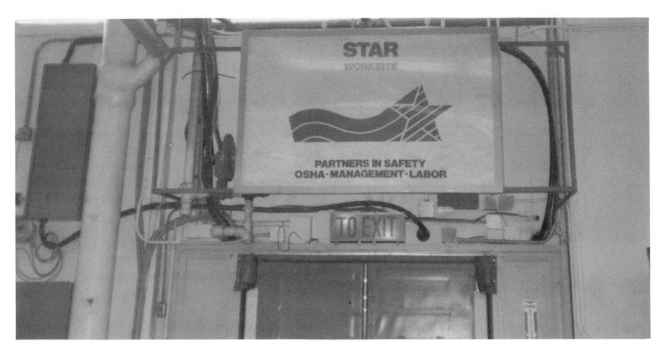

Figure 5-1. Safety recognition organizations publicize safety and renew people's motivation to work safely.

tests are special campaigns that run for a specified time and are launched with advance publicity.

Contests also can be based on many different positive measurements of a safety program:

- Quantity and quality of safety meetings held
- Number of near-accident reports made
- Employee safety suggestions offered
- Job safety analyses completed/revised
- Accident investigation reports turned in complete and on time
- Safety training courses, seminars, programs completed
- Employee safety recommendations made
- Safety inspections held with quality results
- Safety committee involvement maintained

Interdepartmental contests are the most challenging for supervisors. (National intercompany contests usually are directed by safety department personnel.) Since workers are more likely to have a personal interest in the standing of their own department, this type of competition has been most successful. However, when interdepartmental contests are conducted between dissimilar departments in one plant, difficulties created by differences in size, type of operation, and exposure to hazards must be considered. When two or more departments have perfect records, the department with the greatest number of workhours since the last chargeable injury wins.

Whenever possible, every employee on a winning team should receive some sort of recognition or award. This will make everyone a winner and prevent the dissatisfaction that results when only one person receives an award through a lottery or another type of drawing. The best prizes are those that everybody wins—certificates, trading stamps, trophies, plaques—although the prizes themselves may have little intrinsic value.

For contest purposes, the basis for calculating occupational injury and illness incidence rates is given in the publication *A Brief Guide to Record-keeping Requirements for Occupational Injuries and Illnesses* available from the U.S. Department of Labor (see Bibliography).

When departments have wide variations in the number of persons employed and in the kind of work performed, teams of 20 to 50 people can participate in intergroup competition. In order to equalize the variables, each team is made up of a proportionate number of people from high-, medium-, and low-hazard occupations. Workers generally have more interest in the competition when their teams are named after prominent football, baseball, or other sports teams, and the entire competition is named after a league or a sports organization. Team names can be chosen by the group or drawn from a hat. Colored buttons designate team members.

Departmental contests. Less formal contests among departments are also feasible. Contests can be held for good housekeeping, greatest improvement in housekeeping, or for wearing personal protective equipment. Housekeeping contest winners are determined by periodic, unbiased inspections.

Sometimes elaborate point systems are devised so that, in addition to helping win the department trophy, employees can also accumulate points for the department's performance (see Chapter 6, Safety Inspections). Employees can select merchandise from a catalog on the basis of points accumulated. Sometimes trading stamps are used for this purpose.

Recognition. Methods for acknowledging employee cooperation are increasing with the use of personal awards, such as pencils, engraved buttons, billfolds, and similar articles. Some companies also recognize employees who have long-term safety records. They are presented with a more valuable award, such as a watch, and, occasionally, with a testimonial dinner. Since the purpose of contests is to stimulate interest, there is virtually no limit to the ways in which contest and award presentations can be improved by added drama.

Although these activities may sound trivial if you are looking for solid ways to create interest in safety, such contests and awards are highly successful, if properly conducted. One way to make sure that these activities are handled properly is to provide positive recognition. Reward your staff for good safety performances and avoid ridiculing or publicizing poor safety performance.

Safety Posters

Safety posters alert people to safe practices. You can make good use of posters, particularly if you take responsibility for their selection, location, and maintenance.

Poster locations should be selected carefully. Display posters in a prominent location that will not interfere with traffic yet provide high visibility (Figure 5-2). They should be centered at eye level, about 63 in. (1.6 m) from the floor. Place them in well-lighted areas or mount them with their own light, if possible. (Never use a flashing light in a production area.) A good size for the poster board in 22 in. wide × 30 in. long (56 cm × 76 cm). Smaller boards can be used to hold just one poster. Standard National Safety Council posters are available in two sizes: "A" size—8½ × 11 in. (22 cm × 28 cm) and "B" size—17 × 23 in. (43 cm × 58 cm). Poster boards or frames should be attractively painted and covered with glass. One board is usually desirable in a workplace. In washrooms, locker rooms, or lunchrooms, several panels may be used effectively. Materials posted should be displayed separately and kept free of clutter or other notices.

Figure 5-2. Display safety posters where people will see them as they pause in their work routine.

Changing posters and display materials frequently is preferred. Add a few new posters; circulate others. Selection and rotation of posters can be handled by the safety director, safety committee, or human relations department. If you select the posters, choose those that keep employees' interest in safety alive. Posters may also provide notice of forthcoming holidays in addition to safety tips.

Special-Purpose Reminders

The National Safety Council's "POP" (point of problem) posters and safety stickers highlight particular hazards (Figure 5-3). POP posters are 4½ × 5½ in. (10.8 cm × 14.0 cm), and stickers are 2½ × 3¾ in. (6.4 cm × 9.5 cm). Both are self-sticking. Occasionally, a message must be permanently displayed, such as a poster describing correct use of fire equipment or respirators or a warning of radioactive material. These posters should be mounted under glass or on heavy board to make them permanent. However, they should be reviewed regularly, to make sure they are not out of date.

Figure 5-3. Using a wide variety of National Safety Council posters will assure that the message is interesting and noticed.

Suggestion Systems

People often have ideas that can help the department or the company to improve work methods or to reduce work hazards. Unfortunately, these ideas are frequently lost, because there is no effective way to present them.

A well-organized suggestion system encourages employees to contribute ideas and stimulates their thinking about problem solving. Employees may be rewarded by receiving a percentage of the savings resulting from an increase in production or decrease in accidents and injuries. Solutions to safety problems can be rewarded according to the importance of the idea that mitigated injuries.

In an effective suggestion system, a committee receives and evaluates employees' ideas. Some companies establish suggestion departments or use the services of professional organizations.

The company should provide special forms for submitting suggestions and set out special boxes for their collection. Suggestions should be gathered frequently and receipt acknowledged promptly. Management must accept or reject each suggestion as soon as possible. When a suggestion is not accepted, management should explain why to the employees.

Supervisors are usually not eligible to win awards for suggestions. However, you can help employees develop their ideas. Emphasize that suggestions will not be interpreted as criticisms and that no discrimination against the person making the suggestion will occur. You can greatly encourage workers' participation by the ways in which you administer the suggestion system in your own department.

A successful suggestion system must have a clear, precise operating plan that is scrupulously observed. Employees must know that the plan is fair, impartial, and potentially profitable. The results of useful suggestions should be publicized, as well as the names—and photographs, when available—of those whose ideas were accepted.

First Aid Courses

Companies that have conducted good first aid courses have found safety to be a conspicuous by-product. Some companies include first aid in their employee training programs. The value of first-aid training is most obvious in operations that are far removed from professional medical help. In addition, workers find that the skills they learn can be used in off-the-job emergencies.

First-aid training should be standard Red Cross courses or the Bureau of Mines courses. OSHA requirements spell out the need for personnel trained in first aid if there is no infirmary, clinic, hospital, or physician nearby or reasonably accessible to the company (Figure 5-4).

OFF-THE-JOB ACCIDENT PROBLEMS

Off-the-job (OTJ) safety activities help build good employee relations and develop good public relations between the company and the community. These activities are substantial evidence that a company believes in the dignity and worth of each employee. A comprehensive OTJ safety program deserves the vigorous support of every company—large or small—no matter what product or service it produces or how "safe" the company considers itself.

An off-the-job safety problem may be simpler than on-the-job safety, but it is a bigger problem (refer to Table 5-1). In an average year, accidents away from work account for more than 70 percent of all deaths and more than 55 percent of all injuries to workers. As company safety programs become more effective, deaths and injuries at work account for a smaller percentage of total injuries.

These figures are for all industries. In some industries, disabling injuries occur 10 to 20 times more often away from the job than on it. Although many companies are building outstanding safety records, their employees continue to suffer disabling injuries and death when away from work.

The reason employees are safer on the job is

TABLE 5-1. Off-the-job accidental deaths and disabling injuries are far more numerous and more costly to employers than many people expect.

Accident Class	Deaths In 1989	Disabling Injuries In 1989
All Accidents	94,500	9,000,000
Motor-Vehicle	46,900	1,700,000
Public nonwork	42,800	1,500,000
Work-related	3,900	200,000
Home-related	200	Less than 10,000
Work-related	10,400	1,700,000
Nonmotor-vehicle	6,500	1,500,000
Motor-vehicle	3,900	200,000
Home-related	22,500	3,400,000
Nonmotor-vehicle	22,300	3,400,000
Motor-vehicle	200	Less than 10,000
Public	19,000	2,400,000

Reprinted from *Accident Facts—1990 Edition.* Chicago: National Safety Council, 1990.

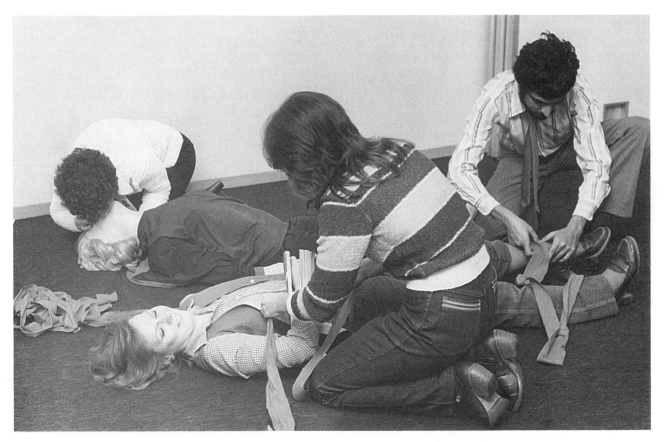

Figure 5-4. Training employees in first aid will help decrease both on-the-job and off-the-job injuries.

because companies continually practice four basic principles of accident prevention:

1. Every effort is made to match the person to the job.
2. Employees are trained and motivated to do their jobs the safe way.
3. Tools, protective equipment, machines, and work areas are kept in first-class condition.
4. Materials are handled according to safe procedures, industrial hygiene is practiced, and machines are safeguarded.

However, when employees leave work for home, they also leave behind a carefully constructed safety network. They are exposed to general hazards on streets and highways, at home, and at play. They are responsible for their own safety. Whether or not they sustain injuries depends on their abilities to supervise themselves and their families.

Companies, on the other hand, have an ongoing responsibility to educate, motivate, and remind employees about the importance of following safe practices on and off the job. This major responsibility is another area in which you must set an example.

Concern with off-the-job safety also represents a sound investment in a company's future. Firms that tackle the total problem of accident prevention claim that safety training is extremely successful, because it reduces the incidence of injuries on and off the job.

The Supervisor's Role in Off-the-Job Safety

Off-the-job safety activities affect the general well-being of employees, the operation of the company's safety program, and the efficiency of your own department. The methods used to promote these activities are the same as those used to promote on-the-job safety.

- As a supervisor, you must set an example for your workers. They will notice if you believe in and practice safety all the time, wherever you may be. You should demonstrate its importance to the well-being of your own family as well as to the health of employees.
- You should determine whether an OTJ injury has kept a person away from work. Encourage

reporting of all accidents, including near misses. These reports can supply topics for safety discussions.

- Part of departmental and committee meetings can be devoted to discussing OTJ safety. You can also talk to people individually before and after meetings.
- Bulletin boards can display vacation and holiday safety hints and newspaper or magazine clippings that may interest employees.
- You can use your company position to support public and home safety activities. Local safety councils, churches, fraternal organizations, and youth groups need safety-minded volunteers to speak at their meetings.

Cost of Off-the-Job Accidents

Off-the-job accidents cost people about $66 billion a year, roughly twice the $33.4 billion price tag of on-the-job accidents. (Figures include property damage, as well as personal losses, wage losses, claims, medical and hospital costs, and administrative expenses.) These accidents affect production output and upset schedules, because skilled workers have been lost.

If replacement workers must be hired, personnel costs, wages spent for training, and reduced production must be added to the OTJ costs. Companies also have to account for tools damaged or materials spoiled by new and inept workers. Finally, wages paid to noninjured persons for nonproductive time, such as court appearances, hospital visits, and clean-up, resulting from an off-the-job accident must be considered.

Nonmonetary costs of OTJ injuries also take a toll. Personal trauma to injured workers is often beyond measure. They must deal with emotional and physical suffering, loss of earning power, and an upset home environment.

In some cases, returning to work does not mean the worker has fully recovered. Mental anguish may be as great a factor as physical impairment in reducing the worker's efficiency. The anxiety such people often suffer can lead to an alarming neglect of safety and result in an increase of on-the-job accidents.

Clearly, attention to OTJ safety has many benefits for workers and companies alike. Safety-conscious employees work more days and are more efficient at work. They spoil fewer materials, do less damage to tools and machines, and have higher rates of production.

Measuring Accident Experience

The off-the-job incidence formula is the standard method of measuring and comparing OTJ accident experience (from ANSI Z16.3–1989—see Bibliography).

$$\frac{\text{Number of OTJ injuries} \times 200,000}{312 \times \text{average number of employees} \times \text{number of months}} = \text{OTJ incidence rate}$$

The formula is based on 312 exposure hours per employee per month. Here is why. An employee normally works eight hours a day, five days a week. The eight hours per day for sleeping are excluded. This gives a person eight hours of exposure per weekday and 16 hours on Saturday and Sunday. This adds up to 72 exposure hours per week multiplied by $4\frac{1}{3}$ weeks per month, for a total of 312 exposure hours per month. No adjustment need be made for overtime since it will be offset by holidays, vacations, and incidental absences.

For example, a plant with 10,000 employees has five OTJ injuries during one month. Each injury causes only one day's absence from work. The OTJ incidence rate for that month will be:

$$\frac{5 \times 200,000}{312 \times 10,000 \times 1} = 0.32$$

To qualify for reporting, an OTJ injury must result in a person's losing at least one full day, either a working day or a holiday, weekend, or vacation. No medical opinion is needed. Fatal or permanent injuries must be included. No scheduled time charges are made for permanent or partial disabilities or fatalities. Only the amount of lost-time injuries and fatalities are reported.

OTJ injuries are classified under three categories: transportation, home, and public.

Transportation. This category includes injuries caused by or resulting from accidents at places other than home and that involve a moving automobile, truck, bicycle, bus, streetcar, motorcycle, railroad, boat, airplane, or pedestrian. This classification also includes injuries that occur when boarding, alighting, or moving within vehicles.

Home. These are injuries occurring at home or in the yard caused by vehicles, firearms, machinery, tools, fire, explosion, exposure to heat or cold, electricity, toxic materials, falls, slips, improper lifting, hot objects or materials, sharp objects, striking an object, overexertion, or animal, insect, or other causes.

6

Safety Inspections

After reading this chapter, you will be able to:

- Understand the importance of conducting continuous and formal safety inspections using a checklist
- Explain the objectives and purposes of an inspection program
- Know how to plan and conduct a safety inspection in your department
- Understand how to inspect work practices in your area
- Know how to complete inspection reports and how to follow up on recommendations made

When should a safety inspection be conducted? Without giving it too much thought, some supervisors would answer, "The third Friday of each month." A much better reply would be, "I conduct safety inspections every time I go through my department." When a safety inspection has become part of your routine, you will have integrated the safety responsibilities of your job with your other duties. One aspect of a supervisory job should not be more important than another.

Safety inspections should be part of every phase of production and a regular element of your standard operating procedure. This type of inspection is called continuous, and it requires that supervisors and their employees constantly be on the lookout for hazards on the job.

FORMAL INSPECTIONS

In addition to continuous inspections, you should make formal inspections once a month, using a checklist (Figure 6-1). (Note: In some operations, you may choose to make inspections more frequently.) These formal inspections can be the foundation for a strong loss control program.

There are three types of scheduled inspections:

- *Periodic inspections.* These include inspec-

tions of specific items that are made weekly, monthly, semi-annually, or at other intervals.
- *Intermittent inspections.* This type of inspection is performed at irregular intervals. Occasionally, an accident in another department that involves equipment similar to the machinery used in your department would lead to an intermittent (special) inspection of your equipment.
- *General inspections.* These inspections are designed to include all areas that do not receive periodic inspection, including parking lots, sidewalks, and fences.

Objective and Purposes

The objectives of an inspection program are to:

- Maintain a safe work environment through hazard recognition and removal
- Ensure that people are following proper safety procedures while working
- Determine which operations meet or exceed acceptable safety and government standards
- Maintain product quality and operational profitability

The basic purpose of safety inspections is to ensure compliance with standards and to serve as a tool to evaluate supervisors' safety performance ac-

Figure 6-1. Supervisors should make formal inspections at least once a month. Areas to check include housekeeping (top), dates of service or maintenance (left), and improper procedures, such as leaving a cover off an electrical box (right). See also the checklist in Figure 6-3.

tivities. In the process of conducting a safety inspection, you may detect potential hazards that require immediate correction or precautions to prevent accidents. Just as inspections of the manufacturing process are important to quality control, safety inspections are vital to loss control (accident prevention).

A safety inspection program should answer the following questions:

- What items need to be inspected?
- What aspects of each item need to be examined?
- What conditions need to be inspected?
- How often must items be inspected?
- Who will conduct the inspection?

Prompt correction of substandard or hazardous conditions detected in an inspection demonstrates to everyone that management is seriously concerned with accident prevention. Also, if you discover that workers are not following safety procedures while performing their jobs, you can take appropriate actions to educate or retrain employees in safety policies and guidelines.

Responsibility for Inspections

Primary safety inspections are your responsibility as first-line supervisor. Because supervisors spend most of their time in their respective departments, they are the people who should be continuously monitoring working conditions.

Completion of a good safety inspection requires that you have: (a) knowledge of your organization's accident experience, (b) familiarity with accident potential and with the standards that apply to your area, (c) ability to make intelligent decisions for corrective action, (d) diplomacy in handling personnel and situations, and (e) clear understanding of your organization's operations—its work flow, systems, and products.

During routine inspections, it is important to check to see that (a) employees are complying with safety rules, (b) no physical hazards exist, (c) aisles and passageways are clear and proper clearances are maintained, and (d) material in-process is properly stacked or stored.

These spot checks emphasize your commitment to safety. Regular formal inspections should also be conducted as frequently as company policy indicates or as conditions dictate.

Working with an Outside Inspector

In addition to OSHA inspections, which are unannounced, sometimes outside professionals (consultants or your company's insurance carrier) are asked to complete an inspection. When this occurs, you should know when the inspectors will arrive so that you can be prepared to answer their questions and offer assistance. Your staff and employees should also be informed of the inspection.

Upon arriving, the inspector should contact you first, so that you can give him or her any information necessary for the inspection. This is especially important when conditions have changed temporarily because of construction, maintenance, equipment downtime, or employee absence.

You may want to accompany the inspector, if there are no rules prohibiting this. However, sometimes the inspector may want to work alone (Figure 6-2).

An inspector is always required to make independent observations, whether or not you participate in the tour. Even if you do not accompany the person, you must be consulted after the inspection to discuss safety recommendations the inspector may propose. The two of you must agree on the importance of each recommendation. Obviously, an inspector should not focus on numerous trivial items just to make a report look good, but any observations about hazardous conditions must be reported.

Once you fully understand the problems and what changes need to be made, you should initiate corrections quickly. Although the inspector's written report must include all hazardous items, it can

Figure 6-2. The inspector may prefer to work alone; the supervisor should be available to answer any questions.

also include a note that you have agreed to make the adjustments as soon as possible. This way the record is clear and will serve as a reminder for the inspector to check these conditions again during the next inspection.

Avoid viewing the report as a personal criticism. Remember, the purpose of inspections is fact-finding, not fault-finding. The process should always be conducted with a high level of professionalism and not be allowed to degenerate into issues of personality. Inspectors should exhibit firm, friendly, and fair attitudes during their work.

You may want the inspector's help in making recommendations for the purchase of new equipment, reassignment of space, or transfer of certain jobs from one department to another. When these suggestions deal with safety, they should be included in the inspector's notes and, possibly, in the report.

Once you have informed your employees about an upcoming inspection, be sure they understand its fact-finding purpose and do not regard it as a personal investigation of their work habits. Explain that the inspector may need to observe them closely while they are working so that he or she can gain a clear understanding of how tasks are performed. An inspector should always ask the employee's permission to watch him or her at work. If this is all done appropriately, the inspection should go smoothly and can be worthwhile for everyone involved.

INSPECTION PLANNING AND PROCEDURES

Timing and Preparation

Inspections should be scheduled when there is a maximum opportunity to view operations and work practices with a minimum of interruptions. Although the areas and routes for inspection should be planned in advance, vary the time and the day on which you conduct a formal inspection so that you are able to check the widest possible variety of conditions.

It is a good idea to review all accidents that have occurred in the area prior to conducting an inspection. In addition to obtaining the regular checklists or formats used, you should have a copy of previous inspection reports. Reviewing these reports makes it possible to determine whether earlier recommendations to remove or correct hazards were followed.

What to Inspect

Many different types of inspection checklists are available for your use. Lists vary in length from hundreds of items to only a few. Each type has its

particular purpose (Figure 6-3 and the following listing). These are some of the items that need to be inspected:

1. Environmental factors (illumination, dusts, gases, sprays, vapors, fumes, noise)
2. Hazardous supplies and materials (explosives, flammables, acids, caustics, toxic materials or by-products)
3. Production and related equipment (mills, shapers, presses, borders, lathes)
4. Power source equipment (steam and gas engines, electrical motors)
5. Electrical equipment (switches, fuses, breakers, outlets, cables, extension and fixture cords, grounds, connectors, connections)
6. Hand tools (wrenches, screwdrivers, hammers, power tools)
7. Personal protective equipment (hard hats, safety glasses, safety shoes, respirators)
8. Personal service and first aid facilities (drinking fountains, wash basins, soap dispensers, safety shoes, eyewash fountains, first aid supplies, stretchers)
9. Fire protection and extinguishing equipment (alarms, water tanks, sprinklers, standpipes, extinguishers, hydrants, hoses)
10. Walkways and roadways (ramps, docks, sidewalks, walkways, aisles, vehicle ways)
11. Elevators, electric stairways, and manlifts (controls, wire ropes, safety devices)
12. Working surfaces (ladders, scaffolds, catwalks, platforms, sling chairs)
13. Materials handling equipment (cranes, dollies, conveyors, hoists, forklifts, chains, ropes, slings)
14. Transportation equipment (automobiles, railroad cars, trucks, front-end loaders, helicopters, motorized carts and buggies)
15. Warning and signaling devices (sirens, crossing and blinker lights, klaxons, warning signs)
16. Containers (scrap bins, disposal receptacles, carboys, barrels, drums, gas cylinders, solvent cans)
17. Storage facilities and areas, both indoor and outdoor (bins, racks, lockers, cabinets, shelves, tanks, closets)
18. Structural openings (windows, doors, stairways, sumps, shafts, pits, floor openings)
19. Buildings and structures (floors, roofs, walls, fencing)
20. Grounds (parking lots, roadways, and sidewalks)
21. Loading and shipping platforms

Figure 6-3. Supervisor's Facility and Administrative Inspection Checklist.

Building/Department _____

This checklist is intended only as a guide in reviewing general facility and administrative items. Only unsatisfactory items and their location need be identified by a check (√). Those items identified as unsatisfactory should be targeted for corrective action.

FACILITY AND OPERATIONS

Machinery and Equipment	Check if Action Required	Location/Comments/Action Required
Equipment in safe operating condition	_____	_____
General safeguarding provided and in place	_____	_____
Operators properly attired (no loose clothing, jewelry)	_____	_____
Point of operation safeguarding provided and functioning properly	_____	_____
Proper tools provided for cleanup and adjustments	_____	_____
Other:	_____	_____

Materials Handling and Storage

	Check if Action Required	Location/Comments/Action Required
Manual materials handling equipment in good condition	_____	_____
Powered materials handling equipment in good condition	_____	_____
Hazardous and toxic materials handled, stored, and transported in accordance to regulatory requirements	_____	_____
Storage areas properly illuminated	_____	_____
Cylinders transported and stored in upright position; properly secured	_____	_____
Shipping/receiving areas in good condition	_____	_____
Racking and other storage procedures followed	_____	_____
Wheel chocks and restraint devices available/functioning properly	_____	_____
Other:	_____	_____

Hand and Portable Power Tools

	Check if Action Required	Location/Comments/Action Required
Correct tools provided	_____	_____
Hand tools and power equipment in good condition	_____	_____
Guards are in place, adjusted properly	_____	_____
Grinding wheel tool rest is within ⅛" of wheel	_____	_____
Stored tools are locked and/or secured	_____	_____
Electric tools GFCI protected	_____	_____
Electric tools and receptacles grounded	_____	_____
Other:	_____	_____

Fire Protection

Portable fire extinguishers:
- Provided as required _____ _____
- Inspected as marked _____ _____
- Location identified _____ _____
- Locations are readily accessible _____ _____

Alarm system tested (as required) _____ _____

Fire doors in good operating condition _____ _____

Exits marked and accessible _____ _____

Fire detectors working _____ _____

Other: _____ _____

Electrical

Outlet boxes covered _____ _____

Electric cords properly placed _____ _____

Outlet circuits properly grounded _____ _____

Portable electric tools:
- GFCI protected _____ _____
- Double insulated _____ _____
- Grounded as required _____ _____

Switches in clean, closed boxes _____ _____

Switches properly identified _____ _____

Circuit fuses, circuit breakers identified _____ _____

Motors are clean, free of oil, grease, and dust _____ _____

Approved extension cords in good condition _____ _____

Other: _____ _____

Housekeeping/Maintenance

Work areas maintained in clean and orderly condition _____ _____

Floors, aisles, work areas free of obstruction, slipping and tripping hazards _____ _____

Washrooms and change facilities clean and well maintained _____ _____

Tools, equipment, and materials properly stored when not in use _____ _____

Waste materials stored in appropriate containers and disposed of in a safe manner _____ _____

Scheduled maintenance:
- General ventilation systems _____ _____
- Local exhaust systems (paint booths, welding areas, etc.) _____ _____
- Machinery (lubrication, belts, servicing, etc.) _____ _____

Other: _____ _____

Personal Protective Equipment

Equipment (determined by exposure)
- Head protection
- Eye protection
- Ear protection
- Foot protection
- Clothes
- Hand protection
- Respiratory protection

Personal protective equipment procedure in place

Other:

Administrative

Training records:
- Safety and health orientation
- Hazard communication (Right-to-Know)
- Safe operating procedures (SOP)
- Confined space entry procedures
- Lockout/tagout
- Evacuation emergency response
- Equipment/vehicle operation
- Fire protection equipment use
- Other:

Plans:
- Disaster preparedness
- Chemical emergencies/spills
- Fire/evacuation
- Emergency medical
- Equipment maintenance
- Other:

Records/reports:
- Injury/illness
- Accidents/incidents
- MSDS's
- Inspection summaries
- Noise surveys
- Equipment service/logs
- Other:

Other:
- OSHA required postings
- Emergency phone listings

- Required labeling   _____   _____
- Defective equipment procedure in place   _____   _____

Completed by: _____

Date: _____

Route to: Maintenance
           Engineering
           Other

---

22. Outside structures (small, isolated buildings)
23. Miscellaneous—any items that do not fit in the preceding categories

Generally, longer checklists are keyed to OSHA standards; such checklists are useful in determining the particular standards or regulations that apply to individual situations. Once the relevant standards are identified, a checklist can be tailored to your needs. In some organizations checklists are computerized for easy follow-up.

Checklists not only serve as reminders of what to look for, but they also serve to document what has been covered in past inspections. They provide direction and permit easy, on-the-spot recording of all findings and comments. If an inspection is interrupted, checklists provide a record of what has already been covered. If you do not have a printed checklist, carry a blank notebook to jot down items during your continuous inspections.

Good checklists also help in follow-up inspections. Remember, however, that a checklist is merely an aid to the inspection process, not an end in itself. If you simply check off items in the list, you are not conducting a safety inspection. A hazard observed during an inspection—even one not included on your list—must be recorded and corrected.

Some items—such as floors, stairways, housekeeping procedures, fire hazards, electrical installations, and chains, ropes, and slings—need particular attention because they represent high-risk areas in a department or company. Be sure to inspect these items often.

Floors. Floors, regardless of their construction, should be carefully inspected, especially slippery floors or those in areas subject to heavy traffic. Here are several items to note:

- Is the surface damaged or wearing out too rapidly?
- Is shrinking present?
- Are there slippery areas?
- Are there holes or unguarded openings?

- Are there indications of cracks, sagging, or warping?
- Are replacements necessary because of deterioration?

Stairways. Always bear in mind that stairs are never to be used for storage. Stairways should be checked to determine whether:

- Treads and risers are in good condition and of uniform width and height
- Handrails are secure and in good condition
- Lighting on stairs is sufficiently bright

Housekeeping. General housekeeping throughout the facility must be checked regularly. Make sure that aisles are marked off with painted lines and kept free of all materials.

Fire protection. Because fire is one of the greatest hazards to an industrial plant, you must pay special attention to fire hazards. Conduct periodic inspections of all fire protection equipment.

Such inspections should review sprinkler systems, alarms, extinguishers, standpipes, hoses, and other equipment. Inspect all exits from the building and all emergency lighting systems. Be sure that the fire protection equipment is in the right place and is not blocked or obscured.

Electrical installations. Electrical installations should be in compliance with the _National Electrical Code_, ANSI/NFPA 70-1990, published by the National Fire Protection Association (see also Chapter 14, Electrical Safety).

Chains, ropes, and slings. Have a qualified expert regularly inspect chains, wire and fiber ropes, and other equipment subject to severe strain in handling heavy equipment and materials. Maintain careful records of each inspection.

INSPECTING WORK PRACTICES

Another important purpose of an inspection is to observe work practices. Are your employees following specific safety procedures and training instructions when doing their jobs?

The best guide to use in determining how a job should be done is a job safety analysis (see Chapter 4, Employee Safety Training). It is strongly recommended that you use these forms when observing your people at work.

Here are some examples of questions you should consider during an inspection.

1. Are machines or tools being used without proper authorizations?
2. Is equipment being operated at unsafe speeds?
3. Are guards and other safety devices being removed or rendered ineffective?
4. Are defective tools or equipment being used? Are tools or equipment being used unsafely?
5. Are employees using their hands or bodies instead of tools or push sticks to manipulate or move items?
6. Is overloading or crowding occurring; are workers failing to stack materials properly?
7. Are materials being handled in unsafe ways; for example, are employees lifting loads or materials improperly?
8. Are employees repairing or adjusting equipment while it is in motion, under pressure, or electrically charged?
9. Are employees failing to use (or using improperly) personal protective equipment and/or other safety devices?
10. Are unsafe, unsanitary, or unhealthy conditions being created by the improper personal hygiene of employees, such as their poor housekeeping, smoking in unauthorized areas, or use of compressed air for cleaning clothes?
11. Are employees standing or working under suspended loads, scaffolds, shafts, or open hatches?

Frequency of Inspections

How often should you conduct inspections? The answer to this question is determined by five factors:

1. *What is the loss severity potential of the problem?* The greater the loss severity potential, the more frequently an item or process should be inspected. A frayed wire rope on an overhead crane block has the potential to cause a much greater loss than a defective wheel on a wheelbarrow. The rope obviously needs to be inspected more frequently than the wheel.
2. *What is the potential for injury to employ-*

ees? If the item or a critical part should fail, how many employees would be endangered and how frequently? The greater the probability for injury to employees, the more often the item should be inspected. For example, a stairway used continuously needs to be inspected more frequently than one that is seldom used.

3. *How quickly can the item or part become hazardous?* The answer will depend on the nature of the part and the condition to which it is subjected. Equipment and tools that get heavy use usually become damaged, defective, or wear out more quickly than those used rarely. Also, an item in one location may be exposed to greater potential damage than an identical item in a different location. The more quickly tools and equipment can become hazardous, the more frequently you should inspect them.

4. *What is the past record of failures?* Maintenance and production records and accident investigation reports can provide (a) valuable information about how frequently items have failed and (b) a description of the results in terms of injuries, damage, delays, and shutdowns. The more frequently a process or equipment has failed in the past and the greater the consequences, the more often that item needs to be inspected.

5. *Are there required inspections?* Some equipment in your department may have to be inspected at regular intervals as mandated by regulation or by a manufacturer's recommendation. When inspections on such equipment are performed, be certain that they are documented properly.

Communicating the Results

It is important to discuss the results of the inspections with your staff or workers. If poor work practices and bad habits have developed, you must advise employees of your observations immediately and explain the correct ways that work should be done.

Always remember to communicate the good news with the bad. Many supervisors fail to mention the positive actions and practices that take place in their departments. It is important to encourage people who follow good work practices. Comments such as, "I'm glad to see that you always check the condition of your tools before using them," give workers positive reinforcement to follow safety procedures.

In a similar vein, if you have found it necessary to correct a person's actions, follow up by acknowledging whenever you see improvement in his or her work patterns. It is always good policy to "communicate the positive" to your people.

INSPECTION REPORTS

A clearly written report must follow each inspection. The report should specify the name of the department or area inspected (giving the boundaries or location, if needed) and the date and time of the inspection.

One way to begin the report is to copy hazards identified on the last report that were not completely corrected. Number each item consecutively. After listing the hazard, specify the recommended corrective action and establish a definite date by which it should be corrected. Record the name of the person responsible for removing the hazard and the date by which the problem should be resolved.

Recording Hazards

The report should contain a description of each hazard uncovered during an inspection. Identify machines and specific operations by their correct names. Describe locations precisely by name or number, and give details about specific hazards. Instead of noting "poor housekeeping," for example, the report should give more details: "Empty pallets left in aisles, slippery spots on the floor from oil leaks, a ladder lying across empty boxes, scrap piled on the floor around machines." Instead of merely noting "guard missing," the report should read, "Guard missing on shear blade of No. 3 machine, SW corner of Bldg. D."

Follow-up

You should follow up inspection reports to correct any problems that have been discovered. If the problem can be resolved by your people, assign someone to correct it promptly. For example, if materials are not properly stacked, ask one or more of your people to stack them safely. If, on the other hand, the condition cannot be corrected by your employees, write a maintenance work-order request to remedy the situation.

You should not only seek to correct hazards but also to take care of the underlying causes. For example, wiping up an oil spill on the floor removes the immediate problem but does not remove the cause. If the oil leak came from a forklift truck, it is essential that the truck be repaired to prevent further oil spills in your own or other departments.

In some cases, taking intermediate action may be the best course. If permanent correction of the problem will take time, consider temporary measures that will help to prevent an accident. Roping off the area, tagging or locking out equipment, or posting warning signs are examples of intermediate actions. They may not be ideal, but they are steps in the right direction. They can prevent injuries or damage while you are working on permanent solutions to the problem.

If in your follow-up work you discover other dangerous conditions, they should be reported to the appropriate person or management immediately. Along with your report, include recommendations for removing or correcting the conditions, even if you can suggest only intermediate steps.

Some of the general categories into which your recommendations might fall include:

- Devise a better process
- Relocate a process
- Redesign a piece of equipment or tool
- Provide personal protective equipment
- Improve training procedures
- Improve maintenance procedures

SUMMARY OF KEY POINTS

Key points covered in this chapter include:

- Safety inspections should be part of every phase of production and a regular element of supervisors' operating procedures. In addition to continuous inspections, supervisors should conduct formal inspections at least once a month, using a checklist. Formal inspections can be periodic, intermittent, or general.
- The objectives of an inspection program are to maintain a safe work environment through hazard recognition and removal, to ensure that people are following proper safety procedures while working, to determine which operations meet or exceed acceptable safety and government standards, and to maintain product quality and operational profitability.
- The purposes of safety inspections are to ensure compliance with standards and to serve as a tool to evaluate a supervisor's safety performance record.
- Safety inspections should determine what items and conditions need to be inspected, how often inspections should be made, and who will conduct the inspections. Supervisors are responsible for safety inspections and must know the organization's accident experience, be familiar with the hazards and safety standards, take corrective action, han-

dle personnel and situations diplomatically, and understand departmental operations thoroughly.

- At times, supervisors may work with outside inspectors. The supervisors should provide all necessary information, be prepared to answer inspectors' questions, and go over any recommendations for improving safety in the department.

- Inspections should be planned during a time when there is a maximum opportunity to view operations and work practices with a minimum of distractions. Reviewing previous inspection reports can help a supervisor determine whether earlier recommendations were followed. Good checklists can help the supervisor conduct a thorough inspection of a job or area.

- Another important purpose of an inspection is to observe work practices. How often such inspections are conducted depends on the loss severity potential of a problem, potential of injury to employees, how quickly an item or part can become hazardous, past record of equipment failures, and whether regular inspections are required.

- The results of an inspection should be communicated to workers and management immediately and corrective action taken. It is important to communicate what workers are doing right as well as what they are doing wrong.

- Clearly written reports must follow each inspection. Reports should specify the name of the department or area, date and time of inspection, specific hazards found, recommended actions or corrections, and date by which the problem should be resolved.

- Supervisors should follow up inspection reports to make sure not only the problems but their underlying causes have been corrected. In some cases, intermediate steps can help prevent an accident if a permanent solution of the problem will take time.

7

Accident Investigation

After reading this chapter, you will be able to:

- Understand the purpose and objectives of accident reporting
- Develop emergency procedures for your department in case of accidents
- Understand the proper steps to take when investigating an accident
- Know how to preserve evidence and how to select and interview accident witnesses
- Understand how to fill out an accident report

Accident investigation is an effective technique for preventing recurring or future accidents. If anything positive results from an accident, it is the opportunity to determine the causes and how to eliminate them. Thorough accident investigation can point out the problem areas within an organization. When these problems are resolved, the result is a safer and healthier work environment.

However, accident investigation involves more than merely filling out forms. Well-managed organizations insist on quality accident investigations, just as they insist on efficient, quality production. Part of your performance evaluation as a supervisor is based on how well you handle this vital part of your job.

platform that is missing toeboards might not hurt anyone. However, it just as easily could have fallen on someone's foot and inflicted a serious injury. Regardless of the outcome, this kind of accident should be reported and the causes investigated. When the causes are removed, serious accidents can be prevented. Immediate action regarding all accidents can prevent future mishaps.

Figure 7-1 shows that accidents are not just events that cause injuries but can be "near misses" or events resulting in property damage as well. As the supervisor, you should have a personal interest in every accident that occurs in your department. In the interest of prevention, you must know and understand the underlying reasons why every accident occurred.

ACCIDENT REPORTING

The basis of a good accident investigation program is an effective, thorough accident system. Many supervisors make the common mistake of reporting and investigating only serious accidents. To help establish an effective accident reporting system, make sure you cover the subject in your new-employee orientation program. Emphasize that *all* accidents must be reported at once—whether they result in personal injuries, illnesses, property damage, or simply near misses. For example, a tool falling off a

FINDING CAUSES

The purpose of accident investigation is to determine the causes and recommend corrective actions to eliminate or minimize these events. Accident investigation should be aimed at fact-finding rather than fault-finding; otherwise, the investigation may do more harm than good. As you investigate, don't put the emphasis on identifying who could be blamed for the accident. This approach can damage your credibility and generally reduce the amount and accuracy of information you receive from work-

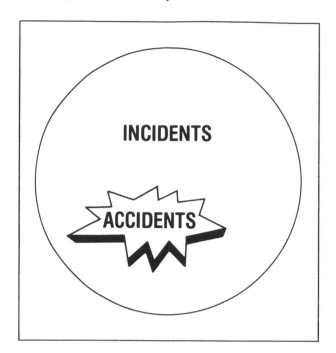

Figure 7-1. Accidents are a part of a group of events that adversely affect the completion of a task. All of the events in this group are incidents; those that do not result in injury and/or property damage are sometimes called "near misses," but they represent significant warnings of potential accidents.

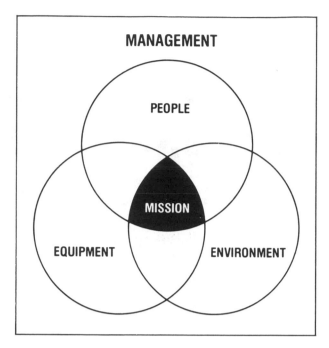

Figure 7-2. This diagram shows a basic system in which people, equipment, and the environment are managed to accomplish a mission. Sometimes they work together in unplanned ways to produce accidents.

ers. This does not mean you ignore oversights or mistakes on the part of employees, nor does it mean that personal responsibility should not be determined, when appropriate. It means that the investigation should be concerned with only the facts. In order to do a quality job of investigating accidents, you must be objective and analytical.

There are many factors in and causes of accidents. The theory of multiple causation states that it is the random combination of various factors that results in accidents. If this is true, additional information about accidents is always needed and helpful.

A report must do more than just identify unsafe acts or hazardous conditions, however. These are only symptoms of, or contributing factors to, accidents. For example, an accident report might explain that the incident was caused by an oily patch on the floor. Instead of merely noting there was oil on the floor, the supervisor must determine how the oil got there. If there was an equipment lubrication leak, had the condition been previously reported? If poor maintenance was the cause, had that been reported? If so, when? If not, why not? These are the kinds of questions that must be asked and answered.

Generally, the methods by which procedures are carried through provide clues to other problems. Accident factors are usually symptoms of other

basic or underlying problems. Inadequate maintenance, poorly designed equipment, untrained employees, and lack of policy enforcement or standard procedures (management control) are all causative factors (Figure 7-2).

EMERGENCY PROCEDURES

A serious accident may not have happened yet in your organization, but it is essential that emergency procedures be established before one does occur. A clear plan for handling accidents can prevent a serious situation from getting further out of control and actually save lives, protect property, and ensure a timely investigation. Management is responsible for developing such a plan, and you are responsible for creating a plan specifically for your operation.

Your emergency procedures plan should specify what should be done in case of an accident. The plan should list key names and phone numbers and assign specific tasks to various workers so that everyone knows exactly what to do in an emergency. You may want to make emergency procedures a topic at your safety meetings.

As the first-line supervisor, you will usually be the first management representative to arrive at the scene of an accident. One of your first steps is to notify the proper authorities and personnel in the company. These include your boss, safety personnel, nurses, and first-aiders. Planning these steps in advance will help make your actions more effective at the time of any accident.

Taking Immediate Action

The safety and health of employees and visitors must be your primary concern when an accident occurs. If an injury or illness results, make sure the affected employee gets immediate medical attention. Take all the necessary steps to provide for emergency rescue. In addition, take any actions that will prevent or minimize the risk of any further accidents happening as a result of the initial emergency.

Securing the Accident Site

After rescue and damage control are complete, secure the accident site for the duration of the investigation. You may need to barricade or isolate the accident scene with ropes, barrier tape, cones, and/or flashing lights that can be used to warn people or otherwise restrict access to the area. In extreme cases, you may want to post guards to make sure no one enters the accident scene.

Nothing should be removed from the accident site without the approval of the person in charge. The site should be maintained, as much as possible, just as it was at the time of the accident to help investigators examine evidence.

Preserving Evidence

Time is of the essence when investigating accidents. The quicker you get to the scene of an accident, the less chance any details will be lost. A prompt and careful investigation of the scene will help to answer the questions of who, where, what, when, how, and why. For example, witnesses will remember more facts, chemical spills will not have had time to evaporate, and dust particles will still remain.

Preserving evidence at the accident scene ultimately makes the investigative process much less frustrating (Figure 7-3). Observing and recording fragile or perishable evidence—such as instrument readings, control panel settings, and details of weather and other environmental conditions—can greatly improve investigation procedures. Evidence can be preserved on film, recorded on tape, diagrammed, or sketched. Detailed notes must accompany any pictures or drawings.

Photography

The camera is one of the most valuable tools for studying accidents. A self-developing camera is often preferred by those with no photographic experience. You can study pictures in detail at your leisure and may notice items you might have overlooked initially. General and specific scenes should be photographed in order to make comprehensive visual records. No one can predict in advance which data will be most useful, so take photographs from many different angles. An old saying for accident photographers is "overshoot and underprint." This means take every possible photo you might need, and make the necessary enlargements after you've studied the proof set. In addition, have someone make accurate and complete sketches and diagrams of the accident scene.

When photographing objects involved in the accident, be sure to identify and measure them to show the proper perspective. Place a ruler or coin next to the object when making a close-up photograph. This technique will help to demonstrate the object's size or perspective. Accurate measurements of the accident area, equipment involved, and materials at hand can be vital to accident investigations, with or without photographs.

EFFECTIVE USE OF WITNESSES

If witnesses are found and are interviewed promptly, they can serve as your best source of information about an accident. The following sections offer some useful tips for finding and interviewing witnesses.

Identifying Witnesses

Avoid restricting your search for witnesses to those who saw the accident happen. Anyone who heard or knows something about the event can offer useful information. Ask witnesses to identify and document the names of others who were in the area, so that everyone can be contacted.

Interviewing Process

Witnesses should be interviewed one at a time and as soon after the accident as possible. The accuracy of people's recall is highest immediately following an event. A prompt interview minimizes the possibility of a witness subconsciously adjusting his or her story. If too much time is allowed to elapse, many things can cloud a person's memory. For example, hearing the opinions of others or reading stories about the accident can influence a witness. A person with a vivid imagination can "remember" situations that did not actually occur. Make sure you set aside time at the beginning of your accident investigation to interview as many witnesses as you can.

Figure 7-3. The supervisor should record all details of the accident scene, including sketches of the scene, as soon after the accident as possible.

Whenever possible, interviews should be conducted at the accident site. This gives witnesses an opportunity to describe and point out what happened. Also, being at the scene can spark a person's memory. Tactful, skilled investigation will usually elicit greater cooperation from employees. Remind the witnesses that you are interested in the facts of the accident, not in placing blame. Such reassurance can dispel people's fear that they may incriminate themselves or others.

How to interview. When conducting a witness interview, establish a relaxed atmosphere. Be a good listener. Witnesses should be allowed to tell their stories without interruption or prompting. More detailed information can be sought after the full story has been told. Interruptions can derail a person's train of thought, influence his or her answers, and inhibit responses. The ways in which you phrase your questions are very important. Ask open-ended questions and avoid leading or putting words in the witness' mouth. For example, asking the question, "Where was he standing?" is preferable to "Was he standing there?" Remember that your pur-

pose is to determine the facts of the accident. You want to gain as much information from the witnesses as possible. As you ask questions, make sure that the specifics of who, what, where, how, and why are included.

Take notes and record employee statements for later review; however, do so as unobtrusively as possible. In some cases, it may be best to wait until the employee has finished explaining what happened before making notes or recording details. After the witness is through, always repeat the information as you heard it. This feedback technique reduces misunderstandings and often leads to further clarifications.

Employees who have been directly involved with the accident should be contacted first, followed by eyewitnesses and those who were nearby. Always use discretion in situations where employees have been injured. Fellow workers might be in shock, unable to speak coherently, or have no memory of the accident. Sometimes you may have to postpone interviews until the employee or a witness who has been involved is in a more stable physical or emotional condition.

It is a good idea for you to solicit ideas from the employees interviewed regarding ways to prevent a recurrence of the accident. Many times they will have the best suggestions. This will also help them to feel involved in the investigation and give you an opportunity to recognize their willingness to participate.

The interview should end on a positive note. Thank each individual for his or her time, for the information supplied, and for the ideas offered.

Reenacting the accident. You may want to ask employees to show you "what they mean or how it happened." In some cases, reenacting an accident can provide valuable information about how and why it occurred (Figure 7-4). Expert investigators have learned to use this technique with caution, however, so the reenactment doesn't cause another accident!

Before asking people to reenact the scene, follow these steps:

1. Ask employees to explain what happened first. This preliminary explanation should give you additional insight about the accident.

2. Make sure that the witnesses thoroughly understand that they are to go through only the *motions* of the accident. They should not repeat the mistakes that caused the original event.

Selecting a location. When it is not possible to interview people at the accident site, chose another location free from distractions and away from other witnesses. Privacy is paramount. When people are discussing what they saw take place, they can influence other witnesses. You want to avoid having people revise their stories. Likewise, the person being interviewed should not be subjected to pressure or influenced by anyone. If possible, do *not* use your office to conduct interviews; employees may find the supervisor's office intimidating, inhibiting, and distracting.

ACCIDENT INVESTIGATION REPORTS

The purpose of accident reporting is to alert and inform management and other concerned people about the circumstances surrounding an accident. The report should record in clear, concise language all the appropriate details of the accident and of the subsequent investigation. Write down all causal factors that might have led to the event. These factors will be in one or more of the following categories: equipment, environment, personnel, and management.

Here are some typical questions to help you identify the accident causes: Was there a safety procedure to detect the hazardous condition? Was the correct equipment, material, or tool readily available? If so, was it used according to established procedure?

Filling Out a Report

The following instructions apply to using the Accident Investigation Report shown in Figure 7-5.

This report is designed primarily for investigation of accidents involving injuries. However, it also can be used to investigate occupational illnesses arising from a single exposure (for example, dermatitis caused by splashed solvent or a respiratory condition caused by the release of a toxic gas). In cases of property damage accidents, simply write "D.N.A" (does not apply) across items 1 through 18 and fill out the rest of the form.

Figure 7-4. Interview witnesses alone and in a private place. If a worker shows you how the accident happened, be sure power is off and all safety procedures are followed so that the accident cannot happen again.

ACCIDENT INVESTIGATION REPORT

CASE NUMBER

COMPANY _____ ADDRESS _____

DEPARTMENT _____ LOCATION (if different from mailing address) _____

1. NAME of INJURED	2. SOCIAL SECURITY NUMBER	3. SEX ☐ M ☐ F	4. AGE	5. DATE of ACCIDENT

6. HOME ADDRESS	7. EMPLOYEE'S USUAL OCCUPATION	8. OCCUPATION at TIME of ACCIDENT

	9. LENGTH of EMPLOYMENT	10. TIME in OCCUP. at TIME of ACCIDENT

11. EMPLOYMENT CATEGORY

9. LENGTH of EMPLOYMENT
☐ Less than 1 mo.   ☐ 6 mos. to 5 yrs.
☐ 1-5 mos.   ☐ More than 5 yrs.

10. TIME in OCCUP. at TIME of ACCIDENT
☐ Less than 1 mo.   ☐ 6 mos. to 5 yrs.
☐ 1-5 mos.   ☐ More than 5 yrs.

11. EMPLOYMENT CATEGORY

☐ Regular, full-time   ☐ Temporary   ☐ Nonemployee

☐ Regular, part-time   ☐ Seasonal

12. CASE NUMBERS and NAMES of OTHERS INJURED in SAME ACCIDENT

13. NATURE of INJURY and PART of BODY

14. NAME and ADDRESS of PHYSICIAN	16. TIME of INJURY	17. SEVERITY of INJURY

14. NAME and ADDRESS of PHYSICIAN

16. TIME of INJURY
A. _____ A.M. P.M.

B. Time within shift

C. Type of shift

17. SEVERITY of INJURY
☐ Fatality
☐ Lost workdays—days away from work
☐ Lost workdays—days of restricted activity
☐ Medical treatment
☐ First aid
☐ Other, specify _____

15. NAME and ADDRESS of HOSPITAL

18. SPECIFIC LOCATION of ACCIDENT

ON EMPLOYER'S PREMISES? ☐ Yes   ☐ No

19. PHASE OF EMPLOYEE's WORKDAY at TIME of INJURY

☐ During rest period   ☐ Entering or leaving plant

☐ During meal period   ☐ Performing work duties

☐ Working overtime.   ☐ Other _____

20. DESCRIBE HOW the ACCIDENT OCCURRED

_____

_____

_____

_____

21. ACCIDENT SEQUENCE. Describe in reverse order of occurrence events preceding the injury and accident. Starting with the injury and moving backward in time, reconstruct the sequence of events that led to the injury.

A. Injury Event _____

B. Accident Event _____

C. Preceding Event #1 _____

D. Preceding Event #2, #3, etc. _____

_____

22. TASK and ACTIVITY at TIME of ACCIDENT

A. General type of task _____

B. Specific activity _____

C. Employee was working:

☐ Alone ☐ With crew or fellow worker ☐ Other, specify _____

23. POSTURE of EMPLOYEE

_____

24. SUPERVISION at TIME of ACCIDENT

☐ Directly supervised   ☐ Not supervised

☐ Indirectly supervised ☐ Supervision not feasible

25. CAUSAL FACTORS. Events and conditions that contributed to the accident. Include those identified by use of the Guide for Identifying Causal Factors and Corrective Actions.

26. CORRECTIVE ACTIONS. Those that have been, or will be, taken to prevent recurrence. Include those indentified by use of the Guide for Identifying Causal Factors and Corrective Actions.

PREPARED BY _____

TITLE _____

DEPARTMENT_____ DATE _____

Developed by the National Safety Council

APPROVED _____

TITLE _____ DATE _____

APPROVED _____ _____

TITLE _____ _____ DATE _____

Figure 7-5. Complete an Accident Investigation Report after you have gathered all possible evidence. See the text for information on how to complete this form.

All questions on this form should be answered. If no answer is available, or the question does not apply, the investigator should indicate this on the form. Answers should be complete and specific. Supplementary sheets can be used for other information, such as drawings and sketches, and should be attached to the report. A separate form should be completed for each employee who is injured in a multiple-injury accident.

The report form meets the recordkeeping requirements specified in OSHA Form 101. The individual entries are explained below.

Department. Enter the department or other local identification of the work area to which the injured is assigned (for example, maintenance shop or shipping room). In some cases, this may not be the area in which the accident occurred.

Location. Enter the location where the accident occurred if different from the employer's mailing address.

1. Name of injured. Record the last name, first name, and middle initial.

2. Social Security Number.

3. Sex.

4. Age. Record the age of the injured at the last birthday, not the date of birth.

5. Date of accident or initial diagnosis of illness.

6. Home address.

7. Employee's usual occupation. Give the occupation to which the employee is normally assigned (for example, assembler, lathe operator, or clerk).

8. Occupation at time of accident. Indicate the occupation in which the injured was working at the time of the accident. In some cases, this may not be the employee's usual occupation.

9. Length of employment. Check the appropriate box to indicate how long the employee has worked for the organization.

10. Time in occupation at time of accident. Record the total time the employee has worked in the occupation indicated in item 8.

11. Employment category. Indicate injured's employment category at the time of the accident (for example, regular, temporary, or seasonal).

12. Case numbers and names of others injured in same accident. For reference purposes, the names and case numbers of all others injured in the same accident should be recorded here.

13. Nature of injury and part of body. Describe exactly the kind of injury, or injuries, resulting from the accident and the part, or parts, of the body affected. For an occupational illness, give the diagnosis and the body part, or parts, affected.

14. Name and address of physician.

15. Name and address of hospital.

16. Time of injury. In part B, indicate in which hour of the shift the injury occurred (for example, 1st hour). In part C, record the type of shift (for example, rotating or straight day).

17. Severity of injury. Check the highest degree of severity of injury. The options are listed in decreasing order of severity.

18. Specific location of accident. Indicate whether the accident or exposure occurred on the employer's premises. Then record the exact location of the accident (for example, at the feed end of No. 2 assembly line or in the locker room). Attach a diagram or map if it would help to identify the location.

19. Phase of employee's workday at time of injury. Indicate what phase of the workday the employee was in when the accident occurred. If "other," be specific.

20. Describe how the accident occurred. Provide a complete, specific description of what happened. Tell what the injured and others involved in the accident were doing prior to the accident; what relevant events preceded the accident; what objects or substances were involved; how the injury occurred and the specific object or substance that inflicted the injury; and what, if anything, happened after the accident. Include only facts obtained in the investigation. Do not record opinions or place blame.

21. Accident sequence. Provide a breakdown of the sequence of events leading to the injury. This breakdown enables the investigator to identify additional areas where corrective action may be taken.

In most accidents, the accident event and the injury event are different. For example, suppose a bursting steam line burns an employee's hands or a chip of metal strikes an employee's face during a grinding operation. In these cases the accident event—the steam line bursting or the metal chip flying up—is separate from the injury event—the steam burning the employee's hands or the chip cutting the employee's face. The question is designed to draw out this distinction and to record other events that led to the accident event.

There also may be events preceding the accident event that, although not accident events themselves, contributed to the accident. These preceding events can take one of two forms. They can be something that happened that should not have happened, or something that did not happen that should have happened. The steam line, for example, may have burst because of excess pressure in the line (preceding event #1). The pressure relief valve may have been corroded shut, preventing the safe release of the excess pressure (preceding event #2). The corrosion may not have been discovered and corrected because a regular inspection and test of the valve was not carried out (preceding event #3).

To determine whether a preceding event should be included in the accident sequence, the investigator should ask whether its occurrence (if it should not have happened) or nonoccurrence (if it should have happened) permitted the sequence of events that brought about the accident and injury events.

Take enough time to think through carefully the sequence of events leading to the injury and to record them separately in the report. The information found in the Finding Causes section, earlier in this chapter, can be used to help identify management system defects that contributed to the events in the accident sequence. By identifying such defects, management may help to prevent other types of accidents in addition to the one under investigation. (See item 25, below.)

For example, the failure to detect the faulty pressure relief valve would lead to a review of all equipment inspection procedures. This review could prevent other accidents that might have resulted from failure to detect faulty equipment in the inspection process.

Additional sheets may be needed to list all of the events involved in the accident sequence.

22. Task and activity at time of accident. In parts A and B, first record the general type of task the employee was performing when the accident occurred (for example, pipe fitting, lathe maintenance, or operating a punch press). Then record the specific activity in which the employee was engaged when the accident occurred (for example, oiling shaft, bolting pipe flanges, or removing material from the press). In part C, check the appropriate box to indicate whether the injured employee was working alone, with a fellow worker, or with a crew.

23. Posture of employee. Record the injured's posture in relation to the surroundings at the time of the accident (for example, standing on a ladder, squatting under a conveyor, or standing at a machine).

24. Supervision at time of accident. Indicate in the appropriate box whether, at the time of the accident, the injured employee was directly supervised, indirectly supervised, or not supervised. If appropriate, indicate whether supervision was not feasible at the time.

25. Causal factors. Record the causal factors (events and conditions that contributed to the accident) that were identified by use of the Finding Causes section earlier in this chapter, discussed under item 21, above.

26. Corrective actions. Describe the corrective actions taken immediately after the accident to prevent a recurrence, including the temporary or interim actions (for example, removed oil from floor) and the permanent actions (for example, repaired leaking oil line). Record other recommended or requested corrective actions.

NOTE: Users may add other data to the form to fulfill local or corporate requirements. Types of data that might be added include:

- Information on accident patterns that are typical of a particular industry or organization. For example, an establishment with many confined-space accidents might wish to add some questions on accidents of that type.
- Information required for special studies. For example, a study tracing the effectiveness of a specific corrective action might be added.
- More detailed severity information, such as the cost of the accident.
- Management data for use in performance reviews and in determining training needs.
- Exposure data for use in calculating incidence rates or injuries associated with certain activities. These data would be estimates or the actual number (or percent) of hours that the employees devote to the activity in a week, month, or year. Incidence rates based on the hours of exposure to specific activities can then be calculated. The use of incidence rates yields a fairer comparison of activities than methods comparing only the total number of cases associated with each activity.

SUMMARY OF KEY POINTS

Key points covered in this chapter include:

- The purpose of accident investigation is to determine the causes and recommend corrective actions to eliminate or minimize these events. All accidents should be investigated, not only those that cause serious injury or property damage. The investigation should emphasize finding facts, not finding fault.
- Supervisors should devise emergency procedures for handling accidents in their departments. The plan should specify what should be done in case of an accident, list key names and phone numbers, and assign specific tasks to workers.
- When an accident occurs, supervisors should see that injured workers get immediate medical attention, then secure the accident site

for the duration of the investigation. Evidence can be preserved on film, recorded on tape, diagrammed, or sketched.

- Witnesses are often the best source of information about an accident. Supervisors should speak with anyone who was in the accident area, not only with those who actually witnessed the event.

- Witnesses should be interviewed one at a time as soon after the accident as possible to ensure accurate recall. The supervisor may ask witnesses to reenact how an accident happened. Interviews should be conducted in a sensitive manner and in a comfortable location to avoid intimidating witnesses or influencing their stories.

- Accident reports are designed to inform management and other concerned people about the circumstances surrounding an accident. Causal factors leading to an accident generally fall into one or more of the following categories: equipment, environment, personnel, and management. These reports must be filled out as completely and thoroughly as possible.

Industrial Hygiene

After reading this chapter, you will be able to:

- Describe various chemical, physical, ergonomic, or biological health hazards commonly encountered on the job
- Recognize an environmental health hazard in one or more of these four categories in your work area
- Understand basic methods of controlling various harmful environmental hazards and stresses
- Understand the concept and purpose of Threshold Limit Values
- Know how to establish standard operating procedures to help ensure a safe, healthy work environment

In addition to safety responsibilities, supervisors—together with management and safety personnel—must make sure that the work area is free from conditions that could be detrimental to health. Consequently, the more you know about industrial hygiene, the better supervisor you will be. This chapter provides information to help you recognize an environmental health hazard in your work area. You can then request assistance from industrial hygienists, who work with medical, safety, and engineering personnel to eliminate or safeguard against such hazards.

Industrial hygienists define their work as "the recognition, evaluation, and control of environmental conditions that may have adverse effects on health, that may be uncomfortable or irritating, or that may have some undesired effect upon the ability of individuals to perform their normal work." It is possible to group these environmental conditions or stresses into four general categories—chemical, physical, ergonomic, and biological stresses. So that you can recognize potential industrial hygiene problems, each category is discussed in some detail in this chapter.

CHEMICAL STRESSES

Chemical compounds in the form of dusts, fumes, smoke, aerosols, mists, gases, vapors, and liquids may cause health problems by inhalation (breathing); by absorption (through direct contact with the skin); or by ingestion (eating or drinking).

Inhalation. The major hazard of employee exposure to chemical compounds is inhalation of airborne contaminants. Contaminants inhaled into the lungs can be classified as gases, vapors, and particulate matter. Particulate matter can be further classified as dust, fumes, smoke, aerosols, or mists.

Absorption. Absorption through the skin can occur quite rapidly if the skin is cut or abraded. Unfortunately, many compounds that exist either in liquid or gaseous form, or both, can be absorbed through intact skin. Some are absorbed through the hair follicles while others penetrate by dissolving into the fats and oils of the skin.

Examples of chemical compounds that can be hazardous by skin absorption are alkaloids; phenols; lead acetate; lead oleate; salts of lead, arsenic, and mercury; nitrobenzene; nitrotoluene; aniline; and nitroglycerine. Other bad actors are triorthocresylphosphate, tetraethyl lead, and parathion and related organic phosphates. Compounds such as toluene and xylene that are good solvents for fats may also be absorbed through the skin, although they are not as hazardous as those mentioned previously because of their lower toxicity. (Toxicity is the ability of a substance to produce disease or physical harm.)

Ingestion. Ordinarily, people do not knowingly eat or drink harmful materials. However, toxic compounds capable of being absorbed from the gastrointestinal tract into the blood—for example, lead oxide—can create serious exposure problems if people working with these substances are allowed to eat or smoke in their work areas. Also, careful and thorough wash-ups are required before eating and at the end of every shift. Workers should change their clothes before leaving work to avoid contaminating their home environment.

Physical Classification of Airborne Materials

Because inhalation of airborne compounds or materials is a common problem, as supervisor you should know the physical classifications of these substances.

Dusts. These are solid particles generated by handling, crushing, grinding, rapid impact, detonation, and decrepitation (breaking apart by heating) of organic or inorganic materials, such as rock, ore, metal, coal, wood, and grain. Dust is a term used in industry to describe airborne solid particles that range in size from 0.1 to 25 μm (μm = 1/10,000 cm = 1/25,000 in.; μm is the abbreviation for micrometer).

A person with normal eyesight can detect individual dust particles as small as 50 μm (micrometers or microns) in diameter. Dust particles below 10 μm in diameter cannot be seen without a microscope. High concentrations of suspended small particles look like haze or smoke.

Dusts settle to the ground under the influence of gravity. The larger the particle, the more quickly it settles. Particles larger than 10 μm in diameter settle quickly while those under 10 μm remain suspended in air for much longer. These smaller particles, called "respirable dusts," can penetrate into the inner recesses of the lungs. Nearly all the particles larger than 10 μm in diameter are trapped in the nose, throat, trachea, or bronchi from which they are either expectorated or swallowed.

Some larger-sized particles can also cause difficulty, however. Ragweed pollen, which ranges from 18–25 μm in diameter, can trigger an allergic reaction known as hay fever when particles enter the upper respiratory tract. Other allergenic dusts, as well as some bacterial and irritant dusts, can also cause respiratory problems in workers exposed to them.

Dust may enter the air from various sources. It may be dispersed when a dusty material is handled—for example, when lead oxide is dumped into a mixer, when talc is dusted on a product, or where asbestos-containing acoustical and/or fireproofing materials are being removed (Figure 8-1). When solid materials are reduced to small sizes in such processes as grinding, crushing, blasting, shaking, and drilling, the mechanical action of the grinding or shaking device can disperse the dust formed. Also, when dusty materials are transported, dust may be dispersed to other plant areas, unless good controls are used to remove or reduce this hazard.

Fumes. Fumes are formed when volatilized solids, such as metals, condense in cool air. The solid particles that make up fumes are extremely fine, usually less than 1.0 μm. In most cases, the hot material reacts with the air to form an oxide.

Examples are lead oxide fumes from smelting and iron oxide fumes from arc welding. Fumes also can be formed when a material such as magnesium metal is burned or when welding or gas cutting is done on galvanized metal. Gases and vapors are not fumes, even if newspaper reporters often (incorrectly) call them that.

Smoke. This hazard is created when carbon or soot particles less than 0.1 μm in size result from the incomplete combustion of such carbonaceous materials as coal or oil. Smoke generally contains liquid droplets as well as dry particles. Tobacco, for instance, produces a wet smoke composed of minute tarry droplets. The size of the particles contained in tobacco smoke is about 0.25 μm.

Aerosols. Liquid droplets or solid particles fine enough to be dispersed and to remain airborne for some time are called aerosols. If inhaled, these can irritate or injure workers' mucus membranes; eyes, noses, and throats; and lungs.

Mists. Mists are suspended liquid droplets generated by chemicals condensing from the gaseous to the liquid state or by a liquid breaking into a dispersed state by splashing, foaming, or atomizing. Mist is formed when a finely divided liquid is suspended in the atmosphere. Examples are the oil mist produced during cutting and grinding operations, acid mists from electroplating, acid or alkali mists from pickling operations, spray-paint mist from spraying operations, and condensation of water vapor into fog or rain.

Gases. Normally gases are formless fluids that occupy the space or enclosure in which they are confined and that can be changed to the liquid or solid state only by the combined effect of increased pressure and decreased temperature. Gases spread out, or diffuse, into the surrounding atmosphere easily and readily. Examples are welding gases, internal combustion engine exhaust gases, and air.

Vapors. The gaseous forms of substances that appear normally in the solid or liquid state (at room temperature and pressure) are called "vapors." The

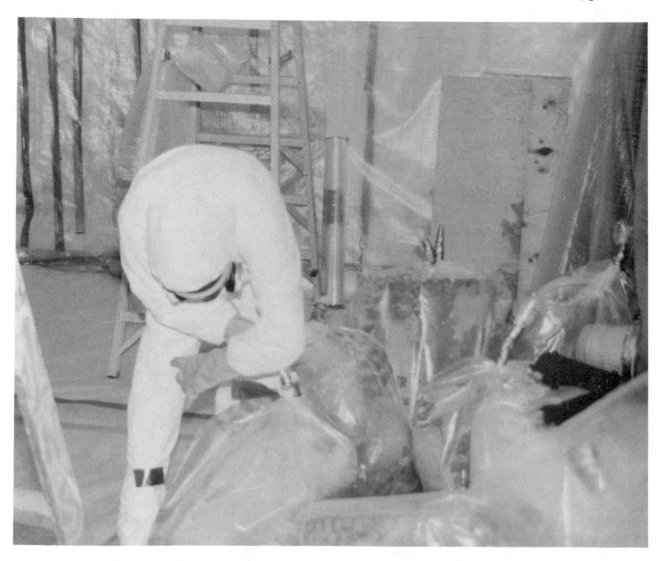

Figure 8-1. To prevent dispersion of asbestos-containing dusts, a worker in an asbestos removal enclosure double-bags all materials, taping each bag securely before disposal. Note that the worker is wearing protective clothing and a filter respirator.

vapor can be changed back to the solid or liquid state either by increasing the pressure or by decreasing the temperature. Evaporation is the process by which a liquid is changed into the vapor state and mixed with the surrounding atmosphere. Solvents that boil at relatively low temperatures—for example, acetone—will vaporize (evaporate) readily at room temperature.

Hazards involved. The hazard associated with breathing a gas, vapor, or mist usually depends upon the solubility of the substance. For example, if the compound is very soluble—such as ammonia, formaldehyde, sulfuric acid, or hydrochloric acid—it is rapidly absorbed in the upper respiratory tract and does not penetrate deeply into the lungs. Consequently, the nose and throat become so irritated

that a person is driven out of the exposure area before he or she is in much danger from the toxicity of the gas. Nevertheless, exposures even for brief periods to high concentrations of these compounds can produce serious health effects.

Compounds that are not soluble in body fluids cause considerably less pain than the soluble ones, but they can penetrate deeply into the lungs. Thus, a serious hazard can be present but not be immediately recognized. Examples of such gases are nitrogen dioxide and phosgene. The immediate danger from these compounds in high concentrations is acute edema or, possibly later, pneumonia or circulatory impairment.

However, numerous chemical compounds do not follow the general solubility rule. They are not especially soluble in water and yet are irritating to the eyes and respiratory tract. They can also cause lung damage and even death under the right conditions. One example is acrolein, which belongs to the chemical family called aldehydes.

Particulates

To evaluate particulate exposures properly, you must know the chemical composition, particle size, dust concentration in air, method of dispersion, and many other factors described in this section. With the exception of certain fibrous materials, dust particles must usually be smaller than 5 μm in order to enter the alveoli or inner recesses of the lungs. Although a few particles up to 10 μm in size may enter the lungs occasionally, nearly all the larger particles are trapped in the nasal passages, throat, larynx, trachea, and bronchi, from which they are expectorated or swallowed and enter into the digestive tract.

Dusts. Most industrial dusts consist of particles that vary widely in size, with the small particles greatly outnumbering the large ones. Consequently, with few exceptions, when dust can be seen in the air around an operation, there are probably more invisible dust particles than visible ones present. The main hazard to personnel occurs when dust becomes airborne. Also, airborne dusts can be flammable and potentially explosive.

A number of occupational diseases result from exposures to such nonmetallic dusts as asbestos. Although specialized knowledge and instruments are needed to determine the severity of a hazard, you should be trained to recognize a hazard and to ask for expert help in controlling it.

A process that produces dust fine enough to remain suspended in the air should be regarded as hazardous until proven safe. An air-monitoring survey of airborne chemicals present in the workplace will determine employee exposure levels and the overall relative safety. Processes that can generate excessive exposure include abrasive blasting or machining, bagging and handling of dry materials, ceramic coating, dry mixing, metal forming, grinding, and metalizing. Consider, for example, free silica, which can cause a damaging lung disease, silicosis. Silica dust is produced in hard-rock mines and by quarrying and dressing granite. Grinding castings that contain mold sand also represent a silica hazard.

Methods of drilling rock with power machinery produce more dust than old-fashioned hand methods. This dust, however, is controlled by applying water to the drill bit so that the dust forms a slurry instead of being suspended in air.

Processes in which materials are crushed, ground, or transported are also potential sources of dust. They should either be controlled by use of wet methods or should be enclosed and ventilated by local exhaust. Points where conveyors are loaded or discharged, transfer points along the conveying system, and heads or boots of elevators should be enclosed and, usually, exhaust-ventilated.

You must be on the alert to see that someone does not cancel the effectiveness of built-in dust controls by tampering with them or by using them improperly. You must also insist that required respiratory equipment be worn by those workers who need supplementary protection.

Fumes. Welding, metalizing, and other hot operations produce fumes, which may be harmful under certain conditions. For example, arc welding volatilizes metal that then condenses—as the metal or its oxide—in the air around the arc. In addition, the rod coating is in part volatilized. Because they are extremely find, these fumes are readily inhaled.

Highly toxic materials, such as those formed when welding structures painted with red lead or when welding galvanized metal, may produce severe symptoms of toxicity rather rapidly. Fumes should be controlled with good local exhaust ventilation or by protecting the welder with respiratory equipment (Figure 8-2).

When pouring brass, zinc volatilizes from the molten mass and oxidizes in the surrounding air to produce a zinc fume, which (in high concentrations) may produce the rather nonspecific disease known as "metal fume fever." If lead is present, it also will become airborne. As a result, brass foundry workers may incur either chronic lead poisoning if the lead is present only in minute amounts or acute poisoning if it is present in substantial amounts.

Most soldering operations, fortunately, do not require temperatures high enough to volatize an appreciable amount of lead. However, some of the lead in the molten solder is oxidized by contact with the air at the surface. If this oxide, often called dross, is mechanically dispersed into the air, it may produce a severe lead-poisoning hazard.

In operations where this condition might happen—for example, soldering or lead battery making—preventing occupational poisoning is largely a matter of scrupulous housekeeping to prevent the lead oxide from dispersing into the air. You can enclose melting pots, dross boxes, and similar operations, and provide exhaust ventilation.

Gases. Gases are used or generated in many industrial processes that often produce toxic waste gases. For example, welding in the presence of chlorinated solvent vapors (from an open tank or degreaser) can produce phosgene, a very toxic gas that

Figure 8-2. This exhaust hood over the die head of a plastics extruder effectively reduces harmful fumes and gases close to the source.

causes respiratory distress and damage. Propane-operated forklift trucks or any process or equipment that burns fuel or other organic materials has the potential to generate carbon monoxide. Carbon monoxide prevents the body from absorbing oxygen, causing headache, nausea, confusion, dizziness, and in severe cases, coma and death. In addition, carbon monoxide and other toxic gases like hydrogen sulfide and methane may be found in confined spaces where organic materials have deteriorated.

You should sample the air for toxic chemicals and test for oxygen and flammable gases before allowing workers to enter any confined space. Many gases are heavier than air and thus may remain present even in confined spaces open to the atmosphere.

Many gases are odorless and colorless, which makes detection unlikely unless appropriate air-sampling equipment is used.

Air sampling may need to be conducted around workers and process equipment to make certain that hazard controls are working. For example, you may need to sample for ozone and other hazardous gases to document a welder's exposure. Industrial hygiene, safety, or engineering personnel should provide you with information on the gases that might be used or generated in your department.

Respiratory hazards. When a respiratory hazard exists or is suspected, the actual airborne concentration of the air contaminant(s) must be measured by an industrial hygienist. While conditions are sometimes similar in different plants within an industry, the degrees of respiratory hazard must be assessed by scientifically valid methods, such as air sampling. Some typical industrial processes and the respiratory hazards that might result are listed in Table 8-1.

You must also be aware of the hazards of oxygen deficiency. Oxygen deficiency results when the atmosphere in question contains less than the normal amount of oxygen found in the atmosphere, about 21 percent. An environment is immediately hazardous to life and health when the oxygen level is 16 percent or lower.

The degree of hazard involved is also an important factor when analyzing respiratory conditions. Some hazards, such as gases and vapors, can produce an immediate threat to life and health when present in high concentrations. On the other hand, oxygen deficiency, by its very nature, is automatically dangerous to life and health.

Respiratory Protection

It is important that you understand the basic concepts of respiratory protection if you are to protect your workers properly. A competent industrial hygienist or safety professional should decide what kind of respiratory protection is needed. However, the more you know about respiratory hazards, and the types and selection of respirators, the better prepared you will be to make sure that employees are protected. Selection of the proper respirator is discussed in Chapter 9, Personal Protective Equipment. See the National Safety Council's Data Sheet 734, *Respiratory Protective Equipment*.

Liquid Chemicals: Solvents

Liquid chemicals are typically used as feed stock, fuel or fuel additives, pesticides, lubricants, detergents and cleaning agents, or as degreasing or processing solvents. Solvents are perhaps the most widespread class of chemicals in manufacturing. Many of these solvents evaporate readily in the air; therefore, their use can pose real exposure problems.

Solvents are usually further categorized as aqueous or organic. *Aqueous solvents* are those that readily dissolve in water. Many acids, alkalis, or detergents, when mixed with water, form aqueous solvent systems.

The term solvent, however, is commonly used to mean *organic solvents*. Many of these chemicals do not mix easily with water but do dissolve other organic materials, such as greases, oils, and fats. Important types of organic solvents include aliphatic, cyclic, aromatic, halogenated, esters, ketones, alcohols, ethers, glycols, aldehydes, hexane,

gasoline, turpentine, benzene, trichloroethylene, freons, ethyl acetate, acetone, formaldehyde, methanol, ethyl ether, and ethylene glycol.

Organic solvents generally have some effect on the central nervous system. They may cause nervous system depression, in which the victim experiences short-term (acute) dizziness, feelings of intoxication and nausea, and a decrease in muscular coordination. Higher levels of exposure may cause loss of consciousness, coma, and, in some cases, death.

Other effects of solvents vary. Some can cause long-term damage to the liver or other organs, or affect the worker's reproductive ability. A few have been found to cause cancer. Some research suggests that long-term exposure to low levels of certain solvents may cause long-lasting effects, including chronic headaches, inability to concentrate, memory losses, nervous disorders such as nervous tics, and reduced nervous system and muscular functioning.

As the supervisor, you can assist the hygienist and safety professional in controlling exposure to such chemicals. Substantial exposures, fortunately, can be controlled—spray-painting booths can be ventilated and degreasing tanks can be exhausted. These will not ordinarily present serious hazards as long as the equipment works. It is the job of the supervisor and engineering, human relations, and safety personnel to make sure that controls and personal protective equipment are properly maintained and are used at all times. Even small exposures on jobs that come up infrequently or that involve small amounts of solvents not covered by standard operating procedures can produce significant exposure hazards.

The point to remember is not how much solvent is used at the job site, but the actual degree of exposure by inhalation or by skin absorption. A close check must be kept on all minor uses of solvents. These chemicals should be issued only after you have determined that they can be used properly and safely. Industrial health authorities or your safety professional can provide information about which solvents to use and for what purposes. It is up to you to see that this information is used and that employees follow the recommendations.

Selection and handling. Getting the job done without hazard to employees or property is dependent upon the proper selection, application, handling, and control of solvents and an understanding of their properties. A good working knowledge of the nomenclature and effects of exposure to solvents is helpful in making a proper assessment of damage or harm.

Keep in mind, however, that nomenclature can be misleading and confusing. Consider, for example,

TABLE 8-1. Potentially Hazardous Operations and Air Contaminants

Process Types	Contaminant Type	Contaminant Examples
Hot operations		
Welding	Gases (g)	Chromates (p)
Chemical reactions	Particulates (p)	Zinc and compounds (p)
Soldering	(dusts, fumes, mists)	Manganese and compounds (p)
Melting		Melting oxides (p)
Molding		Carbon monoxide (g)
Burning		Ozone (g)
		Cadmium oxide (p)
		Fluorides (p)
		Lead (p)
		Vinyl chloride (g)
Liquid operations		
Painting	Vapors (v)	Benzene (v)
Degreasing	Gases (g)	Trichloroethylene (v)
Dipping	Mists (m)	Methylene chloride (v)
Spraying		1,1,1-trichloroethylene (v)
Brushing		Hydrochloric acid (m)
Coating		Sulfuric acid (m)
Etching		Hydrogen chloride (g)
Cleaning		Cyanide salts (m)
Dry cleaning		Chromic acid (m)
Pickling		Hydrogen cyanide (g)
Plating		TDI, MDI (v)
Mixing		Hydrogen sulfide (g)
Galvanizing		Sulfur dioxide (g)
Chemical reactions		Carbon tetrachloride (v)
Solid operations		
Pouring	Dusts (d)	Cement
Mixing		Fibrous glass
Separations		
Extraction		
Crushing		
Conveying		
Loading		
Bagging		
Pressurized spraying		
Cleaning parts	Vapors (v)	Organic solvents (v)
Applying pesticides	Dusts (d)	Chlordane (m)
Degreasing	Mists (m)	Parathion (m)
Sand blasting		Trichlorethylene (v)
Painting		1,1,1-trichloroethane (v)
		Methylene chloride (v)
		Quartz (free silica, d)
Shaping operations		
Cutting	Dusts (d)	Asbestos
Grinding		Beryllium
Filing		Uranium
Milling		Zinc
Molding		Lead
Sawing		
Drilling		

(Reprinted with permission from *Occupational Exposure Sampling Strategy Manual*, NIOSH Pub. No. 77-172.)
Abbreviations: d = dust, g = gases, m = mists, p = particulates, v = vapors.

the two solvents benzine and benzene. The are spelled and pronounced nearly the same, yet their toxicity varies widely. The distinction is especially important because benzene is considered a carcinogen (a substance that tends to produce cancer).

Another nomenclature problem is the similarity of names for the various kinds of solvents in the chlorinated hydrocarbon family. Perchloroethylene, trichloroethylene, trichloromethane, and dichloromethane are a few examples of such solvents that have similar-sounding names, yet each possesses its own characteristic hazards.

Hazard communication. Many state regulations and the federal hazard communication standard require that management provide information about chemical hazards to the workforce. Your role as a supervisor will probably include giving some of this information to your employees. Many of these regulations require:

- An inventory and assessment of chemical hazards in the workplace
- Development and use of labels that describe the hazards of chemicals and the protective measures to use
- Material Safety Data Sheets (MSDSs) that detail chemical hazard and precaution information
- Training on identifying hazards, including specific chemicals or groups of chemicals with which employees work
- A written program that describes how the company intends to accomplish these tasks and provides documentation that workers have been trained

Many companies have developed programs in which in-house or outside experts are used to inform employees about general classes of chemicals. Supervisors and other departmental personnel are often expected to provide information on specific chemical hazards found in the department, and the measures that should be used to control the hazards. Such measures could include engineering controls (ventilation or isolation of a hazardous substance); work-practice controls such as housekeeping programs; or personal protective equipment programs. The company should train you to recognize the hazards and to tell your workers how they can best protect themselves.

If you find a new chemical for which you have not been trained, or which lacks labels, you will need to discuss it with your industrial hygienist, safety professional, or chemist, as well as with the administrator of your hazard-communication program. You should alert your purchasing department if you have received an unlabeled container. Purchasing will need to notify suppliers that materials cannot be accepted without appropriate labels and MSDS.

Degree of hazard severity. The severity of hazard in the use of organic solvents depends on the following facts:

- How the solvent is used
- Type of job operation (determines how the workers are exposed)
- Work pattern
- Duration of exposure
- Operating temperature
- Exposed liquid surface
- Ventilation efficiency
- Evaporation rate of solvent
- Pattern of air flow
- Concentration of vapor in workroom air
- Housekeeping

The solvent hazard, therefore, is determined not only by the toxicity of the solvent itself, but by the conditions of its use—who, what, how, where, and how long. Precautionary labeling and/or MSDS should indicate the major hazards and safeguards (Figure 8-3).

For convenience, job operations employing solvents may be divided into three categories:

1. *Direct contact.* This is a consequence of hand operations. Emergency repair of equipment, spraying or packaging volatile materials without ventilation, cleanup of spills, and manual cleaning using cloths or brushes wetted with solvent are examples of jobs in which employees may have direct contact with the solvent.

Remember that many solvents can penetrate the skin and be absorbed into the body. Solvents are a leading cause of industrial dermatitis or skin diseases. It also is important to note that many solvents can penetrate various glove materials. Thus, properly selected gloves and clothing must be used to prevent solvents from soaking through and coming into direct contact with the skin.

2. *Intermittent or infrequent contact.* This occurs when the solvent is contained in a semi-closed system where exposure can be controlled. Examples are paint spraying in an exhaust-ventilated spray booth, vapor degreasing in a tank with local lateral slot exhaust ventilation, and charging reactors or kettles in a batch-type operation in which

METHANOL

POTENTIAL HAZARDS

HEALTH HAZARDS
Poisonous; may be fatal if inhaled, swallowed or absorbed through skin.
Contact may cause burns to skin and eyes.
Runoff from fire control or dilution water may cause pollution.

EMERGENCY ACTION

Keep unnecessary people away; isolate hazard area and deny entry.
Stay upwind; keep out of low areas.
Self-contained breathing apparatus and chemical protective clothing which is specifically recommended by the shipper or producer may be worn but they do not provide thermal protection unless it is stated by the clothing manufacturer. Structural firefighter's protective clothing is not effective with these materials.
Isolate for 1/2 mile in all directions if tank car or truck is involved in fire.
CALL CHEMTREC AT 1-800-424-9300 FOR EMERGENCY ASSISTANCE. If water pollution occurs, notify the appropriate authorities.

FIRE
Small Fires: Dry chemical, CO_2, Halon, water spray or standard foam.
Large Fires: Water spray, fog or standard foam is recommended.
Move container from fire area if you can do it without risk.
Dike fire control water for later disposal; do not scatter the material.
Cool containers that are exposed to flames with water from the side until well after fire is out. Stay away from ends of tanks.

FIRE OR EXPLOSION
Flammable/combustible material; may be ignited by heat, sparks or flames.
Vapors may travel to a source of ignition and flash back.
Container may explode in heat of fire.
Vapor explosion and poison hazard indoors, outdoors or in sewers.
Runoff to sewer may create fire or explosion hazard.

Withdraw immediately in case of rising sound from venting safety device or any discoloration of tank due to fire.

SPILL OR LEAK
Shut off ignition sources; no flares, smoking or flames in hazard area.
Do not touch spilled material; stop leak if you can do it without risk.
Water spray to reduce vapors; but it may not prevent ignition in closed spaces.
Small Spills: Take up with sand or other noncombustible absorbent material and place into containers for later disposal.
Large Spills: Dike far ahead of spill for later disposal.

FIRST AID
Move victim to fresh air and call emergency medical care; if not breathing, give artificial respiration; if breathing is difficult, give oxygen.
Remove and isolate contaminated clothing and shoes at the site.
In case of contact with material, immediately flush skin or eyes with running water for at least 15 minutes.
Keep victim quiet and maintain normal body temperature.
Effects may be delayed; keep victim under observation.

W.H. BRADY CO., SIGNMARK® DIV. CATALOG NO. 93565

Figure 8-3. An MSDS must list the potential hazards, required safeguards, and emergency actions.

the worker is exposed only at infrequent intervals.

3. *Minimal contact.* This is characterized by remote operation of equipment totally isolated from the work area. This type of operation includes directing chemical plant operations from a control room, mechanical handling of bulk-packaged materials, and other operations where the solvent is contained in a closed system and is not discharged to the atmosphere in the work area.

It is unfortunate that the term "safety" solvent has been applied to some proprietary cold degreasers, because the term is not precise and subject to various interpretations. For example, a "safety" solvent may be considered by some users as nondamaging to the surfaces being cleaned. Other users may consider it to be free from fire or toxicity hazards. Depending on the conditions of use, neither of these criteria may be met by a so-called "safety" solvent.

These solvents are prepared as mixtures of halogenated hydrocarbons and petroleum hydrocarbons to be used for cold degreasing. Although the halogenated hydrocarbons are effective grease and oil solvents and generally are not flammable, they are relatively expensive and may be toxic under adverse conditions. The petroleum hydrocarbons are effective solvents and are inexpensive, but they are flammable (flash points below 100 F) and can also be toxic.

To combine the best qualities of each solvent, manufacturers mix them in an attempt to produce a cold degreaser that has a flash point higher than that of the flammable petroleum hydrocarbon by itself. Such a mixture, however, can present both fire and toxicity hazards, depending on the evaporation rates of the solvents used. If the flammable liquid is more volatile than the nonflammable solvent, the vapors from the mixture can be highly flammable. Conversely, if the nonflammable solvent is more volatile, it can evaporate to leave a flammable liquid.

You should always keep these considerations in mind when using such solvents. Never forget that, under the right conditions, the vapor from a "safety" solvent can be just as deadly and flammable as the vapor from an extremely toxic solvent. If there is a chance that workers might be exposed to these hazardous substances in the workroom air, check the area by sampling the air.

PHYSICAL STRESSES

The second category of environmental factors or stresses involves physical agents. This category includes such hazards as noise, ionizing and nonionizing radiation, visible radiation, temperature extremes, skin problems, and pressure extremes. You should be alert to these hazards because they can have immediate or cumulative effects on employee health.

Noise

Noise—defined as unwanted sound—is a form of vibration that can be conducted through solids, liquids, or gases. The effects of noise on people include the following:

- Psychological effects—noise can startle, annoy, and disrupt concentration, sleep, or relaxation.
- Interference with verbal communication, and as a consequence, interference with job performance and safety.
- Physiological effects—noise-induced hearing loss, aural pain, or even nausea (when the exposure is severe). Some research even links long-term overexposure to noise to circulatory problems and heart attack.

Noise measurement. A source that emits sound waves produces minute changes in air pressure. We generally measure these pressure changes with sound-level meters and noise dosimeters. The human ear can hear sound over a wide range of pressures, with the ratio of highest to lowest pressures about 10,000,000 to 1.

To deal with the problem of this huge pressure range, scientists have developed the decibel (dB) scale, which is logarithmic. Decibels are not linear units like miles or pounds. Rather, they are points on a sharply rising curve. Thus, 10 decibels is 10 times greater than one decibel; 20 decibels is 100 times greater (10×10); 30 decibels is 1,000 times greater ($10 \times 10 \times 10$); and so on. The rustle of leaves is rated at 20 dB; a typical office has a background noise level of about 50 dB. A vacuum cleaner runs at about 70 dB, while a typical milling machine from 4 feet away is rated at 85 dB. The sound of a newspaper press is about 95 dB, a textile loom is 105 dB, a rock band is about 110 dB, a large chipping hammer is 120 dB, and a jet engine registers about 160 dB.

The pitch or frequency of sound is also important in noise measurement because the ear hears some frequencies better than others. Our ears tend not to hear very low or very high frequencies as well as those in the middle range. Thus, noise-measuring instruments are designed to mimic the response of the human ear on what is called the "A" weighting scale. As a result, the standards for noise are written for the dBA scale. The OSHA Noise Standard requires that noise-monitoring instruments be capable of measuring noise between 80 and 130 dBA. Noise above 115 dBA, however, is dangerous to hearing and exposure should not be permitted. Workers should be exposed to noise above 90 dBA only if they use hearing protection or have their exposure time reduced, as discussed below.

Measurement of noise in the workplace can be quite complex and usually requires the services of a trained safety professional or industrial hygienist. They can make decisions regarding engineering controls or other measures to deal with noise hazards. They will also furnish data to the administrator of the company's hearing-conservation program, who can use it to decide what types of hearing protection are needed. Details and definitions of noise, and its measurement and control, are given in the National Safety Council's *Fundamentals of Industrial Hygiene* and *Noise Control: A Guide for Employees and Employer*.

Factors in hearing loss. If the ear is subjected to high levels of noise for a sufficient time, some hearing loss may occur. A number of factors can influence the effect of noise exposure. Among these are:

- Variation in individual susceptibility
- Total energy of the sound
- Frequency distribution of the sound
- Other characteristics of noise exposure, whether it is continuous, intermittent, or made up of a series of impacts
- Total daily time of exposure
- Length of employment in the noisy environment

Criteria have been developed to protect workers against hearing loss. OSHA has established a regulation for Occupational Noise Exposure, which sets allowable noise levels based on the number of hours of exposure. This standard limits unprotected exposure to eight hours at 90 dBA, four hours at 95 dBA, and so on (Table 8-2). Because noise may affect some workers' hearing more than others, it is good practice to require hearing protection when workers are exposed to levels above 90 dBA, regardless of the length of exposure.

The OSHA standard requires employers to reduce noise exposures with administrative and engineering controls, where feasible. The standard also

TABLE 8-2. Permissible Exposures

Duration per Day Hours	Sound Level dB(A)*
8	90
6	92
4	95
3	97
2	100
1½	102
1	105
¾	107
½	110
¼	115

*Sound level in decibels as measured on a standard level meter operating on the A-weighting network with slow meter response.

requires initial monitoring and re-monitoring and whenever changes in production, processes, or controls increase noise exposure. Some individuals, however, may experience hearing loss below 90 dBA. To further protect worker hearing, OSHA enforces its Hearing Conservation Amendment to the noise standard, which requires employers to:

- Provide annual audiometric tests to all employees exposed to noise over 85 dBA
- Offer optional hearing protection to workers exposed above 85 dBA and to make protection mandatory where noise exposures exceed 90 dBA
- Ensure that workers with an existing hearing loss wear protection when exposed to noise levels above 85 dBA
- Provide annual training in:
  - Effects of noise on hearing
  - Proper selection, fitting, use, and care of hearing protection, and training in the advantages and disadvantages of each type
  - Explanation of the purpose and method of the hearing test

The regulation is very specific concerning the ways a hearing test is conducted. To be successful, the test requires the close cooperation of you, the supervisor. If possible, you should schedule employees so that their hearing can be tested before they are exposed to noise on the job. If this is not possible and employees must be called off the job, make certain that the workers selected for hearing tests wear ear plugs or muffs properly before their test. This will help to prevent confusing a temporary detrimental effect on hearing with a more serious long-term effect. Tests must be conducted in a noise-free environment, such as a testing booth (Figure 8-4).

Audiograms. The first audiogram is called the baseline audiogram, because it is the basis of comparison for subsequent yearly audiograms also required to assess hearing. Employees must be notified if a significant shift in hearing is discovered.

As a supervisor, you may be asked to ensure that employees wear their hearing protection, both before audiometric testing and on a regular basis. One of the best ways to achieve employee cooperation is to set an example by wearing protection yourself wherever it is required. The same goes for managers, visitors, and other employees who visit noisy areas.

Hearing protection. The most commonly used hearing protection equipment are ear plugs, canal caps, and ear muffs. Ear plugs are available in both preformed and disposable types. A variety of styles in both plastic and foam or other material is available. Ear plugs can also be custom-made to fit an individual employee's ear.

Canal caps are typically cone-shaped caps attached to a headband. They are most often used where the employee goes into noisy areas for brief periods, as they allow for easy insertion and removal.

Ear muffs fit over the entire outer ear; how effective they are depends on the tightness of the seal between the ear and muff. The earpiece of eyeglasses, for example, can create a significant noise "leak" if it passes between the muff and the wearer's skin.

Each type and brand of protection has been given a "Noise Reduction Rating" (NRR), which describes how much an employee's exposure can be reduced by wearing the device. Not all types offer the same protection. In general, custom-molded plugs offer the greatest NRR. Ear plugs offer the next best NRR, with muffs and caps offering somewhat less. However, it is important to note that not all brands of one type of protector offer the same noise reduction.

Noise reduction ratings were developed in laboratory settings; in the actual work situation, ear plugs, muffs, or caps may not provide the expected degree of protection. Therefore, many health and safety personnel select hearing protection devices that provide an extra margin of safety against the noise in the area. They try to provide hearing protection that will reduce an employee's exposure below 85 dBA. (However, hearing protection often is not the sole answer to noise exposure. At sound levels over 90–100 dBA, engineering controls must be considered to reduce exposure.)

In addition, in selecting protection, the company and employee should also consider the advantages and disadvantages of the different devices for their workplace. For example, it might not be ap-

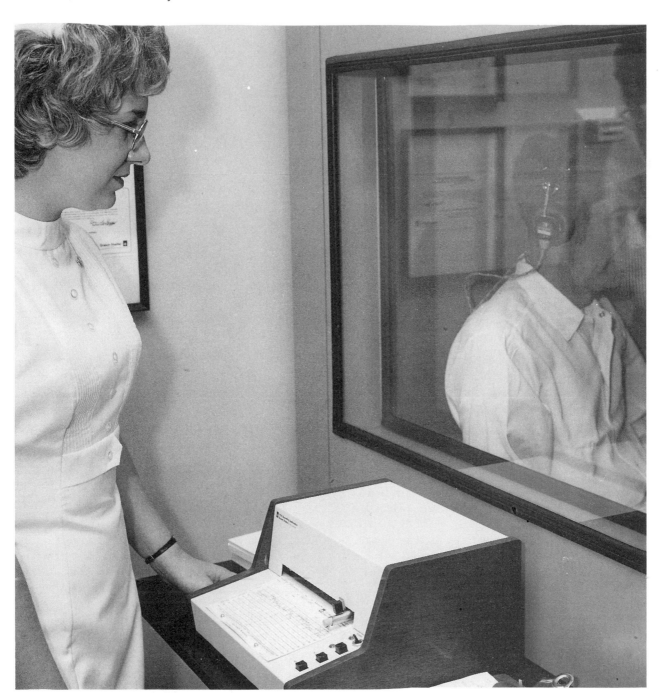

Figure 8-4. A hearing test should be done in an enclosed, soundproof booth by a trained technician.

propriate for a mechanic or someone whose hands are routinely soiled by work to wear foam or other insert plugs. The workers will transfer grit and grime to the outside of the plug which can contribute to ear canal irritation and possible infection.

Hearing protectors must be comfortable when worn properly, or employees will likely find ways to avoid wearing them at all. Even ear plugs, which seem like simple devices, need to be seated properly each time they are worn. They must be inserted fully into the ear canal, not just "propped" at the opening. The administrator of the hearing-conservation program should be responsible for ensuring that employees receive the proper training and are comfortable with the hearing protection selected. It is up to you as supervisor to ensure that employees wear hearing protection when needed. Check for use of hearing protection when you do routine inspections, but also be a good role model—wear yours!

Ionizing Radiation

Although ionizing radiation is a complex subject, a brief discussion is included here because it is one of the more dangerous physical agents. Radiation is a form of energy. Familiar forms of radiant energy include light, which enables us to see, and infrared, which we feel as heat. The relationship among the different forms of electromagnetic radiation is shown in Figure 8-5.

Ionizing radiation refers only to those types which involve ionization (a change in the normal electrical balance in an atom). Only the forms of radiation discussed below possess sufficient energy to change this electrical balance, which can in turn produce damage in living cells. Other forms of radiation that are not ionizing may also cause damage but usually to a lesser extent and by a different mechanism.

Certain chemicals, when exposed to radiation, may form hazardous compounds or more hazardous forms of an already toxic substance. For example, when oxygen is electrified, ozone is formed. Ozone is highly irritating to the nose, throat, and lung tissues. Therefore, when working with chemicals and ionizing radiation, it is important to know what reactions between the two can take place.

Gamma radiation. This type of radiation, produced from radioactive materials and X-radiation, is highly penetrating and can damage body tissues. The amount of damage depends, in part, on the energy of the radiation and also on the relative sensitivity of the tissue.

Alpha and beta radiation. These types of ionizing radiation take the form of particles. Alpha particles are heavy and thus do not travel far in air. Although alpha particles are usually stopped by the skin, if they are inhaled or ingested, they can do considerable internal damage. Beta particles are lighter and thus travel farther. If a beta source is left on the skin, it can damage the tissue. Proper shielding material is necessary to protect employees from beta radiation.

High-energy protons and neutrons. In the handling of some types of radioactive materials and in the operation of high-powered particle accelerators, high-speed protons and high-speed and thermal neutrons can be created. A harmful dose is dependent on both the number of these particles and

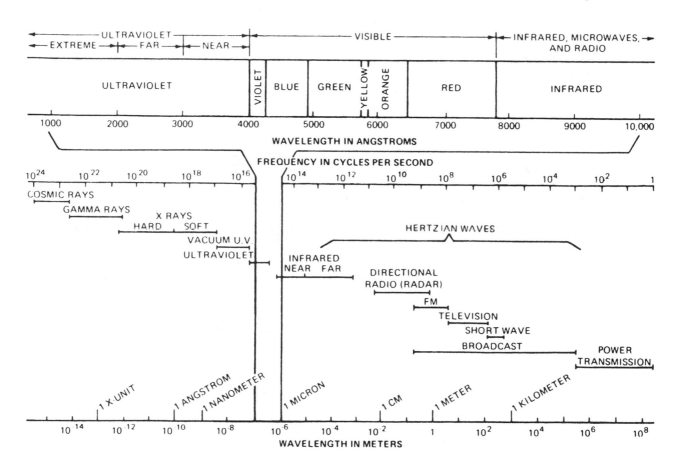

Figure 8-5. The electromagnetic spectrum, encompassing the ionizing radiations and the nonionizing radiations (expanded portion and right). Top portion expands spectrum between 10^{-7} and 10^{-8} m. Note: cycles per second (cps) = hertz (Hz).

their energy distribution. Exposure potential must be evaluated by a health physicist (radiation specialist).

X-radiation. This hazard is produced by high-potential electrical discharge in a vacuum and should be anticipated in evacuated electrical apparatus operating at a potential of 10,000 volts or more. Although X-radiation is certain to be produced in such apparatus, it is dangerous only if it penetrates the jacket of the apparatus and enters the inhabited part of the workroom.

If a worker is overexposed to the radiation of a comparatively low-voltage X-ray tube, dermatitis of the hand is generally the first result. It is characterized by rough, dry skin, a wartlike growth, and dry, brittle nails. Continued exposure and more penetrating X-rays can cause bone destruction.

Beta sources and X- and gamma radiation are found in industry in some gages to measure the thickness of various materials or in beta-source thickness gages. Sealed gamma sources and X-ray generators are used in industry as various types of level gages, for example, to measure the level of a liquid inside a tank. They are also used to X-ray the welds at joints.

These types of radiation all cause injury by ionizing the tissue in which they are absorbed. Such injuries differ widely in location and extent, but the determining factors are the ability of the particular radiation to penetrate through the tissues and the amount of ionization produced by a given amount of the radiation. Alpha radiation, for instance, is classed as 20 times more biologically active than X-rays because the alpha particles are heavy and can greatly disrupt and ionize the tissue molecules with which they come in contact.

Radiation protection. There are two general methods for preventing injuries from any penetrating radiation. First, workers can be separated from the hazard by distance or by shielding with appropriate materials to reduce the radiation received to below the maximum permissible dose. Second, the time of exposure can be limited so that workers will not receive a harmful dose.

Rules for working with ionizing radiation are set by industrial hygienists or health physicists. They must be strictly followed at all times to avoid the chance of radiation injuries and illnesses.

Nonionizing Radiation

Electromagnetic radiation has varying effects on the body, depending largely on the particular wavelength of the radiation involved. Following, in approximate order of decreasing wavelength and increasing frequency, are some hazards associated with different regions of the nonionizing electromagnetic radiation spectrum.

Low frequency. The longer wavelengths—including power frequencies, broadcast, radio, and short-wave radio—can produce general heating of the body. The health hazards of low-frequency radiation were previously thought to be minimal. New data suggest that exposure to radiation at electric power frequencies may be associated with adverse health effects, including cancer. However, scientists have not yet discovered how or why these effects occur. Protection from electric field exposure can be provided by shielding and by limiting access to the affected area.

Microwaves. These are wavelengths of 3 m to 3 μm (100 to 100,000 megahertz, MHz). They are found in radar, communications, and diathermy applications. Microwave intensities may be sufficient to cause significant heating of tissues.

The effect is related to wavelength, power intensity, and time of exposure. Generally, longer wavelengths will produce a greater temperature increase in deeper tissues than will shorter wavelengths. However, for a given power intensity, workers are less aware of the heat from longer wavelengths because it is absorbed beneath the skin.

An intolerable rise in body temperature, as well as localized damage, can result from exposure to wavelengths of sufficient intensity and time. In addition, flammable gases and vapors may ignite when they are inside metallic objects bombarded by a microwave beam.

Power intensities from microwaves are given in units of watts per square centimeter (w/cm²). Areas with a power intensity greater than 0.01 w/cm² should be avoided. In such areas, dummy loads should be used to absorb the energy output while equipment is being operated or tested. If a dummy load cannot be used, adjacent populated areas should be protected by adequate shielding.

Infrared radiation (IR). This type of radiation does not penetrate much below the superficial layer of the skin, so its major effect is to heat the skin tissues immediately below this layer. Longer wave infrared is completely absorbed in the surface layer of the skin. Mid-range infrared can cause skin burns and increases skin pigmentation. Short-wavelength IR radiation from furnaces can cause eye problems known as "glass blower's cataract" or "heat cataract." Exposed workers need to wear appropriate eye protection gear with darkened filtered lenses (see Chapter 9, Personal Protective Equipment).

Ultraviolet (UV) and visible radiation. These two types of radiation also do not penetrate appreciably below the skin; their effects are essentially to heat the surface. However, high-intensity visible radiation can severely damage the retina of the eye, even causing blindness in a relatively short exposure time. This is the reason behind the insistent warnings to the public not to stare directly at the sun during periods of partial solar eclipses. The automatic defense mechanism of closing the eyes or looking away usually prevents damage. It should be emphasized that this defense mechanism is not effective against visible light emitted by lasers.

The most common exposure to ultraviolet radiation is from direct sunshine. Ultraviolet radiation is responsible for "suntan" and "sunburn." People who continually work outdoors in the full light of the sun may develop tumors on exposed areas of the skin. These tumors occasionally become malignant.

The effects of ultraviolet light are generally more of a problem because ultraviolet can produce a severe burn with no warning, and excessive exposure can result in significant damage to the eye lens. Because ultraviolet radiation in industry is found around electrical arcs, they should be shielded by materials opaque to UV light.

Electric welding arcs and germicidal lamps are the most common strong producers of ultraviolet radiation in industry. The ordinary fluorescent lamp generates a good deal of ultraviolet light inside the bulb, but the UV is essentially absorbed by the glass bulb and its fluorescent coating.

Some industrial materials also influence the effects of ultraviolet radiation from the sun on human skin. After exposure to compounds such as cresols, for example, the skin is exceptionally sensitive to the sun. Even a short exposure in the late afternoon when the sun is low is likely to produce a severe sunburn. Other compounds minimize the effects of UV rays. Some of these are used in certain protective creams, for example, sunblocks found in various suntan lotions.

Visible Radiation or Lighting

Visible radiation, which falls about midway in the electromagnetic spectrum, should concern you as a supervisor, because it can affect both the quality and accuracy of employees' work. Good lighting invariably results in increased product quality with less spoilage and increased production. Good lighting contributes to sanitary, clean, and neat operations (Figure 8-6).

Proper lighting. Good lighting is the result of several factors: the amount and color of the light, direction and diffusion, and the nature of illuminated surfaces. A dark-gray, dirty surface may reflect only 10 or 12 percent of a light source while a light-colored, clean surface may reflect more than 90 percent.

Lighting should be bright enough for easy viewing of work tasks and directed so that it does not create glare. The level should be high enough to permit efficient sight. The *Accident Prevention Manual for Industrial Operations* and *Fundamentals of Industrial Hygiene* list the illumination levels and brightness ratios recommended for various occupational tasks by the Illuminating Engineering Society. The IES's *Practice for Industrial Lighting* is designated as American National Standard ANSI/IES RP7.

One of the biggest problems associated with lighting is glare—brightness within the field of vision that causes discomfort or interferes with proper vision. The brightness can be caused by either direct or reflected light. To prevent glare, keep the source of light well above the line of vision, or shield it with opaque or translucent material. You can dull reflections by having such surfaces matte (dull) finished rather than polished or smooth.

The video display terminals of computers are especially likely to be affected by glare. You can reduce glare by dimming overhead lights and using a task light instead, by placing the terminal parallel to windows, or by installing an antiglare screen.

Another visual problem occurs when an object is too bright for its surrounding field. A highly reflecting white paper in the center of a dark, nonreflecting surface, or a brightly illuminated control handle or dial on a dark or dirty machine are two examples. The contrast can irritate the eyes. To prevent this condition, keep surfaces light or dark with little differences in surface reflectivity. Instead, use color contrasts to set objects apart from their backgrounds.

Although it is generally best to provide even, shadow-free light, some jobs require contrast lighting. In these cases, keep the general or background light well diffused and free of glare and add a supplementary source of light directed to cast shadows as desired. Keep in mind that too much contrast in brightness between the work area and its surroundings can be as distracting and harmful as direct or reflected glare.

Lasers. Lasers emit beams of coherent light of a single color (or wavelength). Contrast this with conventional light sources that produce random, disordered light-wave mixtures of various wavelengths. The word *laser* is an acronym for Light Amplification by Stimulated Emission of Radiation. A laser is made up of light waves that are nearly parallel to each other (collimated), and that are all traveling in the same direction. Electrons of atoms capable of emitting excess energy in the form of visible light are "pumped" full of enough energy to force them into higher-energy-level orbits. These electrons

Figure 8-6. Good lighting promotes increased safe production and good house-keeping.

quickly return to their normal-energy-level orbits but, in so doing, they emit a photon of visible light. The light waves that are given off are directed to produce the coherent laser beam. The maser, the laser's predecessor, emits microwaves instead of light. Some companies call their lasers "optical masers."

Lasers are extremely versatile, and new applications for this invention are being developed every year. They are used for welding microscopic parts, for welding heat-resistant exotic metals, for communications instruments, for all kinds of guidance systems, for various surgical procedures, in chemistry laboratories, and in numerous other applications.

Biological hazards. The eye is the organ most vulnerable to injury induced by laser energy. This is because the cornea and lens focus the parallel laser beam onto a small spot on the retina. Make sure workers are instructed never to observe a laser beam or its reflection directly or to look at it with an optical aid, such as a binocular or a microscope.

Also, the work area should contain no reflective surfaces (such as mirrors or highly polished furni-

ture) as even a reflected laser beam can be hazardous. Suitable shielding to contain the laser beam should be provided. Other safety measures may also be necessary, such as storing combustible solvents and other materials in proper containers, shielding them from laser beams or induced electric sparks, and safeguarding all potential electrical hazards. The fact that infrared radiation of certain lasers may not be visible to the naked eye contributes to the potential hazard. Eyes must be protected (Figure 8-7). Lasers generating in the ultraviolet range of the electromagnetic spectrum produce corneal burns rather than retinal damage, because of the way the eye handles ultraviolet light.

Other factors that influence the degree of eye injury induced by laser light include the following:

- Pupil size—the smaller the pupil diameter, the smaller the amount of laser energy permitted to the retina
- Power of the cornea and lens to focus the incident light on the retina
- Distance from the source of energy to the retina

Figure 8-7. No one type of safety glass gives protection from all laser wavelengths so safety glasses suitable for protection against laser beams must be selected for the specific wavelength used.

- Energy and wavelength of the laser
- Pigmentation of the subject
- Place on the retina where light is focused
- Divergence of the laser light
- Presence of scattering media in the light path

Control of exposure to laser radiation should include use of protective devices built into the laser equipment, training employees in eye hazards, and use of protective eyewear as necessary. To develop a program of controls for the more hazardous lasers might require the help of a laser specialist.

Temperature Extremes

General experience shows that extremes of temperature affect the amount of work that people can do and the manner in which they do it. In industry, people are more often exposed to hazards associated with high temperatures than with low temperatures.

The body is continuously producing heat through its metabolic processes. Since these processes are designed to operate efficiently within only a narrow temperature range, the body must dissipate excess heat as rapidly as it is produced. Body temperature is regulated by a complex set of thermostatic controls that react quickly to significant changes in internal and external temperatures.

Sweating. Sweating is the most important of the temperature-regulating and heat-dissipating processes. Almost all parts of the skin are provided with sweat glands that excrete a liquid (mostly water and a little salt) to the surface of the body. This process goes on continuously, even under conditions of rest.

In an individual who is resting and not under stress, the sweating rate is approximately 1 liter per day, which is evaporated in the air as rapidly as it is excreted. Under the stress of heavy work or high temperature, the sweating rate may increase to as much as 4 liters (approximately one gallon) in 4 hours. Up to 10 to 12 gm (150 to 190 grains) of salt per day will be lost with the water. Both the water and salt loss must be replaced promptly to ensure good health. Such stress is especially dangerous for those with heart problems because the loss of salt and water can affect the heart's ability to function.

If the sweat evaporates as rapidly as it forms, and if the heat stress does not exceed the maximum sweating rate of which the body is capable, an individual can maintain the necessary constant temperature. The rate of evaporation depends on the moisture content and temperature of the surrounding air (heat and humidity), and the rate of air movement. The industrial hygienist must consider all these factors when determining the effect of the thermal environment on humans.

Radiant heat. Radiant heat is electromagnetic energy that does not heat the air it passes through. It affects the body's ability to remain in equilibrium with its surroundings. For instance, an object, such as a glass window, may be far below the body temperature of a nearby worker. As a result, a large amount of heat will be radiated from the person near the window, who may feel chilled even if the air in the immediate environment is fairly warm. Conversely, if an object in the surroundings, such as a furnace wall, is above body temperature, people can receive considerable heat by radiation. They will find it difficult to keep cool even if the room is air conditioned.

Thus, when radiant heat strikes an object (such as a person), it is absorbed and produces the sensation of heat. Merely blowing air around offers no relief from the source. The only protection is to set up a barrier. Placing radiant-reflective shielding between the heat source and the worker is usually the most cost-effective way to reduce employee exposure. Also, usage of heat-reflective personal protective clothing may be considered when exposures are short and intermittent.

Heat conduction and convection. If heat can be conducted through the clothing and dissipated into the air, this will cool the body to some degree. Conduction usually is not an important means of cooling, however, because the conductivity of clothing and the heat capacity of air are usually low.

Conduction and convection become an important means of heat loss when the body is in contact with a good cooling agent, such as water. For this reason, when people are exposed to cold water, they become chilled much more rapidly than when exposed to air at the same temperature.

Air movement cools the body by conduction and convection. More importantly, moving air removes the layer of saturated air around the body (which is formed rapidly by evaporation of sweat) and replaces it with a fresh layer, capable of accepting more moisture.

Effects of high temperature. The effects of high temperature are counteracted by the body's attempt to keep the internal temperature down through increasing the heart rate. The capillaries in the skin dilate to bring more blood to the surface so that the rate of cooling increases and, gradually, the body temperature stabilizes.

If the thermal environment is tolerable, these measures will soon restore the body's equilibrium to the point where the heart rate and the body temperature remain constant. If this equilibrium is not reached by the time the body temperature is about 102 F (38.9 C), corresponding to the sweating rate of about 2 liters per hour, there is imminent danger of heatstroke. Intermittent rest periods for people exposed to extreme heat reduces this danger.

Heatstroke or sunstroke. This is not necessarily the result of exposure to the sun. It is a caused by exposure to an environment in which the body is unable to cool itself sufficiently. Sweating stops, and the body can no longer rid itself of excess heat. As a result, the body soon reaches a point where the heat-regulating mechanism breaks down completely and the internal temperature rises rapidly.

The symptoms are hot, dry skin (which may be red, mottled, or bluish); severe headache; visual disturbances; rapid temperature rise; and confusion, delirium, and loss of consciousness. The condition is recognizable by a flushed face and high temperature. The victim should be removed from the heat immediately and cooled as rapidly as possible, usually by being wrapped in cool, wet sheets. Studies show that for people admitted to hospital emergency rooms for heatstroke, the higher the body temper-

ature upon admission, the higher the mortality rate. Since the condition can be fatal, medical help should be obtained immediately.

Heatstroke is a much more serious condition than heat cramps or heat exhaustion, discussed next. An important factor is excessive physical exertion. The only method of control is to reduce the temperature of the surroundings or to increase the ability of the body to cool itself so that body temperature does not rise. Heat shields, discussed under the next section, Preventing heat stress, are of value here.

Heat cramps. These may result from exposure to high temperature for a relatively long time, particularly if accompanied by heavy exertion, and excessive loss of salt and moisture from the body. Even if the moisture is replaced by drinking plenty of water, an excessive loss of salt may provoke heat cramps or heat exhaustion.

Heat cramps are characterized by the cramping of the muscles of either the skeletal system or the intestines. In either case, the condition may be relieved in a few hours under proper treatment, although soreness may persist for several days.

Heat exhaustion. This may result from sustained physical exertion in a hot environment, especially if a worker has not been acclimatized to heat, and has not replaced the water lost in sweating. Its symptoms are a relatively low temperature, pallor, weak pulse, dizziness, profuse sweating, and cool, moist skin.

Preventing heat stress. Most heat-related health problems can be prevented or, at least, the risk can be reduced. Following a few basic precautions should lessen heat stress.

Acclimatization. Acclimatize workers to heat by giving them short exposures, followed by gradually longer periods of work in the hot environment. New employees and workers returning from an absence of two weeks or more should have a five-day period of acclimatization. This period might begin with 50 percent of the normal workload and time exposures the first day and gradually build up to 100 percent on the fifth day.

Mechanical cooling. A variety of engineering controls, including general ventilation and spot cooling by local exhaust ventilation at points of high heat production, may be helpful. Evaporative cooling, mechanical refrigeration, and cooling fans are other ways to reduce heat, along with eliminating steam leaks. Equipment modifications, the use of power tools to reduce manual labor, and using personal cooling devices or protective clothing (Figure 8-8) are other ways to reduce heat exposure for workers.

Procedures. Work practices can also be tailored

Figure 8-8. Jobs with high radiant heat loads often require reflective clothing. This first-response firefighting team is properly equipped with heat-protective clothing.

to prevent heat disorders. Companies can provide a period of acclimatization for new workers and those returning from vacation. They can make plenty of cool drinking water available, as much as a quart per worker per hour (Figure 8-9). Companies should train first aid workers to recognize and treat heat stress and should post the names of trained staff known to all workers. Always consider individual workers' physical conditions when determining their fitness for working in hot environments. Older workers, obese workers, and workers on certain

types of medication are at greater risk for heat-stress illnesses.

If you alternate work and rest periods, but with longer rest periods in a cool area, you can help workers avoid heat strain. If possible, schedule heavy work during the cooler parts of the day and provide appropriate protective clothing. Allow workers to interrupt their jobs if they become extremely uncomfortable in the heat.

Rehydration. Employee education is vital so that workers are aware of the need to replace fluids

Figure 8-9. To help prevent heat disorders, cooled drinking water should be easily accessible and workers should be trained to recognize symptoms of heat stress.

and salt lost by sweating, and so they can recognize dehydration, exhaustion, fainting, heat cramps, salt deficiency, heat exhaustion, and heatstroke as heat disorders. Workers should be advised to drink beyond the point of thirst. Adding salt to the diet is not usually recommended or needed for a worker who is acclimatized to heat. For workers on a salt-restricted diet who need to become acclimatized, adding salt during the first two days of exposure may be necessary to replace the amount lost through sweating; however, the worker's physician must be consulted before taking this step. Workers should also be informed of the importance of weighing themselves daily before and after work to avoid dehydration.

Companies may want to supply a cooled carbohydrate electrolyte beverage to encourage rehydration. Such a beverage should have six to eight percent carbohydrate to provide energy and stimulate fluid absorption as well as some sodium to help maintain the desire to drink and offset loss of sodium from sweating. A number of carbohydrate electrolyte beverages are on the market.

Workers should be educated about the negative effects of high-carbohydrate (above 10 percent) and caffinated beverages. The former will tend to decrease fluid absorption, and the latter will increase fluid loss through urination.

Dermatitis

Skin problems in industry account for the single largest cost of all hygiene problems. Individual skin disorders usually can be traced to exposure to some kind of chemical compound, or to some form of physical abrasion or irritation of the skin. Although rarely a direct cause of death, skin disorders cause much discomfort and are often hard to cure (Figure 8-10). Even normally harmless substances can cause irritations of varying severity in some skins.

Sources of hazards. Causes of occupational skin disorders are classified in these ways:

1. Mechanical agents—friction, pressure, trauma.
2. Physical agents—heat, cold, radiation.

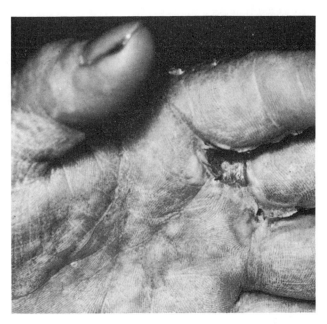

Figure 8-10. Eczematous dermatitis is a form of contact dermatitis caused by working with organic solvents. It is one of the most prevalent types of dermatitis.

3. Chemical agents—organic and inorganic. These are subdivided according to their action on the skin as primary irritants or sensitizers.
4. Biological agents—bacteria, fungi, parasites.
5. Plant poisons—several hundred plants and woods can cause dermatitis. The best known example is poison ivy.

There are two general types of skin reactions: primary irritation dermatitis and sensitization dermatitis. Practically all persons suffer primary dermatitis from mechanical, physical, or chemical agents. Brief contact with a high concentration of a primary irritant or prolonged exposure to a low concentration will result in inflammation. Allergic reaction is not a factor in these conditions. Sensitization dermatitis, on the other hand, is the result of an allergic reaction to a given substance. Once sensitization develops, even small amounts of the material may cause symptoms. Some substances can produce both types of dermatitis. Examples are organic solvents, formaldehyde, and chromic acid.

Occupational acne. Cutting fluids are frequently responsible for occupational acne. When the skin becomes infected, oil pimples or boils form that resemble adolescent acne. Dirty oil, poor materials handling, and addition of germicides in the fluid increase the possibility of occupational acne.

To reduce the possibility of this condition, keep machines and areas around them clean. Change cutting fluids at regular intervals. Have employees use protective creams or gloves, aprons, or face shields. Encourage employees who work in these areas to shower at the end of each shift. Work clothes should be changed before going home. Abrasive cleaners or brushes should be avoided. Instead, workers should use warm water, mild soap, and a soft brush. They should be discouraged from wearing oil-soaked clothing. Some sensitive skins may require special cleaners.

Some types of skin are more susceptible than others, but most people will develop oil acne with sufficient exposure. Certain factors help account for some skin disorders. Any investigation of occupational skin disease should take into consideration:

1. Degree of perspiration
2. Personal cleanliness
3. Preexistence of skin disorders
4. Allergic conditions
5. Diet

Atmospheric Pressures

It has been recognized from the beginning of caisson work, dating back to about 1850, that those who work under greater-than-normal atmospheric pressures are subject to various disorders. The main effect, decompression sickness (commonly known as the bends), results from the release of nitrogen bubbles into the bloodstream and surrounding tissues during decompression. The bubbles lodge at the joints and under muscles, causing severe cramps. To prevent this trouble, decompression is carried out slowly and by stages so that the nitrogen can be eliminated without forming bubbles.

Deep-sea divers are supplied with a mixture of helium and oxygen for breathing. Since helium is an inert diluent and is less soluble in blood and tissue than is nitrogen, it presents a less formidable decompression problem.

Common troubles encountered by workers operating under compressed air are pain and congestion in the ears caused by the inability to ventilate the middle ear properly during compression and decompression. As a result, many workers under compressed air suffer from temporary hearing loss. The cause of this damage is considered to be obstruction of the eustachian tubes that prevents proper equalization of pressure on the middle ear. Occasionally, permanent hearing loss may result.

The effects of low pressure on a worker are much the same as the effects of decompression from a high pressure. If pressure is reduced too rapidly, decompression sickness and ear disturbances similar to, if not identical with, the divers' conditions may result.

Persons working at reduced pressure are also subject to oxygen starvation, which can have serious and insidious effects upon the senses and judgment. There is a considerable amount of evidence showing that exposure for 3 to 4 hours to an altitude of 9,000 ft (2.7 km) above sea level, without breathing an atmosphere enriched with oxygen, can result in severely impaired judgment. Even if pure oxygen is provided, the altitude should be limited to that giving the same partial pressure of oxygen as air at 8,000 ft (2.4 km).

Reduced pressure is not the only condition under which oxygen starvation may occur. Deficiency of oxygen in the atmosphere of confined spaces is commonly experienced in industry. (Normal air contains roughly 21 percent oxygen by volume.) The oxygen content of any tank or other confined space should be checked before workers are allowed to enter. Instruments, such as the oxygen analyzer, are commercially available for this purpose (Figure 8-11).

The first physiologic signs of a deficiency of oxygen (anoxia) are increased rate and depth of breathing. Oxygen concentrations of less than 16 percent by volume cause dizziness, rapid heartbeat,

Figure 8-11. This portable, hand-held oxygen indicator with digital display and remote galvanic sensor operates by diffusion and measures oxygen on a 0–100 percent-by-volume range. (Courtesy MSA.)

and headache. One should never enter or remain in areas where tests have indicated such low concentrations unless wearing self-contained, supplied-air respiratory equipment.

In an oxygen-deficient atmosphere, workers may find it hard to move and have a hazy lack of concern about the imminence of death. In cases of sudden entry into areas containing little or no oxygen, the individual usually has no warning symptoms but immediately loses consciousness and has no recollection of the incident if rescued and revived.

Atmospheric contaminants are discussed in this chapter in the sections Threshold Limit Values and Standard Operating Procedures. Entry into confined spaces also requires compliance with lockout/tagout procedures that should be used to reduce hazards from incoming process lines, agitators, and other equipment. These are discussed in Chapter 11, Machine Safeguarding.

ERGONOMIC STRESSES

The third category of environmental factors is ergonomic. Involved are human reactions to monotony, fatigue, repeated motion, and repeated shock. This subject is discussed in greater detail in Chapter 10, Ergonomics.

The term "ergonomics" literally means the customs, habits, and laws of work. According to the International Labor Office, it is:

> the application of human biological science in conjunction with the engineering sciences to achieve the optimum mutual adjustment of man and his work, their benefits being measured in terms of human efficiency and well-being.

The ergonomics approach goes beyond productivity, health, and safety. It includes consideration of the total physiological and psychological demands of the job upon the workers.

The human body can endure considerable discomfort and stress and can perform many awkward and unnatural movements, but only for a limited time. When unnatural conditions or motions continue for prolonged periods, they may exceed workers' physiological limitations. To ensure a continuously high level of performance, work systems must be tailored to human capacities and limitations.

Biomechanics is that phase of engineering devoted to the improvement of the worker-machine-task relationship in an effort to reduce operator discomfort and fatigue. Biotechnology, a broader term, encompasses biomechanics, human factors engineering, and engineering psychology. To arrive at biotechnological solutions to work-stress problems, the sciences of anatomy, physiology, psychology, anthropometry, and kinesiology need to be brought into play.

Areas of concern are:

- Strictly biomechanical aspects—the consideration of stress on muscles, bone, nerves, and joints
- Sensory aspects—the consideration of eye fatigue, odor, audio signals, tactile surfaces, and the like
- External environment aspects—the consideration of lighting, glare, temperature, humidity, noise, atmospheric contaminants, and vibration
- The psychological and social aspects of the working environment

Information obtained from studying these factors can be translated into tangible changes in work environments. Reducing fatigue and stress with redesigned hand tools, adjustable chairs and workbenches, better lighting, control of heat and humidity, and noise reduction are definitely rewarding improvements.

In the broad sense, the benefits that can be expected from designing work systems to minimize physical stress on workers are:

- More efficient operation
- Fewer accidents
- Lower cost of operation
- Reduced training time
- More efficient use of personnel

See Chapter 10 for more detailed information on ergonomic stresses.

BIOLOGICAL STRESSES

Biological stresses include any virus, bacterium, fungus, or any other living organism capable of causing a disease in humans.

Tuberculosis

Tuberculosis and other bacterial infections are classified as occupational diseases when they are contracted by nurses, doctors, or attendants caring for tuberculosis patients, or during autopsy or laboratory work where the bacilli may be present.

Fungus Infections

A number of occupational infections are common to workers in agriculture and in closely related industrial jobs. Grain handlers who inhale grain dust are likely to come in contact with some of the fungi that contaminate grain from time to time. Some of these fungi can and do flourish in the human lung, causing a condition called "farmer's lung." Fine, easily spread fungi, such as rust and smut, can be contacted once the covering of infected grain is broken and may produce sensitization.

One example, coccidioidomycosis or Valley fever, is a disease caused by a fungus present in the soil of southwestern states and the San Joaquin Valley in California. It causes a lung disease in humans called Valley fever, which may be either mild or severe. Because the spores are found in the soil, Valley fever outbreaks sometimes occur during excavation and construction work.

Byssinosis

Byssinosis occurs in individuals who have experienced prolonged exposure to heavy air concentrations of cotton dust. Flax dust also has been incriminated.

The exact mode of action of the cotton dust is unknown, but one or more of these factors may be important: (a) toxic action of microorganisms adhering to the inhaled fibers, (b) mechanical irritation from the fibers, and (c) allergic stimulation by the inhaled cotton fibers or adherent materials. It takes several years of exposure before manifestations are noticed.

Anthrax

Anthrax is a highly virulent bacterial infection. In spite of considerable effort in quarantining infected animals and in sterilizing imported animal products, this disease remains a problem. With prompt detection and modern methods of treatment, it is less likely to be fatal than it was a few years ago. Vaccination of high-risk workers and use of personal protection can prevent infection.

Q Fever

There have been reports of infection with the rickettsial organism (Q fever) among meat and livestock handlers. It is similar to, but apparently not identical with, tick fever, which has been known for many years. Q fever may come from contacting freshly killed carcasses or droppings of infected cattle. Probably in the latter case, transmission is by inhalation of the infectious dust. Prevention undoubtedly depends upon recognition and elimination of the disease in the animal host. Workers should use personal and respiratory protection when working in dusty and contaminated environments, and should receive immunization.

Brucellosis

Brucellosis (undulant fever) has long been known as an infection produced by drinking unpasteurized milk from cows suffering from Bang's disease (infectious abortion). Since it can also be contracted by handling the animals or their flesh and is also transmitted by swine and goats, it is an occupational ailment of slaughterhouse workers, as well as of farmers, although more common among the latter. Here also, preventive measures are primarily proper testing and control of the animals to eradicate the disease. Again, personal protection when handling infected materials is important.

Erysipelas

The other major bacterial infection to which slaughterhouse workers and fish handlers are subject is erysipelas. This disease seems to be especially virulent when contracted from the slime of fish.

Leptospirosis

Leptospirosis can infect livestock and wild animals. It causes a flu-like illness and, in severe cases, liver damage. Hunters and people working in rice paddies may come in contact with water containing liquid animal waste. Control includes immunization of animals and appropriate personal protection for workers.

Psittacosis

Psittacosis is a risk for poultry workers and poultry-processing workers. This disease causes a mild flu-like illness and conjunctivitis. Sanitation and personal hygiene are important in prevention.

Lyme Disease

Lyme disease is caused by bacteria that live in the gut of deer ticks. It causes a bullseye-shaped rash and flu-like symptoms. If left untreated, it can lead to rheumatoid arthritis. Employees who work outdoors, especially in woods or brush, or who have potential contact with wildlife or who handle stock or domestic animals are most vulnerable. Prevention requires awareness of the signs and symptoms, often difficult to identify at the time of the bite. Precautions should include covering exposed skin with clothing, tucking pant legs into boots, and surveying the body for ticks on leaving the area. Ticks found should be removed carefully, using protective gloves. The tick should not be squeezed or crushed, as this could release the bacteria. First aid should be administered to the wound.

AIDS

AIDS, caused by a Human Immunodeficiency Virus or HIV, is usually contracted by sexual contact, intravenous drug use involving shared contaminated needles, or by infants from their mothers before or during birth. It cannot be transmitted by casual contact with an infected person in the workplace. The virus is present in the blood and other body fluids of infected persons, and can be a hazard for workers who might have contact with these fluids: medical and emergency rescue personnel, hospital and laboratory workers.

The Centers for Disease Control have developed guidelines for workers in health-care and hospital settings. These specify that all patients should be assumed to be infectious for HIV and other blood-borne illnesses (such as hepatitis B). Precautions to avoid the spread of bloodborne disease in the workplace include:

- Proper handling and disposal of all needles in impervious containers labeled as biohazardous
- Prohibiting recapping of needles and other measures to avoid needlestick injuries
- Using personal resuscitation clothing, including impervious gloves, aprons, boots, face shields, and goggles in some situations
- Using resuscitation devices instead of mouth-to-mouth techniques
- Adopting appropriate sterilization, disinfection, and housekeeping methods where potential for exposure to blood or body fluids is involved

Upper Respiratory Tract Infections

Workers exposed to dusts from such vegetable fibers as cotton, bagasse (sugar cane residues), hemp, flax, or grain may develop upper respiratory tract infections ranging from chronic irritation of the nose and throat to bronchitis, complicated by asthma, emphysema, or pneumonia, or a combination. Dust from any of these fibers may also be a source of allergens, histamine, and toxic metabolic products of microorganisms.

THRESHOLD LIMIT VALUES

Guidelines called "Threshold Limit Values" (abbreviated as TLVs) have been established for airborne concentrations of many chemical compounds. Since as supervisor you may be involved in discussing problems with employees and industrial hygienists conducting surveys, you should understand something about the meaning of TLVs and the terminology in which concentrations are expressed.

Concept and Purpose

The basic idea of the TLV is fairly simple. TLVs represent conditions under which it is believed that nearly all workers may be repeatedly exposed without adverse effect. However, because individual susceptibility varies widely, exposure of an occasional individual at (or even below) the threshold limit may not prevent discomfort, aggravation of a preexisting condition, or occupational illness. In addition to the TLVs set for chemical compounds, there are limits for physical agents, such as noise, microwaves, and heat stress.

The TLV may be a time-weighted average figure that would be acceptable for an 8-hour exposure. For some substances, such as an extremely irritating one, a time-weighted average concentration would not be acceptable, so a ceiling value is established. In other words, a ceiling limit means that at no time during the 8-hour work period should the airborne concentrations exceed that limit.

It is important to understand that TLVs are not intended to be fine lines between safe and unsafe conditions. Considerable judgment needs to be used in deciding on an appropriate exposure level in a particular work situation. Many companies choose to set an in-house "action level" of one-half or less of the TLV. When the concentration of the chemical exceeds the "action level," further air monitoring is often conducted, and measures are taken to reduce employee exposure.

As more toxicological information has become available, the wisdom of this practice is clear. Many of the TLVs have been reduced as epidemiological and laboratory studies become available that show health effects at lower exposure levels. The National Institute for Occupational Safety and Health (NIOSH) also has set Recommended Exposure Levels (RELs) for a number of substances; many of these RELs are below the Threshold Limit Values.

Establishing Values

Threshold Limit Values have been established for more than 600 substances, as well as for heat and cold stress, vibration, radiation, and biological exposure indexes. The TLVs are developed and reviewed by members of the American Conference of Governmental Industrial Hygienists, a voluntary organization composed of industrial hygienists, toxicologists, and medical personnel who work in government and educational institutions.

The data for establishing a TLV come from animal studies, human studies, and industrial experience. The limit may be selected for one of several reasons. It may be based on the fact that a substance is very irritating to the majority of people exposed to concentrations above a given level. Or the substance may be an asphyxiant. Other reasons for establishing a limit might be that the chemical compound is an anesthetic, or a fibrogenic, or can cause allergic reactions or malignancy. Some TLVs are established because above a certain airborne concentration, a substance creates a nuisance to workers.

The concentrations of airborne materials capable of causing problems are quite small. Consequently, industrial hygienists use special terminology to define these concentrations. They often talk in terms of parts (of contaminant) per million (parts of air) when describing the airborne concentration of a gas or vapor. If measuring airborne particulate matter, such as a dust or fume, they use the term milligrams (of dust) per cubic meter (of air), or mg/m^3 to define concentrations. For materials composed of fibers, such as asbestos or fiberglass, industrial hygienists may also use the number of fibers present in one cubic centimeter of air being tested, in which case the descriptive term becomes the number of fibers per cubic centimeter (F/cm^3).

As an example of the small concentrations involved, the industrial hygienist commonly samples and measures substances in the air of the working environment in concentrations ranging from 1 to 100 parts per million (ppm). Some idea of the magnitude of these concentrations can be appreciated when one realizes that 1 in. in 16 miles, one cent in $10,000, one oz of salt in 62,500 lb of sugar, one oz of oil in 7,812.5 gal of water—all represent one part per million.

STANDARD OPERATING PROCEDURES (SOP)

One of your primary functions as supervisor is to assist in developing standard operating procedures for jobs that require industrial hygiene controls, and to enforce these procedures once they have been established. Common jobs involving maintenance and repair of systems for storing and transporting fluids or entering confined spaces for cleaning and repairs are controlled almost entirely by the immediate supervisor.

Confined Space Entry Procedures

Confined spaces are generally considered to be areas that are not intended for employee occupancy, are difficult to enter and leave due to tight openings and awkward space configurations, and that present serious hazards to the occupants. These hazards can include oxygen deficiency, presence of flammable and/or toxic chemicals, as well as safety and equipment hazards. Examples of confined spaces include spaces normally entered through a manhole, such as storage tanks, bins, process vessels, furnaces, boilers, silos, dust collectors; open-topped spaces more than 4 ft deep that may not have good ventilation, such as pits, trenches, slumps, or wells; or other structures like septic tanks, underground tunnels, caissons, or large-diameter pipes.

You should have a standard operating procedure for entering confined spaces that has been established by a trained health and safety expert. Even a space that hasn't been in use may have been closed for some time and developed an oxygen deficiency. As discussed earlier, the atmosphere must be tested for oxygen content and for the presence of any flammable gas, toxic particulates, vapors, or gases. The space may have to be force-ventilated before beginning and, perhaps, continuously during the work operation. It may be unsafe to enter without self-contained breathing apparatus.

For example, if a tank were filled with solvent and has not been thoroughly cleaned, the atmosphere within may be saturated with the solvent vapor. Even if the volatile solvent has been flushed with water and steam, the tank may still be hazardous. There are records of severe poisoning of workers who have entered "cleaned" tanks that were found to be relatively free of solvent vapor. In these instances, the scale that was disturbed by the cleaning process released trapped solvents.

For these reasons, it is universal practice to provide approved respiratory equipment for workers entering confined spaces. An industrial hygienist or safety professional will advise you on which is the proper respirator for adequate protection. Remember that air-purifying respirators (those without an air supply) must *never* be used in oxygen-deficient atmospheres. Because the inside of many confined spaces might seem relatively harmless, workers often neglect to use the equipment supplied. It is your job to prevent workers from entering these spaces without approved equipment and attached lifelines, or without a similarly equipped employee outside to assist should any serious difficulties arise (Figure 8-12).

Federal and state regulations may also require the use of entry systems that ensure all entry procedures have been checked and approved by a competent, trained individual. Compliance with lockout/tagout regulations and programs must also be considered. Many confined spaces contain agitators or other equipment that can be very dangerous if left plugged in (see Chapter 11).

Maintaining Pumps and Valves

The other job that can often lead to serious exposures to chemical vapors is maintenance of pumps, valves, and packed joints in solvent lines. Vapor from dripping packing glands may collect to form a fire and explosion hazard as well as a health hazard. Pneumatic conveyors can also pose flammability hazards. When material that can cause dust explosions is transported, care must be taken that the proper air velocity is maintained and that the equipment is bonded electrically and grounded to prevent static electricity from igniting flammable materials.

The lower explosive limit for most flammable solvents is about 1½ percent. The Threshold Limit Value for less toxic solvents is about 1/10 of one percent. Control measures for solvent exposures try to keep the concentration below the Threshold Limit Value for health reasons. Although the fire hazard will be taken care of "automatically" in general areas, a dripping gland could still produce a flammable concentration in its immediate area.

Industrial Hygiene Controls

Finally, as supervisor, you should be aware of the various kinds of controls used to maintain a healthy work environment. The type and extent of these controls depend on the physical, chemical, and toxic properties of the air contaminant, the evaluation made of the exposure, and the operation that disperses the contaminant. The extensive controls needed for lead oxide dust, for example, would not be needed for limestone dust, since limestone is less toxic.

General methods of controlling harmful hazards or stresses include the following:

1. Substitution of a less harmful material for one that is dangerous to health. Make sure a professional carries out this task so that a more hazardous material is not mistakenly substituted.
2. Change or alteration in a process to minimize worker contact.
3. Isolation or enclosure of a process or work operation to reduce the number of persons exposed, or isolation of the workers in clean control booths.
4. Wetting down dust during working opera-

Figure 8-12. An air-supplied respirator. (Courtesy E. D. Bullard Company.)

tions and cleanup to reduce generation of dust.

5. Local exhaust ventilation at the point of generation and dispersion of contaminants.

6. General or dilution ventilation with clean air to provide a safe atmosphere. (Don't use dilution ventilation improperly for toxic chemicals; local exhaust ventilation is the way to control these.)

7. Personal protective devices, such as special clothing, and eye and respiratory protec-

tion. (These are discussed in Chapter 9, Personal Protective Equipment.)

8. Good housekeeping, including cleanliness of the workplace, proper waste disposal, adequate washing and eating facilities, potable drinking water, and control of insects and rodents.

9. Special control methods for specific hazards, such as reduction of exposure time, film badges and similar monitoring devices, continuous sampling with preset alarms,

and medical programs to detect intake of toxic materials.

10. Training and education to supplement engineering controls. (Training in work methods and in emergency response procedures—medical, gas release, etc.—to teach who should and should not respond in specific emergencies.)

As a supervisor, you play an important role in making sure that controls are effective. You must observe and study process changes, process enclosures, use of wet methods, local exhaust ventilation, and general ventilation carefully and systematically to make sure that they are functioning properly, or to determine if maintenance or repair work is needed. Even more important is the part you must play in enforcing the use of personal protective equipment, such as hearing protectors and respirators, in areas where they should be worn. If this is not done, your company can be cited and penalized by regulating agencies.

Good housekeeping, special monitoring devices, and safety and hygiene training and education all call for your full cooperation. In fact, as supervisor, you are a key person where safety and industrial hygiene controls are concerned. Without your help and guidance, these controls are of questionable value in ensuring a safe, healthy workplace.

SUMMARY OF KEY POINTS

The chapter covered the following key points:

- The more supervisors know about industrial hygiene, the better they can fulfill the objectives of their job: to ensure safe production. Supervisors should know how to identify health hazards and take steps to correct them in four key areas of industrial hygiene: chemical, physical, ergonomic, and biological stresses.
- Chemical stresses exist in the form of dusts, fumes, smoke, aerosols, mists, gases, vapors, and liquids. They can cause health problems when workers inhale, absorb (through the hair or skin), or ingest one or more of these substances.
- The major exposure hazard to chemical compounds is through inhalation of airborne contaminants in gases, vapors, and particulate matter (dust, fumes, smoke, aerosols, and mists). Supervisors should know the physical classifications of these substances.
- The hazard associated with breathing an airborne contaminant depends on the nature of the chemical but can cause symptoms ranging from mild irritation of the nose and throat to serious injuries deep in the lungs. Concentration of airborne contaminants must be measured by an industrial hygienist. Supervisors should also be alert to the hazards of an oxygen-deficient environment. Workers can guard against respiratory hazards through the use of respiratory protective equipment and through reducing their exposure to the hazards.

- Absorption of chemicals can occur through breaks in the skin or through direct contact with intact skin. Solvents, classified as aqueous or organic, are among the most widely used liquid chemicals in industry and present the greatest hazard to workers. Solvent hazard is determined not only by the toxicity of the chemical but by the conditions of its use. Job operations involving solvents can be classified as (a) direct contact with the chemical, (b) intermittent or infrequent contact, and (c) minimal contact.

- Adverse effects of solvents vary from mild skin irritation, to more serious disorders of the central nervous system and major organs, to the formation of cancerous tumors. Even low-level exposure over time can cause long-lasting effects on workers' health. Preventive and protective measures include the proper selection, application, handling, and control of solvents; a working knowledge of each chemical's risks to worker health; a sound training program to educate workers in chemical hazards; and an effective program of engineering controls, work-practice controls, and personal protective equipment.

- Although workers may not knowingly eat or drink harmful substances, toxic chemicals such as lead oxide can be absorbed from the gastrointestinal tract into the blood. Employees working with highly toxic materials should not eat or smoke in the work area, should always wash their hands after every job, and should maintain a clean environment.

- Physical stresses include hazards such as noise, ionizing and nonionizing radiation, visible radiation, temperature extremes, skin disorders, and pressure extremes.

- Noise, defined as an unwanted sound, can have harmful psychological and physiological effects on workers as well as interfere with verbal communication on the job. Effects vary based on individual differences among workers and conditions on the job. Supervisors should try to reduce or eliminate the noise hazard as well as take protective and preventive measures to ensure workers' health.

- According to OSHA Noise Standard, noise above 85 to 90 dBA is harmful to workers. When exposed to these (or higher) noise levels, employees must either be given hearing protection equipment, have their exposure time reduced, or both, to ensure proper working conditions. Companies are required to administer annual audiometric tests to all workers exposed to noise over 85 dBA. Hearing protection equipment includes ear plugs, canal caps, and ear muffs. Supervisors should make sure these devices fit workers properly and are correctly used.

- Nonionizing radiation involves a change in the normal electrical balance in an atom. The most common forms encountered on the job include gamma radiation, alpha and beta radiation, high-energy protons and neutrons, and X-radiation. Chemicals also can interact with ionizing radiation to form highly toxic substances. This type of radiation causes injury by ionizing the tissues in which they are absorbed. Injuries can range from mild skin irritations to severe radiation poisoning, to the formation of various types of cancers. Workers can be protected from ionizing radiation by separating or shielding them from the hazard or by reducing their time of exposure.

- Ionizing radiation includes low-frequency radiation, microwaves, infrared radiation, and ultraviolet and visible radiation. These types of radiation heat the tissues in which they are absorbed. Injuries range from mild irritations to severe burns and cancerous tumors as well as possible damage to the eyes. Protective measures include shielding workers from the radiation source and providing adequate protective gear.

- Visible radiation or light can affect both the quantity and accuracy of employees' work. Proper lighting in the work area is the result of the amount and color of light, direction and diffusion, and nature of illuminated surfaces. Lighting should be bright enough to produce easy viewing and directed so that it does not create glare. Supervisors should work to eliminate or reduce glare off all reflective surfaces and VDT terminals. Visible-light tools such as lasers, if viewed directly, can injure the retina of the eyes. Protective measures include shielding laser sources, training workers in eye hazards, and use of protective eyewear.

- Temperature extremes present health hazards when the body cannot heat or cool itself adequately to overcome the extreme. Most work hazards are associated with high temperatures. The body adapts to high temperatures by sweating, radiating heat, and through heat conduction or convection. If the body cannot overcome the extreme, a worker may suffer from heatstroke, heat cramps, or heat exhaustion. Supervisors and workers must know the appropriate emergency medical steps to counteract these potentially life-threatening conditions.

- Workers can be protected against high temperatures by allowing them to acclimatize to a hot environment, providing plenty of fluids, allowing intermittent rest periods in a cooler environment, and providing proper work clothing and protective gear.

- Occupational dermatitis, or skin disorders, are generally caused by mechanical, physical, chemical, or biological agents, and plant poisons. Skin reactions are classified as primary irritation dermatitis and sensitization dermatitis. Mechanical, physical, and chemical agents produce primary irritation dermatitis. Sensitization dermatitis, on the other hand, is the result of an allergic reaction to a given substance. Occupational acne is caused by certain cutting fluids. Protective and preventive measures include scrupulous housekeeping, showering at the end of each shift, changing work clothes before leaving the job, and protective clothing.

- Atmospheric pressures related to high altitude or underwater work can cause decompression sickness, hearing problems, and oxygen starvation. An oxygen-deficient environment can impair workers' judgment or cause sudden loss of consciousness and even death in a short time. Workers should be allowed sufficient time to decompress from underwater or high-altitude work and be equipped with proper independent air supplies when entering oxygen-deficient environments.

- Ergonomics involves creating the best "fit" between worker and job, including such factors as work environment (seating, desks, lighting, noise, and so on); body postures and movements; and psychological and social aspects. The disciplines of biomechanics and biotechnology are used to solve work-stress problems on the job. Minimizing stress on workers can raise efficiency and reduce costs, number of accidents, and training time.

- Biological stresses include any virus, bacterium, fungus, or other living organism capable of causing disease in humans. The most common biological hazards in workplaces include tuberculosis, fungus infections, byssinosis, anthrax, Q fever, brucellosis, erysipelas, leptospirosis, psittacosis, Lyme disease, AIDS, and upper respiratory tract infections. Supervisors must know how to prevent organisms from entering the workplace, what steps to take if an infection occurs, and how to protect workers on the job.

- Threshold Limit Values (TLVs) have been established for airborne concentrations of many chemical compounds. TLVs represent conditions under which it is believed that nearly all workers may be repeatedly exposed without adverse effect. These values have been established for more than 600 substances and may be used to calculate exposure times for employees who work in areas where airborne contamination exists. Because the toxic levels of many substances is not known for certain, NIOSH has also set Recommended Exposure Levels (RELs) that are below the TLVs for some airborne chemicals.

- Supervisors should help to develop and enforce standard operating procedures for jobs that require industrial hygiene controls. The type and extent of controls depend on the physical, chemical, and toxic properties of contaminants, evaluation made of the exposure, and the operation that disperses or removes the contaminant. Jobs for which standard operating procedures are commonly established include (1) maintenance and repair of systems for storing and transporting fluids, (2) entering confined spaces to clean and repair them, and (3) maintaining pumps and valves. Workers should wear protective gear and be furnished with independent air supplies, if necessary.

- Supervisors are key personnel in creating effective industrial hygiene controls. They must make use of good housekeeping measures, special monitoring devices, and safety/hygiene training and education to help ensure a safe, healthy workplace.

Personal Protective Equipment

After reading this chapter, you will be able to:

- Know how to develop an effective personal protective equipment program for your workers to control hazards
- Overcome employees' objections to using personal protective equipment and educate them about the importance of such gear
- Know the types of protective equipment available for the head, face, eyes, ears, respiratory tract, torso, and extremities
- Understand how to select the best protective equipment to fit the hazard involved

When a hazard is identified in the workplace, every effort should be made to eliminate it so that employees are not harmed. Elimination thus becomes the first thrust of control. For instance, if there is a turret lathe, screw machine, or milling operation that produces large quantities of steel chips that fly all over an aisleway, the engineering control to be taken, short of eliminating the operation, is either to enclose the operation or provide a deflector to keep the chips from flying in the aisle. The first approach is to determine whether the hazard can be engineered out of the operation. Another way to reduce or control the hazard is to isolate the process. Can that dip tank be relocated? Can the spray booth be enclosed? Can that noisy machine operation be isolated, separated from the rest of the plant?

A second approach is to control the hazard by administrative control. For example, a milling operation might be done at only certain times of the day, after the shift ends. This reduces the flying-chip hazard to other employees. Now, in spite of these two approaches and because the milling operation must be done and employees will be present, it will be necessary to use personal protective equipment (PPE). For example, safety glasses may be required for the work area.

Too often PPE usage is considered the last thing to do, in the scheme of hazard control. It should not be. Personal protective equipment can provide that added protection to the employee even when the hazard is being controlled by other means. In the milling operation, a deflector virtually eliminates flying chips, but because there may be a slight chance of flying chips, the operators and other affected employees should wear eye protection.

In some situations the only available protection will be the use of PPE and often in emergencies, PPE will be required for the safety of workers. PPE should be considered one aspect of the safety program.

CONTROLLING HAZARDS

People who must work in hazardous areas should use PPE, which can protect a person from head to toe. To develop an effective PPE program, supervisors should:

- Be familiar with required standards and requirements of government regulations
- Be able to identify hazards
- Be familiar with the safety equipment on the market to protect against the specific hazard(s)
- Know the company procedures for paying for and maintaining the equipment
- Develop an effective method for convincing

employees to dress safely and to wear the proper protective equipment

- Review all material safety data sheets (MSDS) that require personal protective equipment for protection against hazardous chemicals and materials
- Consider establishing an industrial hygiene evaluation procedure to determine whether PPE is needed to meet MSDS requirements

As the supervisor, once you know what and when personal protective equipment is needed, you should make sure employees use it. If workers fail to see the reason for using protective equipment, you must impress on them its value and help them to recognize the need for it (Figure 9-1). Getting some workers to use protective equipment may be one of the toughest jobs you face.

To fulfill this responsibility, supervisors must learn about safety equipment. You should be familiar with government regulations, be able to recognize hazards, and know your equipment and your dealer so that you are thoroughly familiar with all types of protective equipment. Get help from your safety and hygiene personnel to keep up to date. Even more, allowing employees to participate in equipment selection will enhance their acceptance of it.

Local safety meetings are a good source of information, because equipment is usually on display and manufacturers' representatives are handy. Hundreds of firms exhibit at the National Safety Congress and Exposition held each fall. Company safety and human relations departments can also be helpful. A wealth of information on safety equipment is published—manufacturers' catalogs, trade literature, and the equipment issue of *Safety & Health*, published each March.

The rule to follow when specifying or buying safety equipment is to insist on the best and to deal only with reputable firms. Do not take a chance on inferior items just because they may be less expensive. Protective equipment should conform to standards where they apply; standards for protective equipment are discussed later in this chapter.

Figure 9-1. During orientation as well as every day supervisors need to motivate employees to wear properly fitted personal protective equipment. Here the supervisor is talking to a new employee about obtaining a better fitting pair of safety glasses.

Overcoming Objections

One of the biggest problems supervisors face is overcoming the objections of some workers who have to wear such protective equipment. Most of the objections you will encounter are quite similar, whether employees are talking about eye, ear, or head protection; respiratory protective equipment; or protective clothing such as aprons, gloves, and even safety shoes.

To understand why employees do not wear the prescribed protective equipment, try to be objective and to see the entire picture. Workers have common concerns regarding personal protection equipment.

First, proper fit and comfort are important. Make sure that safety glasses, ear protectors, respirators, aprons, gloves, and shoes are properly fitted. No one wants to wear something that does not feel comfortable, that is not clean, or that constantly needs adjusting (Figure 9-2).

Second, appearance should be considered. If the piece of equipment does not look attractive—for example, big clumsy safety shoes as opposed to snug steel-toed moccasins or wing-tip dress shoes—workers are turned off. If possible, let workers select their style.

Finally, workers will want to know how easily protective equipment can be cleaned and maintained. This is particularly important in the case of hearing and respiratory protection.

Check with safety or industrial hygiene personnel who can specify the appropriate equipment. They should make sure that it fits correctly, is attractive, and that there is a good cleaning and maintenance program so that the equipment is always sanitary. By using common sense, empathizing with your people, and understanding the basic principles about protective equipment, you can overcome the objections.

Selling the Need for PPE

Closely associated with overcoming objections is selling workers on the need for personal protective equipment. If your people can be made to see the need for such protection, your job is much easier. When looking at personal protective equipment, picture the human body and, starting with the head, analyze all of the hazards or the types of accidents that could possibly occur. For example, the head is vulnerable to injuries such as bumps and abrasions,

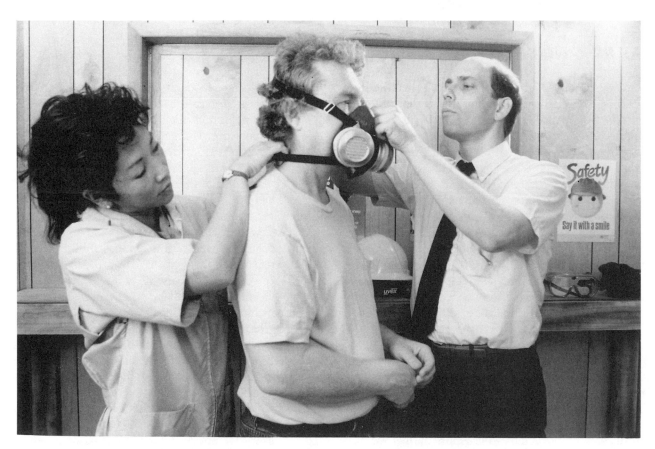

Figure 9-2. Protective equipment should be properly fitted. The fit of this cannister respirator is being checked by a technician.

so think of what types of protection are needed: hard hats, bump caps? Then move on to the eyes, ears, throat and lungs (respiratory protection), arms and legs, torso, hands and toes. In this checklist fashion, you will be reminded of all the hazards that can occur and the protection that is required.

Cost of Equipment

Companies differ in their personal protective equipment payment policies. Usually companies that require extremely clean operations will furnish laundered coveralls, aprons, smocks, and other garments.

Personal items, such as safety shoes and prescription safety glasses, are often sold on a shared-cost basis with employees paying part of the cost. Some firms maintain safety shoe stores for their employees. Other items considered necessary to the job are often supplied free to the employee. A welder's helmet and welder's gloves are good examples. Ordinary work clothes and work gloves are, however, usually purchased by the employee.

Companies should be concerned with controlling costs, but not at the expense of safety. In the long run it can be less expensive to purchase necessary equipment than to pay for workers' injuries caused by the lack of proper equipment.

HEAD PROTECTION

Safety helmets are needed on jobs where a person's head is menaced by falling or flying objects or by bumps (Figure 9-3). American National Standard Z89.1–1986, *Protective Headwear for Industrial Workers*, gives specifications that a protective hat must meet.

Impact resistance is essential. Where contact with energized circuits is possible, only helmets that meet the requirements of Class B, ANSI Z89.1, should be worn. These helmets should have no conductive fittings passing through the shell. Class B hard helmets are tested at 20,000 volts.

A brim around the helmet provides the most complete protection for the head, face, and back of the neck. In a situation where the brim would get in the way, a helmet with a cap visor may have to be worn.

Another type of head protection is known as "bump hats" or "bump caps." These are used only in confined spaces where the hazard is limited to bumping the head on some obstruction. Bump hats should never be used on construction sites, shipyards, or other locations where more dangerous hazards are present. These head gear do not meet the requirements of ANSI Z89.1.

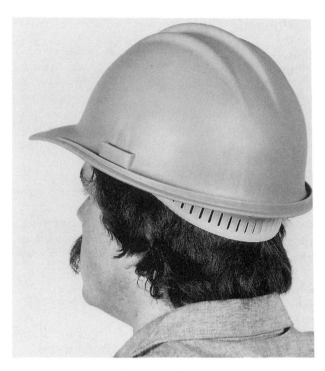

Figure 9-3. Safety helmets displace the impact of blows to the head from falling or flying objects or bumps. This safety helmet also has a rain trough. (Courtesy E.D. Bullard Company.)

Types and Materials

Plastic, molded under high pressure, is most frequently used for safety helmets. It resists impact, water, oil, and electricity. Fiberglass impregnated with resin is preferred because of its high strength-to-weight ratio, high dielectric strength, and resistance to moisture. Helmets with reflective trim are available on special order. They give added protection against traffic accidents that may occur when employees work at night or in darkened areas.

The hard outer shell of the helmet is supported by a suspension—a cradle attached to a headband that keeps the shell away from the head and provides some degree of protection against falling objects. This suspension must be adjusted to ensure a good fit. Keep extra cradles and headbands on hand, and replace a cradle or headband whenever it shows signs of deterioration or soiling.

Bands, cradles, and shells can be washed in warm, soapy water and rinsed, or steam cleaned. Thirty days is the recommended maximum interval between cleanings. Make sure helmets and suspensions are thoroughly cleaned before reissuing them. The suspension should be replaced at least once a year.

Never attempt to repair the shell of a helmet once it has been broken or punctured. A few extra shells should be kept on hand as replacements. Do not let people drill holes in their safety helmets to improve ventilation or let them cut notches in the brims. Such practices destroy the ability of the helmet to protect the wearer.

Auxiliary Features

Liners for safety head gear are available for cold weather use. Do not let workers remove the safety helmet suspension in order to wear the helmet over parka hoods. This practice completely destroys the protection given by the helmet and has led to tragic results.

A chin strap is useful when the wearer may be exposed to strong winds, such as those on oil derricks, for example (Figure 9-4).

An eye shield of transparent plastic can be attached to some helmets. The shield is secured under the brim and lies flat against it when not in use.

Brackets to support welding masks or miners' cap lamps are available on some safety helmets. Other mounting accessories for hearing protection, and eye and face protection are available (Figure 9-5).

Standard colors of safety helmets are white, yellow, red, green, blue, brown, and black. Other colors are available on special order. Colors that go all the way through the laminated shell are permanent. Painting is not recommended because paint may contain solvents that can make the shell brittle.

Often distinctive colors or designs are used to designate the wearer's department or trade, especially in companies or plants where certain areas are restricted to a few selected employees.

Obtaining User Acceptance

In some cases, workers' personal preferences in appearance and dress may clash with safety regulations. For example, hair styles may be incompatible with wearing safety helmets. As the supervisor, you will have to convince people that it is more important (indeed, it may be required) to wear helmets than to have stylish hairdos.

In certain industries, such as in food-processing plants, workers are required not only to wear bump caps but to cover all facial hair—beards and mustaches—with hair nets. This is a requirement of the FDA and is done for sanitary reasons. Here supervisors have an additional job cut out for them; they must explain the reasons for this protection to the workers.

In overcoming objections and convincing people to use PPE, you must be prepared to counter all types of complaints. In the case of safety helmets, for instance, workers may complain that in winter the helmets are too cold. A liner or skull cap large enough to come down over the ears can take care of this objection. However, the suspension must not be removed. Some people may complain that helmets give them headaches. This is a possible effect until the employee becomes accustomed to wearing the helmet. In general, however, there is no physiological reason for a properly fitted helmet to cause a headache.

Although safety helmets are slightly heavier than ordinary hats, the difference is unnoticed after a while, particularly if the headband and suspension are properly adjusted. Workers should not be bothered by the helmets unless there are unique physiological or psychological reasons for their complaints.

FACE PROTECTION

Many types of personal protective equipment shield the face (and sometimes also the head and neck) against impact, chemical or hot metal splashes, heat radiation, and other hazards.

Face shields of clear plastic protect the eyes and

Figure 9-4. This rescue worker needs a safety helmet with a 3-point chin strap to ensure the helmet does not fall off as he climbs. (Courtesy E.D. Bullard Company.)

Figure 9-5. This tree trimmer uses a safety helmet with a 2-point chin strap and attachments for a safety shield and muff-type hearing protectors. (Courtesy E.D. Bullard Company.)

face of a person who is sawing or buffing metal, doing sanding or light grinding, or handling chemicals. The shield should be slow burning and must be replaced if warped or scratched. A regular replacement schedule must be set up because plastic tends to become brittle with age.

The headgear and shield should be easy to clean and be adjustable to the size and contour of the head. Many types permit the wearer to raise the shield without removing the headgear.

Plastic shields with dichroic coating or metal screen face shields deflect heat from a person and still permit good visibility. They are used around blast furnaces, soaking pits, heating furnaces, and other sources of radiant heat.

Helmets

Welding helmets protect the eyes and face against the splashes of molten metal and the radiation produced by arc welding (Figure 9-6). Helmets should have the proper filter glass to keep ultraviolet and visible rays from harming the eyes. A list of shade numbers recommended for various operations is given in Table 9-1. Workers' eyes vary due to age, general health, and the care given to them. Two people may require different shaded lenses when doing the same job. Cracked or chipped filter lenses must be replaced. Otherwise, they will permit harmful rays to reach the welder's eyes.

The shell of the helmet must resist sparks, molten metal, and flying particles. It should be a poor heat conductor and a nonconductor of electricity. Helmets that develop pinholes or cracks must be discarded. Helmets should have headgear that permit workers to use both hands and to raise the helmet to position their work.

Most types of helmets have a replaceable, heat-treated glass or plastic covering to protect the filter lens against pitting and scratching. Some helmets have a lift-front glass holder that permits the welder to inspect the work without lifting or removing the helmet.

Impact goggles should be worn under helmets to protect welders from flying particles when the helmets are raised. The spectacle type with side

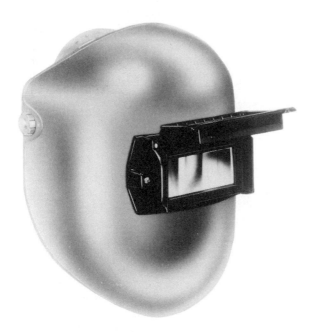

Figure 9-6. This welder's helmet protects against splashes of molten metal and radiation. (Courtesy Sellstrom Manufacturing Co.)

TABLE 9-1. Selection of Shade Number for Welding Filters

Welding Operation	Suggested Shade Number*
Shielded Metal-Arc Welding, up to 5/32 in. (4 mm) electrodes	10
Shielded Metal-Arc Welding, 3/16 to 1/4 in. (4.8 to 6.4 mm) electrodes	12
Shielded Metal-Arc Welding, over 1/4 in. (6.4 mm) electrodes	14
Gas Metal-Arc Welding (Nonferrous)	11
Gas Metal-Arc Welding (Ferrous)	12
Gas Tungsten-Arc Welding	12
Atomic Hydrogen Welding	12
Carbon Arc Welding	14
Torch Soldering	2
Torch Brazing	3 or 4
Light Cutting, up to 1 in. (25 mm)	3 or 4
Medium Cutting, 1 to 6 in. (26 to 150 mm)	4 or 5
Heavy Cutting, over 6 in. (150 mm)	5 or 6
Gas Welding (Light), up to 1/8 in. (3.2 mm)	4 or 5
Gas Welding (Medium), 1/8 to 1/2 in. (3.2 to 12.7 mm)	5 or 6
Gas Welding (Heavy), over 1/2 in. (12.7 mm)	6 or 8

*The choice of a filter shade may be made on the basis of visual acuity and may therefore vary widely from one individual to another, particularly under different current densities, materials, and welding processes. However, the degree of protection from radiant energy afforded by the filter plate or lens when chosen to allow visual acuity will still remain in excess of the needs of eye filter protection. Filter plate shades as low as shade 8 have proven suitably radiation-absorbent for protection from the arc-welding processes.

In gas welding or oxygen cutting where the torch produces a high yellow light, it is desirable to use a filter lens that absorbs the yellow or sodium line in the visible light of the operation (spectrum).

Notes:

1. All filter lenses and plates shall meet the test for transmission of radiant energy prescribed in ANSI Z87.1–1989, *Practice for Occupational and Educational Eye and Face Protection.*

2. All glass for lenses shall be tempered, substantially free from striae, air bubbles, waves, and other flaws. Except when a lens is ground to provide proper optical correction for defective vision, the front and rear surfaces of lenses and windows shall be smooth and parallel.

3. Lenses shall bear some permanent distinctive marking by which the source and shade may be readily identified.

Source: *U.S. Code of Federal Regulations,* Title 29—Labor, Chapter XVII, Part 1910, Occupational Safety and Health Standards, § 1910.252(e)(2)(i), and Part 1926, Safety and Health Regulations for Construction, § 1926.102(a)(5).

shields is recommended as minimum protection from flying materials, adjacent work, or from the popping scale of a fresh weld.

Welders' helpers should have proper welding goggles or helmets to wear while assisting in a welding operation or while chipping flux away after a bead has been run over a joint. The danger of foreign bodies becoming lodged in the eye is great for workers performing this latter operation without protection.

Shields and Goggles

A hand-held shield can be used where the protection of a helmet is not needed, such as for inspection work, tack welding, and other operations requiring little or no welding by the user. Frame and lens construction are similar to that of the helmet.

Welding goggles are available with filter glass shades up to No. 8. If darker shades are required, then complete face protection is needed because of the danger of skin burns. When shades darker than No. 3 must be used, side shields or cup goggles are recommended.

Hoods

Acid-proof hoods that cover head, face, and neck are used by persons exposed to the risk of severe splashes from corrosive chemicals (Figure 9-7). This type of hood has a window of glass or plastic that is securely joined to the hood to prevent acid from

seeping through. Hoods made of rubber, neoprene, plastic film, or impregnated fabrics are available for resistance to different chemicals. Consult manufacturers to find the protective properties of each material.

Hoods with air supply should be worn for work

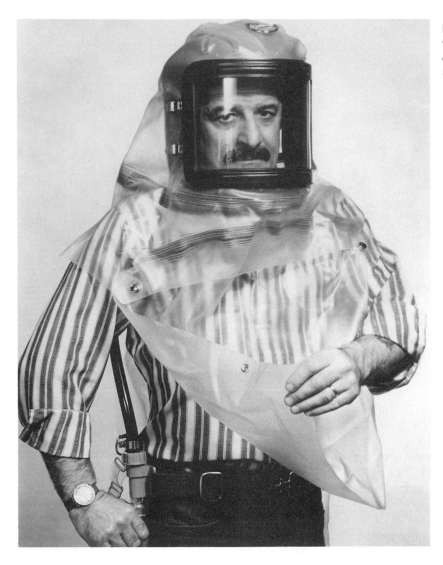

Figure 9-7. This air-supplied hood with plastic window shields against chemical splashes. (Courtesy AO Safety Products.)

around toxic fumes or dusts. These hoods provide a supply of clean, breathing-quality air, which excludes toxic materials and maintains worker comfort. To support the air hose, the worker should wear a harness or belt.

Transparent face shields, supported from a head harness or headband, are used when there is limited exposure to direct splashes of corrosive chemicals. Splash-resistant cup goggles should be worn under the shield for added protection.

EYE PROTECTION

Industrial operations expose the eyes to a variety of hazards, such as flying objects, splashes of corrosive liquids or molten metals, dusts, and harmful radiation. Eye injuries not only disable a person but they often disfigure the face. Per-injury cost is high to both employee and employer.

Flying objects such as metal or stone chips,

nails, or abrasive grits cause most injuries. The National Society to Prevent Blindness lists the other chief causes of eye injury as:

- Abrasive wheels (small flying particles)
- Corrosive substances
- Damaging visible or thermal radiation
- Splashing metal
- Poisonous gases or fumes

Operations in which hardened metal tools are struck together, where equipment or material is struck by a metal handtool, or where the cutting action of a tool causes particles to fly usually require the user and other workers nearby to wear eye protection. The hazard can be minimized by using nonferrous, "soft" striking tools and by shielding the job with metal, wood, or canvas deflectors. Safety goggles or face shields should be worn when woodworking or cutting tools are used at head level or overhead with the chance of particles falling or flying into the eyes (Figure 9-8).

Occasionally, the need for eye protection is overlooked on many potentially hazardous jobs. These include cutting wire and cable, striking wrenches, using hand drills, chipping concrete, removing nails from scrap lumber, shoveling material to head level, working on the leeward side of the job, using wrenches and hammers overhead, and other jobs where particles or debris may fall. Make sure workers on these jobs wear the proper protective eye gear.

Contact Lenses

Because contact lenses are very popular among workers, you should understand the potential safety problems related to them. Where there are appreciable amounts of dust, smoke, irritating fumes, or liquid irritations that could splash into the eyes, contact lenses usually are not recommended. Often a judgment must be made, however, as to the degree of potential hazard. Some workers obtain better vision correction with contact lenses and, therefore, need them. These persons should wear the proper safety spectacles or goggles over their contacts when in an area where any sort of potential eye hazard exists.

One rarely noticed regulation under OSHA prohibits the wearing of contact lenses in contaminated areas with respirators; see *CFR* 1910.134(e)(5)(ii).

The National Society to Prevent Blindness, through its Advisory Committee on Industrial Eye Health and Safety, has issued the following position statement:

Except for work environments in which there are significant risks of ocular injury, employees may be allowed to wear contacts on the job. However, contact lens wearers must conform to management policy regarding contact lens use. When formulating policy on contact lens use, OSHA and NIOSH guidelines should be used. The National Society to Prevent Blindness offers a list of recommendations on contact lens use in the workplace. To obtain these recommendations and for further information, contact the Society, 500 East Remington Road, Schaumburg, IL 60173, (708) 843-2020.

Goggles

Goggles and other kinds of eye protection are available in many styles, along with the protective medium of heat-treated or chemically treated glass, plastic, wire screen, or light-filtering glass (Figure 9-9). Supervisors should be familiar with the various forms of eye protection and should know which ones are the best for each job.

Types and materials. Cover goggles are frequently worn over ordinary spectacles. Lenses that have not been heat- or chemically treated are easily broken. Cover goggles protect against pitting, as well as breaking.

There goggles include the cup type with heat-treated lenses and the wide-vision type with plastic lenses. Both are used for heavy grinding, machining, shipping, riveting, working with molten metals, and similar heavy operations. They offer the advantage of being wide enough to protect the eye socket and to distribute a blow over a wide area.

Spectacles without side shields may be worn only when it is unlikely that particles will fly toward the side of the face. However, spectacles with side shields are recommended for all industrial uses. Frames must be rigid enough to hold lenses directly in front of the eyes. These glasses should be fitted by a specialist.

Goggles with soft vinyl or rubber frames protect eyes against splashes of corrosive chemicals and exposure to fine dusts or mists. Lenses can be heat-treated glass or acid-resistant plastic. For exposures involving chemical splashes, these goggles are equipped with baffled ventilators on the sides. For vapor or gas exposures, they must be nonventilated. Some types are made to fit over spectacles.

Dust goggles should be worn by employees who

Figure 9-8. Fog ban safety goggles with high impact-resistant polycarbonate lenses. (Courtesy General Bandages, Inc.)

AMERICAN NATIONAL STANDARD Z87.1-1989

PROTECTIVE DEVICES

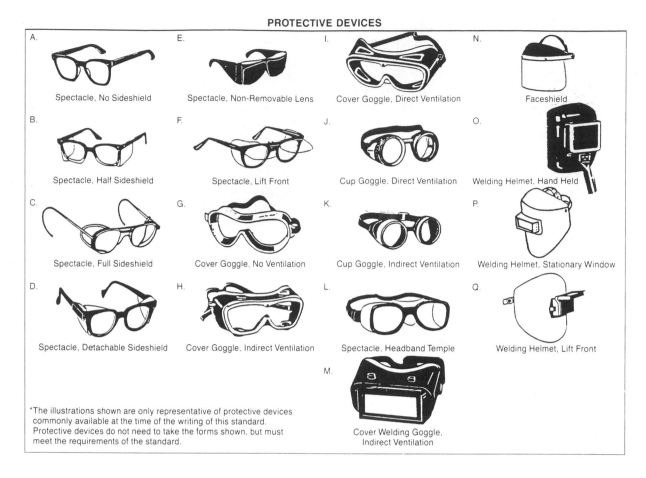

A. Spectacle, No Sideshield
B. Spectacle, Half Sideshield
C. Spectacle, Full Sideshield
D. Spectacle, Detachable Sideshield
E. Spectacle, Non-Removable Lens
F. Spectacle, Lift Front
G. Cover Goggle, No Ventilation
H. Cover Goggle, Indirect Ventilation
I. Cover Goggle, Direct Ventilation
J. Cup Goggle, Direct Ventilation
K. Cup Goggle, Indirect Ventilation
L. Spectacle, Headband Temple
M. Cover Welding Goggle, Indirect Ventilation
N. Faceshield
O. Welding Helmet, Hand Held
P. Welding Helmet, Stationary Window
Q. Welding Helmet, Lift Front

*The illustrations shown are only representative of protective devices commonly available at the time of the writing of this standard. Protective devices do not need to take the forms shown, but must meet the requirements of the standard.

NOTES:

(1) Care shall be taken to recognize the possibility of multiple and simultaneous exposure to a variety of hazards. Adequate protection against the highest level of each of the hazards must be provided.

(2) Operations involving heat may also involve optical radiation. Protection from both hazards shall be provided.

(3) Faceshields shall only be worn over primary eye protection.

(4) Filter lenses shall meet the requirements for shade designations in Table 1.

(5) Persons whose vision requires the use of prescription (Rx) lenses shall wear either protective devices fitted with prescription (Rx) lenses or protective devices designed to be worn over regular prescription (Rx) eyewear.

(6) Wearers of contact lenses shall also be required to wear appropriate covering eye and face protection devices in a hazardous environment. It should be recognized that dusty and/or chemical environments may represent an additional hazard to contact lens wearers.

(7) Caution should be exercised in the use of metal frame protective devices in electrical hazard areas.

(8) Refer to Section 6.5, Special Purpose Lenses.

(9) Welding helmets or handshields shall be used only over primary eye protection.

(10) Non-sideshield spectacles are available for frontal protection only.

Figure 9-9. Goggles and other eye protection must be selected for the specific hazard on the job. Notes refer to Table 9-2. (Reprinted with permission from ANSI Z87.1–1989.)

work around noncorrosive dusts—for example, in cement and flour mills. The goggles have heat-treated or filter lenses; wire-screen ventilators around the eye cup provide air circulation.

Welders' goggles with filter lenses are available for such operations as oxyacetylene welding, cutting, lead burning, and brazing. Table 9-1 is a guide for the selection of the proper shade numbers. These recommendations may be varied to suit individual needs.

TABLE 9-2. Eye and Face Protective Equipment to Be Worn for Various Types of Exposures

AMERICAN NATIONAL STANDARD Z87.1-1989

		ASSESSMENT SEE NOTE (1)	PROTECTOR TYPE	PROTECTORS	LIMITATIONS	NOT RECOMMENDED
I M P A C T	Chipping, grinding, machining, masonry work, riveting, and sanding.	Flying fragments, objects, large chips, particles, sand, dirt, etc.	B,C,D, E,F,G, H,I,J, K,L,N	Spectacles, goggles faceshields SEE NOTES (1) (3) (5) (6) (10) For severe exposure add N	Protective devices do not provide unlimited protection. SEE NOTE (7)	Protectors that do not provide protection from side exposure. SEE NOTE (10) Filter or tinted lenses that restrict light transmittance, unless it is determined that a glare hazard exists. Refer to OPTICAL RADIATION.
H E A T	Furnace operations, pouring, casting, hot dipping, gas cutting, and welding.	Hot sparks	B,C,D, E,F,G, H,I,J, K,L,*N	Faceshields, goggles, spectacles *For severe exposure add N SEE NOTE (2) (3)	Spectacles, cup and cover type goggles do not provide unlimited facial protection. SEE NOTE (2)	Protectors that do not provide protection from side exposure.
		Splash from molten metals	*N	*Faceshields worn over goggles H,K SEE NOTE (2) (3)		
		High temperature exposure	N	Screen faceshields. Reflective faceshields. SEE NOTE (2) (3)	SEE NOTE (3)	
C H E M I C A L	Acid and chemicals handling, degreasing, plating	Splash	G,H,K *N	Goggles, eyecup and cover types. *For severe exposure, add N	Ventilation should be adequate but well protected from splash entry	Spectacles, welding helmets, handshields
		Irritating mists	G	Special purpose goggles	SEE NOTE (3)	
D U S T	Woodworking, buffing, general dusty conditions.	Nuisance dust	G,H,K	Goggles, eyecup and cover types	Atmospheric conditions and the restricted ventilation of the protector can cause lenses to fog. Frequent cleaning may be required.	

				TYPICAL FILTER LENS SHADE	PRO-TECTORS		
O P T I C A L R A D I A T I O N		WELDING:		SEE NOTE (9)			
		Electric Arc	O,P,Q	10-14	Welding Helmets or Welding Shields	Protection from optical radiation is directly related to filter lens density. SEE NOTE (4). Select the darkest shade that allows adequate task performance.	Protectors that do not provide protection from optical radiation. SEE NOTE (4)
		WELDING:		SEE NOTE (9)			
		Gas	J,K,L, M,N,O, P,Q	4-8	Welding Goggles or Welding Faceshield		
		CUTTING		3-6			
		TORCH BRAZING		3-4		SEE NOTE (3)	
		TORCH SOLDERING	B,C,D, E,F,N	1.5-3	Spectacles or Welding Faceshield		
		GLARE	A,B	Spectacle SEE NOTE (9) (10)		Shaded or Special Purpose lenses, as suitable. SEE NOTE (8)	

16

All goggle frames should be of corrosion-resistant material that will neither irritate nor discolor the skin, that can withstand sterilization, and that are flame-resistant or nonflammable. Metal frames should not be worn around electrical equipment or near intense heat.

To permit the widest range of vision, goggles should be fitted as close to the eyes as possible without the eyelashes touching the lens. The lenses should have no appreciable distortion or prism effect. Some specifications are covered by ANSI Z87.1-1989, *Practice for Occupational and Educational Eye and Face Protection* (Figure 9-9 and Table 9-2).

Cleaning after use. Eye protection equipment must be sterilized before being reissued to different employees. The proper procedure is to disassemble the equipment and wash it with soap or detergent in warm water. Rinse thoroughly. Immerse all parts in a solution containing germicide, deodorant, and fungicide. Do not rinse, but hang to dry in air. When dry, place the parts in a clean, dust-proof container—a plastic bag is acceptable. Many companies have stations that dispense cleaning liquid and tissues to encourage frequent cleaning after goggles get smudged or dirty. Carrying cases or storage cabinets also help promote care of eye protection.

Replace defective parts. If a lens has more than just the most superficial scratch, nick, or pit, it should be replaced. Such damage can materially reduce protection afforded the wearer.

Some companies keep extra goggles in the main supply department, while others keep them in individual departments, along with a stock of parts. A third practice is to have an eye protection cart with a trained attendant who makes the rounds of the company to clean, adjust, repair, or replace eye protection on the job.

The following are other important considerations regarding the care of eye protection equipment and overcoming employee complaints. Glasses should never be placed with the lenses facing down because they could become scratched, pitted, or dirty. They should not be kept in a bench drawer or tool box, unless the goggles are in a sturdy case. Cleaning stations or materials should be easily available.

A lens "fog" problem can usually be eliminated by use of one of the many commercial antifog preparations. In hot weather, workers can wear a sweatband to keep perspiration off their goggles. Although it takes a little effort to keep the goggles clean, this should not be an excuse for workers to discard them and risk losing an eye.

EAR PROTECTION

Excessive noise should be reduced by engineering changes and administrative controls, whenever possible, as outlined in Chapter 8, Industrial Hygiene. Ear protection should be used only as a last resort.

Although there is some disagreement as to the maximum intensity of sound to which the human ear can be subjected without damage to hearing, the standards of the Occupational Safety and Health Act should be used as a minimum level of protection. Under OSHA, where the sound levels exceed an 8-hour time-weighted average of 85 dB measured on the A scale, a continuing, effective hearing conservation program shall be administered. This level may be increased slightly as the duration of exposure decreases. (Review the discussion in Chapter 8.)

On the decibel scale, zero is the threshold of hearing for a sensitive ear, and 130 dB is the threshold of pain. The decibel is a ratio-type scale. If the sound pressure is doubled, it increases 6 dB; if halved, it decreases 6 dB. If cut to one-tenth, it decreases 10 dB; if cut to one one-hundredth, it decreases 20 dB. The human ear is not as sensitive as a sound-measuring device; it often interprets a 10 dB drop in intensity as being "half as loud."

It takes specialized equipment and trained personnel to analyze a noise exposure and a specialist or company medical department to recommend the types of ear protection that will be most effective (Figure 9-10). The best ear protection is the one that is accepted by the individual and is worn properly. Properly fitted protectors can be worn continuously by most people and will provide adequate protection against most industrial noise exposures.

Insert Ear Protectors

Insert protectors are, of course, inserted into ear canals and vary considerably in design and material. Materials used are pliable rubber, soft or medium plastic, wax, and cotton. Rubber and plastic are popular because they are inexpensive, easy to keep clean, and give good performance. Wax tends to lose its effectiveness during the workday because jaw movement changes the shape of the ear canal and this breaks the acoustical seal between ear and insert. Wax inserts may be objectionable for use in dirty areas because they must be shaped by hand. They should be used only once. Cotton is a poor choice because of its low attenuating properties and because it must be hand formed.

Still another type of ear plug popular with workers is one that is molded to fit each ear. After being allowed to set, it will hold its form. Because

Figure 9-10. This selection of ear-protective devices includes ear muffs, ear plugs, and Swedish wool.

each person's ear canal is shaped differently, these plugs become the property of the individual to whom they were fitted. The plugs, of course, must be fitted by a trained, qualified professional.

Some pressure is required to fit a rubber or plastic earplug into the ear canal so that the noise does not leak around the edges (as much protection as 15 dB could be lost). As a result, points of pressure develop, and they may cause discomfort. Even though plugs can be fitted individually for each ear, a good seal cannot be obtained without some initial discomfort. However, if the earplugs are made of soft material and kept clean, they should not create any lasting problems. Avoid earplugs composed of hard, rigid materials as they could injure the ear canal.

Skin irritations, injured ear drums, or other harmful side effects of PPE are rare when ear protectors are properly designed, well fitted, and kept clean. They should cause no more difficulty than does a pair of well-fitted safety goggles. If fitting continues to be a problem, casts can be made of the ear canals, and a plastic plug can be custom-made for each.

As supervisor, you should be aware of noise hazards in your department. You should also be able to demonstrate the correct use of earplugs when other forms of control are not practical.

Muff Devices

Cup or muff devices cover the external ear to provide an acoustic barrier. The effectiveness of these devices varies with the size, seal material, shell mass, and suspension of the muff as well as with the size and shape of workers' heads. Muffs are made in a universal type or in specific head, neck, or chin sizes. Hearing protection kits that can be used with hard helmets are also available.

In the past few years, some employees have listened to radios with earphones. You, as the supervisor, should make sure that these are not worn in the shop. In addition to the possibility of damage to the ears from excessive noise levels, they are not safe in a working environment. People may not be able to hear oncoming plant trucks, vehicular traffic, other workers, or any number of industrial noises. Most Air Force bases have banned their use for this reason.

RESPIRATORY PROTECTION

Respiratory equipment can be regarded as emergency equipment, or equipment for occasional use. Of course, if contaminants are present, they should be removed at the source, or the process should be isolated. Since leaks and breakdowns do occur, however, and since some operations expose a person only briefly and infrequently, respiratory equipment should be available. Workers must be instructed and trained in its proper use and its limitations.

Types of Equipment

Respiratory equipment includes air-purifying devices (mechanical filter respirators, chemical cartridge respirators, combination mechanical filter and chemical cartridge respirators, and masks with canisters, Figure 9-11); air-supplied devices (airline respirators, Figure 9-12); and self-contained breathing apparatus (Figure 9-13).

Air-purifying devices remove contaminants from air as it is being breathed. They can be used only in environments containing sufficient oxygen to sustain life. Air-purifying devices are only effective in the limited concentration ranges for which they are designed, and must never be used where contaminant levels exceed the respirator manufacturer's accepted protection factor. These respirators generally consist of a soft, resilient facepiece and some kind of replaceable filtering element. Several types of air-purifying respirators, however, are available as completely disposable units. Various chemical filters can be employed to remove specific gases and vapors, while mechanical filters remove particulate contaminants. Air-purifying devices are never used in environments that are immediately hazardous to life or health.

Mechanical filter respirators must protect against exposure to nuisance dusts and pneumoconiosis-producing dusts, mists, and fumes. Examples of nuisance dusts are aluminum, cellulose, cement, flour, gypsum, and limestone.

Pneumoconiosis comes from three Greek words that mean "lung," "dust," and "abnormal condition." The generally accepted meaning of the word is merely "dusty lung." The kind of dust inhaled determines the type of condition or injury. Many organic dusts are capable of producing lung diseases, but not all these diseases are classified as pneu-

Figure 9-12. This is a type C supplied-air respirator and a 5-minute compressed air self-contained escape breathing apparatus to use in the event of a primary air source failure. (Courtesy National Draeger.)

moconioses because they don't represent a "dusty condition" of the lung.

In rare cases, so much dust has been inhaled that it causes mechanical blockage of the air spaces. Flour dust has been known to do this. Some dusts are essentially inert and remain in the lungs indefinitely with no recognizable irritation. A few (like limestone dust) may gradually dissolve and be eliminated without harm. Another type of filter respirator is approved for toxic dusts, such as lead, asbestos, arsenic, cadmium, manganese, selenium, and their compounds.

Protection against mists—for example, chromic acid, and exposure to such fumes as zinc and lead—is given by mechanical filter respirators specifically approved for such exposures. The filter, usually made of paper or felt, should be replaced frequently. If it becomes clogged, it restricts breathing or becomes inoperative. This condition can happen as frequently as several times a shift.

Figure 9-11. These employees are wearing filter and cartridge respirators while taking an air sample.

Figure 9-13. This compressed air breathing apparatus is designed to provide respiratory protection during entry into and escape from atmospheres immediately dangerous to life or health.

A mechanical filter respirator is of no value as protection against chemical vapors, injurious gases, or oxygen deficiency. To use it under these conditions is a serious mistake.

Chemical cartridge respirators have either a half-mask facepiece or a full-mask facepiece connected to one or more small containers (cartridges) or sorbent, typically activated charcoal or soda lime (a mixture of calcium hydroxide with sodium or potassium hydroxide) for absorption of low concentrations of certain vapors and gases.

The life of the cartridges can be relatively short. For protection against mercury vapors, the nominal container life is 8 hours. After use, the cartridges must be discarded. These respirators must not be used in atmospheres immediately dangerous to life or health, such as those deficient in oxygen.

Gas masks consist of a facepiece or mouthpiece connected by a flexible tube to a canister. Inhaled air, drawn through the canister, is cleaned chemically. Unfortunately, no one chemical has been found that removes all contaminants. Therefore, the canister must be chosen to match the contaminants. Gas mask canisters are color-coded according to the type of exposure (Table 9-3).

Gas masks have definite limitations on their effectiveness. Both gas concentration and length of time influence this. Gas masks, like chemical cartridge respirators, do not protect against oxygen deficiency. When the canister is used up, it should be removed and replaced by a fresh one. Even if they have not been used, canisters should be replaced periodically. Manufacturers will indicate the maximum effective life.

Gas masks must be quickly available for emergencies. For example, in an ice plant where an ammonia leak is likely, masks should be placed either just inside or just outside the exit doors so that they can be reached quickly. Masks should be stored away from moisture, heat, and direct sunlight. They should be inspected regularly.

Hose masks, with or without a blower, should not be used in atmospheres immediately dangerous to life or health.

Supplied-air devices deliver breathing air through a hose connected to the wearer's facepiece. The air source used is monitored frequently to make sure it does not become contaminated—for example, with carbon monoxide.

The air-line respirator can be used in atmospheres not immediately dangerous to life or health, especially where working conditions demand continuous use of a respirator. Each person should be assigned his or her own respirator.

Air-line respirators are connected to a compressed air line. A trap and filter should be installed

TABLE 9-3. Color Assigned to Canister or Cartridge

Atmospheric Contaminant(s) to Be Protected Against	Color Assigned	ISCC-NBS Centroid Color Number	ISCC-NBS Centroid Color Name
Acid gases	White	263	White
Organic vapors	Black	267	Black
Ammonia gas	Green	139	Vivid green
Carbon monoxide gas	Blue	178	Strong blue
Acid gases and organic vapors	Yellow	82	Vivid yellow
Acid gases, ammonia, and organic vapors	Brown	75	Deep yellow brown
Acid gases, ammonia, carbon monoxide, and organic vapors	Red	11	Vivid red
Other vapors and gases not listed above	Olive	106	Light olive
Radioactive materials (except tritium and noble gases)	Purple	218	Strong purple
Dusts, fumes, and mists (other than radioactive materials)	Orange	48	Vivid orange

Notes:
1. A purple (ISCC-NBS Centroid Number 218) stripe shall be used to identify radioactive materials in combination with any vapor or gas.
2. An orange (ISCC-NBS Centroid Number 48) stripe shall be used to identify dusts, fumes, and mists in combination with any vapor or gas.
3. Where labels only are colored to conform with this table, the canister or cartridge body shall be gray (ISCC-NBS Centroid Number 265), or a metal canister or cartridge body may be left in its natural metallic color.
4. The user shall refer to the wording of the label to determine the type and degree of protection the canister or cartridge will afford.

Source: ANSI K13.1–1973.

in the compressed air line ahead of the mask to separate oil, water, grit, scale, or other matter from the air stream. When line pressures are over 25 psi (170 kPa), a pressure regulator is required. A pressure release valve, set to operate if the regulator fails, should be installed.

To get clean air, keep the compressor intake away from any source of contamination, for example, internal combustion engine exhaust. The compressor should have a carbon monoxide alarm in order to guard against the carbon monoxide hazard either from overheated lubricating oil or from engine exhaust. The most desirable air supply is provided by a nonlubricated or externally lubricated medium-pressure blower—for example, a rotary compressor.

If workers need to move from place to place, they may find that the air hose is a nuisance. You should be aware that using the hose will reduce its efficiency. Care must also be exercised to prevent damage to the hose; for example, it should not be permitted to be in oil.

Abrasive blasting helmets are a variety of an air-line respirator designed to protect the head, neck, and eyes against the impact of the abrasive, and to give a supply of breathing air. The air quality requirements are the same as those described for air-line respirators.

The helmet should be covered both inside and out with a tough, resilient material. This increases comfort and still resists the abrasive. Some helmets have an outer hood of impregnated material and a zippered inner cape for quick removal. The helmet should contain a glass window, protected by a 30- to 60-mesh fine wire screen or plastic cover plate. Safety glass, used to prevent shattering under a heavy blow, should be free of color and glass defects.

A lighter weight helmet is approved for less hazardous work, such as spraying paint, buffing, bagging, and working in clean rooms (Figure 9-12).

Self-contained breathing apparatus offers protection for various periods of time by providing a portable air supply, which is usually worn by the user. The four principal types of self-contained breathing apparatus are oxygen rebreathing, self-generating, demand, and pressure-demand.

The wearer of a self-contained breathing apparatus is independent of the surrounding atmosphere; therefore, this kind of respiratory protective equipment must be used in environments where air contaminants are immediately harmful to life. This equipment is frequently used in mine rescue work and in firefighting (Figure 9-13). Because of the extreme hazard, no one wearing self-contained breathing apparatus should work in an oxygen-deficient atmosphere unless other people similarly equipped are standing by, ready to give help.

Protection factors are a measure of the overall effectiveness of a respirator. These factors, based on tests and on professional judgment, range from 5 to 10,000. The maximum use concentration for a respirator is determined by multiplying the Threshold Limit Values (TLVs) of the substance to be protected against by the protection factor. For example, a respirator with a protection factor of 10 for acetic acid would protect a worker against concentrations up to 10 times its TLV. Since the TLV of acetic acid is 10 ppm, the worker would be protected in atmospheres containing acetic acid concentrations as high as 100 ppm. (TLVs were explained in Chapter 8.)

Selecting the Respirator

Air contaminants range from relatively harmless substances to toxic dusts, vapors, mists, fumes, and gases that may be extremely harmful. Respiratory protection is required in certain areas or during certain operations when engineering controls are not available to reduce airborne concentrations of contaminants to a safe level. Engineering controls might not be present because they are technically unfeasible, or the hazardous operation might be done infrequently, making controls impractical. Respiratory protection is also needed while engineering controls are being implemented.

Your first step in selecting a respirator is to determine the chemical or other offending substance in the environment and to evaluate the extent of the hazard. With this information, you can choose the respirator that best protects against the particular hazard. Before respiratory equipment is ordered, discuss the type of exposure you have identified with the company safety or industrial hygiene department, and with manufacturers and dealers.

You must make sure the respirator fits workers properly. In addition, you should explain why it is essential to wear the respiratory protection, how it works, and, in general, why workers should wear the proper respirator for the operation in question.

Only respirators approved by the National Institute for Occupational Safety and Health (NIOSH) and the Mine Safety and Health Administration (MSHA) are acceptable. A NIOSH-MSHA-approved respirator is assigned an approval number prefixed by the letters TC, indicating that it has been tested and certified by NIOSH and MSHA for the air contaminant at the concentration range stated.

This approval ensures that the design, durability, and workmanship of the equipment meet minimum standards. Technicians have tested the respirator for worker safety, freedom of movement, vision, fit, and comfort of facepiece and headpiece; the ease with which the filter and other parts can be replaced; tightness of the seal against dust; freedom from leakage; and resistance to air flow when the wearer is inhaling and exhaling.

Cleaning the Respirator

Respirator facepieces and harnesses should be cleaned and inspected regularly. If several persons must use the same respirator, it should be disinfected after each use in order to comply with OSHA regulations. Methods of disinfection include:

1. Immersion in a weak solution of quarternary ammonium compound, followed by a warm-water rinse. This solution is not normally injurious to skin or to rubber.
2. Washing in warm, soapy water and rinsing for at least one minute in clean water of 120 F (49 C) minimum temperature.

The supervisor or a designated employee should inspect respirators at intervals to check for damage or improperly functioning parts, such as headbands or valve seats.

If the respirator is removed at intervals, dust settles on it. When the respirator is replaced, these dust particles are transferred to the skin and cause irritation. To prevent this, the wearer should keep the respirator on at all times when in a contaminated atmosphere.

BODY PROTECTION

The most common protection for the abdomen and trunk is the full apron. Aprons are made of various materials. Leather or fabric aprons, with padding or stays, offer protection against light impact and against sharp knives and cleavers, such as those used in packinghouses. Asbestos coats and aprons are often used by those who work around hot metal or other sources of intense conductive heat.

An apron worn near moving machinery should fit snugly around the waist. Neck and waist straps should be either light strings or instant-release fasteners in case the garment is caught. There should be a fastener at each end of the strap to prevent severe friction burns should the strap be caught or drawn across the back of the neck. Split aprons should be worn on jobs that require mobility on the part of the worker. Fasteners draw each section snugly around the legs.

Welders are often required to wear leather vests or capes and sleeves, especially when doing overhead welding, as protection against hot sparks and bits of molten metal. On jobs where employees must carry heavy, angular loads, pads of cushioned leather or padded duck are used to protect the shoulders and back from injury.

Safety Belts and Harnesses

Safety belts and harnesses with lifelines attached should be worn by those who work at high levels or in closed spaces and by those who work where they may be buried by loose material or be injured in confined spaces. This discussion, however, does not include vehicular seat belts or linemen's belts.

Normal use involves comparatively light stresses applied during regular work—stresses that rarely exceed the static weight of the user. Emergency use means stopping an individual when he or she falls. Every part of the belt may be subjected to an impact loading many times the weight of the wearer.

A window cleaner's belt, for example, is subjected to a moderate load most of the time it is used. It will be subjected to a severe loading, however, if the worker falls when only one terminal of the belt is attached. A belt for a person who leans back while working should, therefore, have two D-rings, one on each side of the belt, to which a throw rope or lanyard can be attached. The rope is then anchored.

A harness-type safety belt is better at distributing the shock of an arrested fall (Figure 9-14). This shock is distributed over shoulders, back, and waist, instead of being concentrated at the waist. A two-line harness distributes the lifting and lowering stresses better than a one-line harness. The harness permits a person to be lifted with a straight back

Figure 9-14. This full-support body harness is ideal for workers on elevated sites. The fall-arrest line is attached to the D-ring at the center back. Two lines can be attached at the shoulders by D-rings for lifting and lowering.

rather than bent over a waist strap. This makes rescue easier if the victim is unconscious, buried, or must be lifted out through a manhole. Wherever a job requires use of a self-contained breathing apparatus or supplied-air device, a harness and lifeline should also be used to assist in escape.

When workers are at risk of falling some distance, the safety harness should be designed to distribute the impact force over the legs and chest as well as the waist. A shock absorber or decelerating device, which brings the falling person to a gradual stop, lessens the impact load on both the equipment and the person. To prevent a long fall, the line should be tied off overhead and should be as short as movements of the worker will permit.

Ladder safety devices. Fixed ladders more than 24 ft (7.5 m) high must have some device to prevent a climber from falling. Baskets and ladder guards are useful in some cases, but cost or space may require a ladder safety device. This consists of a body belt and a clamp that rides a wire rope or rigid device on the ladder. The clamp engages a safety device that locks on the rope or ladder when the climber slips or misses a step (ANSI A14.3–1984, *Safety Requirements for Fixed Ladders*). Safety requirements for other types of ladders can be found in the following ANSI standards:

Portable Ladders, Wood	ANSI A14.1–1982
Portable Ladders, Metal	ANSI A14.2–1982
Job-Made Ladders	ANSI A14.4–1984
Portable Ladders, Reinforced Plastic	ANSI A14.5–1982

Materials. Belts made of natural or synthetic materials are furnished by most manufacturers. Nylon belts, straps, and harnesses, although widely used, may not be satisfactory under some chemical conditions. Special types of belting are available for certain environments—wax-treated belts resist paint solvents or mildew, neoprene-impregnated belts resist acids and oil products. Be sure to consult with your safety department so you obtain the right equipment.

Safety belts are designed to give a little when stress loaded. If a person does fall and stresses a belt, the belt should be taken out of service and replaced. Several kinds of devices that attach the belt to a lifeline serve to reduce the stress of a fall; these minimize both the injury to the user and the damage to the belt.

Inspection. Safety belts and lines should be examined each time before using. At least once every three months, belts should be checked by a trained inspector.

Web belts should be inspected for worn and torn fibers. When a number of the outer fibers are worn or cut through, the belt should be discarded.

Belt hardware should be examined and worn parts replaced. If the belt is riveted, each rivet should be carefully inspected for wear around the rivet hole. Dirt and dust should be brushed carefully from belts to prevent fiber damage. Broken or damaged stitching may require belt replacement.

Web belts may be washed in warm, soapy water, rinsed with clear water, and dried by moderate heat. If these belts are worn under unusual conditions, or if a dressing is to be used, consult the manufacturer.

The outer surface of rope lines should be examined for cuts and for worn or broken fibers. Manila rope should be discarded if it has become smaller in diameter or has acquired a smooth look. Inner fibers of manila rope should be examined for breaks, discoloration, and deterioration. If the rope shows any of these signs, it should be discarded.

A steel wire rope should be examined for broken strands, rust, and kinks that may weaken it. Ropes must be kept clean, dry, and rust free. They should be lubricated frequently, especially before use in acid atmospheres or before exposure to salt water. After such use, wire rope should be carefully cleaned and again coated with oil.

Lifelines

Manila rope of ¾-in. (1.9 cm) diameter or nylon rope of ½-in. (1.3 cm) diameter is recommended for lifelines. Nylon is more resistant to wear or abrasion than is manila (abaca). It is more elastic (absorbing shocks and sudden loads better) and has high tensile strength (wet or dry). Nylon is tough, flexible, and easy to handle. Because it resists mildew, it can be stored wet.

Breaking strength is always measured in a straight-line pull on the rope. Half-inch (1.3 cm) diameter manila rope breaks at approximately 2,650 lb (1,200 kg), half-inch diameter nylon rope is rated at 6,400 lb (2,900 kg, ±5 percent). Using a safety factor of 5 for manila and 9 for nylon gives safe load strengths of 530 lb (240 kg) and 710 lb (320 kg), respectively.

Manila rope ¾-in. (1.9 cm) in diameter breaks at approximately 5,400 lb (2,450 kg), giving a safe load strength of 1,080 lb (490 kg). When a low load-limit shock absorber is used, ¾-in. manila rope or ½-in. nylon rope is considered strong enough for most lifelines. Without a shock absorber, even ¾-in. manila may not be strong enough to arrest a long fall.

Wire rope should not be used for lifelines where falls are possible, unless a shock-absorbing device is also used. The rigidity of steel greatly increases the impact loading. Steel is, furthermore, hazardous around electricity.

Knots reduce the strength of all ropes. The degree of loss depends upon the type of knot and the amount of moisture in the rope (Table 9-4). More information is given in Chapter 11, Materials Handling.

Flotation Devices

Operations that require working on off-shore rigs, merchant vessels, barges, tugs, and so forth require flotation devices according to U.S. Coast Guard standards. In these cases, you should make certain that this equipment meets all applicable standards, is in good condition, and, most important, is worn when required. Many near-accidents and, indeed, drownings easily could have been prevented had this personal protective equipment been worn.

Protection Against Ionizing Radiation

Ionizing radiation, dangerous because of its serious biological effects, need not be feared if it is respected and if suitable precautionary measures are taken. The National Safety Council's *Fundamentals of Industrial Hygiene* has much information on the topic and many references. (Also see Chapter 8, Industrial Hygiene, for a more detailed discussion of avoiding radiation hazards.)

Standards for protection against radiation, specified in OSHA regulations 1910.96 and .97, reference the Nuclear Regulatory Commission regulations. These are spelled out in the *Code of Federal Regulations*, Title 10, "Energy," Part 20.

Rubber gloves for handling radioactive materials, disposable clothing, suits with supplied air, and approved respiratory devices are some of the specialized equipment needed. Under no circumstances should contaminated clothing be worn into clean areas. Thorough washing with soap and water is usually the best general method for decontamination of the hands and other parts of the body, regardless of the contaminant.

Before you or your employees work with radio-

TABLE 9-4. Approximate Efficiency of Manila Rope Hitches

	Percent
Full strength of dry rope	100
Eye splice over metal thimble	90
Short splice in rope	80
Timber hitch, round turn, half hitch	65
Bowline, slip knot, clove hitch	60
Square knot, weaver's knot, sheet bend	50
Flemish eye, overhand knot	45

Source: Oregon Safety Code for Places of Employment.

active materials, check with the medical and safety departments. You must be absolutely assured in writing that all people will be adequately protected before any tests are made or procedures changed that involve radioactive materials or isotopes. Make sure that medical clearances and other authorizations, in addition to badges and dosimeters, are issued to all personnel assigned to work in areas restricted for radioisotope use.

Employees should be monitored continuously for exposure to radiation. At work, they should wear film badges that are developed and replaced at regular intervals, depending on the radiation intensity level to which they are exposed. The film badge records the total exposure for a given period, but it does not indicate the precise time during the period when the exposure occurred.

For better protection, each person should also wear two pocket dosimeters. These show the dosage received during any part of a work period so that the individual can check at any time for exposure. The dosimeter does not provide a permanent record, but is essential for safety during work periods. It is read at least once, usually at the end of a shift. It can be reset by means of a dosimeter charger. Radiation accidents resulting in overexposure should be reported immediately to the medical director.

Protection Against Chemicals

Protecting workers in and around chemicals is of paramount importance. Many chemical substances to which employees are exposed can be irritating and result in serious burns on the body, hands, arms, and legs. You, as the supervisor, have the responsibility of evaluating operations using chemicals and determining what protective equipment or proper clothing is needed. (See the discussion of chemical hazards in Chapter 8.)

If you need help for this task, contact your safety department, insurance company safety engineer, industrial hygiene department, protective equipment supplier, or a chemist. The best way to protect people is by making sure they wear protective clothing designed specifically for the chemicals involved. Keep in mind, however, that the best course of action may be to replace the chemical with a less toxic one. If the chemical cannot be eliminated, replaced, or isolated, then your only recourse is protective clothing.

Protective Clothing

Ordinary work clothing, if clean, in good repair, and suited to the job, may be considered safe. "Protective clothing" refers to garments designed for specific, hazardous jobs where ordinary work clothes do not give enough protection against such injuries as abrasions, burns, and scratches.

Guidelines. Good fit is important. Most work trousers or slacks are made extra long to provide for shrinkage and to fit tall workers. Trousers that are too long must be shortened to proper length, preferably without cuffs. If cuffs are made, they should be securely stitched down so the wearers cannot catch their heels in them. Cuffs should never be worn near operations that produce flying embers, sparks, or other harmful matter.

Neckties, long or loose sleeves, gloves, and loose-fitting garments (especially about the waist) create a hazard because they are easily caught in moving (or revolving) machine parts. Likewise, all types of jewelry are out of place in a shop—rings, bracelets, and wristwatches can cause serious injury. A finger may be torn off if a ring catches on a moving machine part, or on a fixed object when the body is moving rapidly. Necklaces, key chains, and watch chains also constitute hazards near moving machinery. Metal jewelry worn around electrical equipment, including batteries, can be dangerous.

Clothing soaked in oil or flammable solvent is easily ignited and is a definite hazard. Remember that skin irritations are often caused by continued contact with clothing that has been soaked with solvents or oils (see Dermatitis in Chapter 8).

If oil or dust gets in the hair, the worker should wear a cap to guard against dirt and infections and to keep hair looking nice. A cap may also help protect a person from moving parts of machines that cannot be completely guarded.

Materials. A number of protective materials are used in making clothes to protect workers against various hazards. Supervisors should be familiar with these materials.

Aluminized and reflective clothing has a coating that reflects radiant heat. Aluminized asbestos or glass fiber is used for heavy-duty suits, and aluminized fabric for fire-approach suits (Figure 8-8).

Asbestos has been used as protection against intense conducted heat and flames. When used with a radiation barrier of reflective material, asbestos offers protection in firefighting and rescue work; however, it must always be treated so as not to give off airborne fibers. Use of this material is declining due to health hazards associated with asbestos exposure.

Flame-resistant cotton fabric is often worn by people who work near sparks and open flames. Although the fabric is durable, the flame-proofing treatment may have to be repeated after one to four launderings.

Flame-resistant duck, used for garments worn around sparks and open flames, is lightweight, strong, and long lasting. However, it is not considered adequate protection against extreme radiant heat.

Glass fiber is used in multilayered construction to insulate clothing. The facing is made of glass cloth or of aluminized fabric.

Impervious materials (such as rubber, neoprene, vinyl, and fabrics coated with these materials) protect against dust, vapors, mists, moisture, and corrosives. Rubber is used often because it resists solvents, acids, alkalis, and other corrosives. Neoprene resists petroleum oils, solvents, acids, alkalis, and other corrosives.

Leather protects against light impact and against sparks, molten metal splashes, and infrared and ultraviolet radiation.

Synthetic fibers (such as acrylics and low-density polyethylene) resist acids, many solvents, mildew, abrasion and tearing, and repeated launderings. Because some fabrics generate static electricity, garments should not be worn in explosive or high-oxygen atmospheres, unless they are treated with an antistatic agent.

Water-resistant duck is useful for exposures to water and noncorrosive liquids. When it is aluminum-coated, the material also protects against radiant heat.

Wool may be used for clothing that protects against splashes of molten metal and small quantities of acid and small flames.

Permanent press fabrics are used in a wide variety of protective clothing. However, care must be taken to determine whether or not they are flammable.

PROTECTING EXTREMITIES

Workers' extremities are highly vulnerable to injury in most work environments. Protective clothing and gear can reduce the number and severity of injuries workers suffer each year. As supervisor, you should educate and train all employees in the importance of using such protection. Injuries to or losses of fingers, toes, even parts of an arm or leg can happen with frightening suddenness. Make sure workers understand that their best protection is prevention. This section discusses protective equipment for arms, hands, and fingers and for feet and legs.

Arms, Hands, Fingers

Fingers and hands are exposed to cuts, scratches, bruises, and burns. Although fingers are hard to protect (because they are needed for practically all work), they can be shielded from any common in-

juries with such proper protective equipment as the following:

1. Heat-resistant gloves protect against burns and discomfort when the hands are exposed to sustained conductive heat.
2. Metal mesh gloves, used by those who work constantly with knives, protect against cuts and blows from sharp or rough objects.
3. Rubber gloves are worn by electricians. They must be tested regularly for dielectric strength.
4. Rubber, neoprene, and vinyl gloves are used when handling chemicals and corrosives. Neoprene and vinyl are particularly useful when petroleum products are handled.
5. Leather gloves are able to resist sparks, moderate heat, chips, and rough objects. They provide some cushioning against blows. They are generally used for heavy-duty work. Chrome-tanned leather or horsehide gloves are used by welders (Figure 9-15).
6. Chrome-tanned cowhide leather gloves with steel-stapled leather patches or steel staples on palms and fingers are often used in foundries and steel mills.
7. Cotton fabric gloves are suitable for protection against dirt, slivers, chafing, or abra-

Figure 9-15. Welding gloves with self-extinguishing fleece lining. (Courtesy Elliot Glove Company.)

sion. They are not heavy enough to use in handling rough, sharp, or heavy materials.

8. Heated gloves are designed for use in cold environments, such as deep freezers, and can be part of a heated-clothing system. Other types designed for such work are insulated with foam, and most such gloves are waterproof.

Specially made electrically tested rubber gloves, worn under leather gloves to prevent punctures, are used by linemen and electricians who work with high-voltage equipment. A daily visual inspection and air test (by mouth) must be made. Make sure that rubber gloves extend well above the wrist so that there is no gap between the coat or shirt sleeve and glove. The shortest glove is 14 in. (42 cm). (See National Safety Council Industrial Data Sheet 598, *Flexible Insulating Protective Equipment for Electrical Workers.*)

People are willing to wear gloves when the dangers of not wearing them are apparent. With gloves, workers do not have to worry about cutting their hands, and they can grip materials better. As a result, production increases. Still, if employees become lax, you should remind them often to wear their gloves.

In some instances, however, gloves may be a hazard, particularly when worn around certain machining operations, such as a drill press. The material may become caught in machine parts. Use of gloves in these areas should be prohibited.

Hand leathers or hand pads are often more satisfactory than gloves for protecting against heat, abrasion, and splinters. Wristlets or arm protectors are available in the same materials as gloves.

Feet, Legs

Foot protection. About a quarter of a million disabling occupational foot injuries take place each year. This points to the need for foot protection in most industries, and the need for supervisors to see that their workers wear this gear. All safety shoes have toes reinforced with a toe cap. The three classifications are shown in Table 9-5. Safety shoes also have other protective features, as well as reinforced toes.

Many shoe manufacturers and jobbers cooperate with companies to set up shoe sale departments. They provide experienced people to see that shoes are properly chosen for the hazard involved and are properly fitted. Some companies keep shoes on hand for sale; others conduct business by mail. An employer can arrange for the purchase of shoes at local outlets. Some dealers have a mobile shoe

TABLE 9-5. Minimum Requirements for Safety-Toe Shoes

Classification	Compression Pounds	Impact Foot Pounds	Clearance Inches
			Men
75	2,500	75	16/32
50	1,750	50	16/32
30	1,000	30	16/32
			Women
			15/32
			15/32
			15/32

Note: 1 pound force = 4.45 newtons
1 foot pound force = 1.36 joules
1 inch = 2.54 centimeters

Source: American National Standard Z41–1983, *Safety-Toe Footwear.*

service—a truck equipped as a shoe store and manned by an experienced fitter. The responsibility for proper care of safety shoes ordinarily rests with the employees. Letting employees select their shoe style encourages cooperation (Figure 9-16).

Metal-free shoes, boots, and other footwear are available for use where there are specific electrical hazards or fire and explosion hazards.

"Congress" or gaiter-type shoes are used to protect people from splashes of molten metal or from welding sparks. This type can be removed quickly to avoid serious burns. These shoes have no laces or eyelets to catch molten metal.

Reinforced or inner soles of flexible metal are built into shoes worn in areas where there are hazards from protruding nails and when the likelihood of contact with energized electrical equipment is remote, as in the construction industry.

For wet work conditions, in dairies and breweries, rubber boots and shoes, leather shoes with wood soles, or wood-soled sandals are effective. Wood soles have been so commonly used by workers handling hot asphalt that they are sometimes called "pavers' sandals."

Safety shoes with metatarsal guards should always be worn during operations where heavy materials, such as pig iron, heavy castings, and timbers, are handled. They are recommended whenever there is a possibility of objects falling and striking the foot above the toe cap. Metal foot guards are long enough to protect the foot back to the ankle, and may be made of heavy gage or corrugated sheet metal. They are usually built onto the shoe.

Supervisors must often overcome workers' objections to wearing foot protection. Safety shoes used to be hot and heavy, and people often complained that they were uncomfortable. Current designs now make some safety shoes as comfortable, practical, and attractive as ordinary street shoes. The steel cap weighs about as much as a pair of rimless eyeglasses or a wristwatch. The toe box is

Figure 9-16. Safety shoes are available in women's sizes and styles.

insulated with felt to keep the feet from getting too hot or cold. Some shoes have inner soles of foam latex with tiny "breathing cells."

Some people object to wearing safety shoes because they do not cover the smallest toes. However, studies show that 75 percent of all toe fractures happen to the first and second toes. In most accidents, the toe box takes the load of the impact for the entire front part of the foot.

Another objection commonly voiced is that if the toe box were crushed, the steel edge would cut off the toes. Actually, accidents of this type are rare. In the majority of cases, safety shoes give sufficient protection. Freak accidents may happen, against which there is no sure protection. But, in any event, a blow that would crush the toe cap would certainly smash someone's toes if he or she were not wearing foot protection.

Leg protection. Leggings that encircle the leg from ankle to knee and have a flap at the bottom to protect the instep, protect the entire leg. The front part may be reinforced to give impact protection. Such guards are worn by persons who work around molten metal. Leggings should permit rapid removal in case of emergency. Hard fiber or metal guards are available to protect shins against impact.

Padding can protect parts of the legs from injury. Knee pads protect employees whose work, like cement finishing or tile setting, requires much kneeling. Ballistic nylon pads are often used to shield the thighs and upper leg against injury from chainsaws.

Foot and leg protectors are available in many different materials. The type selected depends on the work being done. Where molten metals, sparks, and heat are the major hazards, heat- or flame-resistant materials or leather is best. Where acids, alkalis, and hot water are encountered, natural or synthetic rubber or plastic, resistant to the specific exposure, can be used.

SUMMARY OF KEY POINTS

This chapter covered the following key points:

- Supervisors must be familiar with government regulations regarding personal protective equipment (PPE), be able to recognize hazards requiring use of PPE for worker safety, know the best type of equipment for the particular hazard workers face, and educate employees in proper equipment use and maintenance.

- Supervisors can overcome employee objections to PPE by making sure the equipment fits properly, looks appealing, and is easy to clean and repair or replace. Employees must be sold on the need for PPE and how the equipment can protect various parts of the body.

- Protective equipment may be purchased by the company, jointly paid for by the company and employees, or furnished by employees. Regardless, the best rule of thumb is to buy the best equipment and to deal only with reputable firms.

- Head protection is provided by pressure-molded, high-impact helmets worn to prevent head injuries from falling or flying objects or from bumps against stationary objects. These helmets can be insulated against electric shock and contain a hard outer shell with an inner suspension, or cradle, and headband. Helmets should be kept clean and replaced if damaged. Auxiliary features include liners, chin straps, eye shields, brackets, and standard or special-order colors. Helmets should be adapted to fit each worker comfortably and to eliminate any excuse for not wearing the head gear.

- The most common causes of eye injuries are flying objects, particles from abrasive wheels, corrosive substances, visible or thermal radiation, splashing metal, and poisonous gases or fumes. Workers can be protected either by shielding them from the job hazard and/or by providing safety goggles and face shields. Employees wearing contact lenses should be particularly careful in contaminated areas. Goggles are the most common eye protection gear, and supervisors should know which types are the best for each job. All eye-wear gear should be sterilized after use and defective parts replaced. Workers should be trained in the proper use, handling, and care of protective eye-wear.

- Face protection should shield the face (and sometimes head and neck) from impact, chemical or hot metal splashes, heat radiation, and other hazards. This type of equipment includes helmets, shields and goggles, and hoods (covering the head, face, and neck). Some hoods are provided with independent air supplies for work around toxic fumes or dusts. Face protection gear should be selected for the particular hazard, be easy to clean, and be adjustable to the size and contour of in-

dividual heads. Supervisors should set up regular replacement schedules, because plastic gear tends to become brittle with age.

- Hearing protection equipment must be provided to workers exposed to noise levels above 90 dBA on an 8-hour time-weighted average to reduce the noise to an acceptable level. This equipment includes insert ear protectors and muff devices. Insert protectors are inserted into ear canals and can be made of pliable rubber or plastic, wax, and cotton. These protectors must be designed and fitted properly and kept clean or they can cause ear infections and hearing loss. Cup or muff devices cover the external ear; their effectiveness depends on the size, shape, seal material, shell mass, suspension of the muff, and how well they fit the worker's head. Workers should be thoroughly trained in the use and care of protective hearing equipment.

- Respiratory equipment can be regarded as emergency equipment or gear for occasional use on the job. This equipment includes air-purifying devices, mechanical filter respirators, chemical cartridge respirators, gas masks, hose masks, supplied-air devices, abrasive blasting helmets, and self-contained breathing apparatus. To select the proper respiratory equipment, supervisors must first determine the chemical or other hazard in the environment, evaluate the extent of the hazard, and choose the respirator that best protects workers against that hazard. Workers should be trained in proper use and maintenance of respiratory equipment, and the gear should be inspected and tested frequently.

- The most common protection for abdomen and trunk are full aprons, safety belts and harnesses, lifelines, flotation devices, suits and equipment designed for radiation and chemical hazards, and protective clothing. Safety belts and harnesses can either contain the worker or help to distribute the shock of an arrested fall. Belts, harnesses, and lifelines are made of various natural and synthetic materials and must be cleaned and inspected frequently. Flotation devices are required for all operations on water, such as off-shore oil rigs, merchant vessels, barges, tugs, and so on. Equipment must meet all standards and be worn when required.

- Suits and equipment for radiation and chemical hazards include gloves, disposable clothing, suits with supplied air, and respiratory devices. Employees exposed to radiation hazards must be monitored continuously to ensure safety levels. Protective clothing can be made of aluminized and reflective material, asbestos, flame-resistant cotton or duck, fiberglass, impervious materials, leather, and various synthetic materials. This clothing is designed for specific, hazardous jobs where ordinary work clothes do not give enough protection.

- Protective equipment for arms, hands, and fingers are generally gloves, wristlets, and arm protectors made of various materials designed to protect these extremities against cuts, scratches, bruises, and burns. Supervisors should know which type of glove is best for a particular hazard: heat-resistant, metal mesh, rubber, leather, chrome-tanned cowhide, cotton, or thermal gloves. Gloves should be inspected regularly and kept clean.

- Safety shoes are the best protection against foot injuries, while leggings and knee, thigh, and leg pads offer protection for the legs. All safety shoes have toes reinforced with a toe cap. Supervisors should see that the right safety shoes are chosen for the particular hazard and that they fit each worker properly. Safety shoes include metal-free footwear, "congress" or gaiter-type shoes, reinforced shoes, boots and other footwear for wet work, and shoes with metatarsal guards. Leggings must not only protect workers' legs but permit rapid removal in case of emergency. Hard fiber or metal guards protect shins, while nylon or cotton padding protects knees, thighs, and upper legs. Supervisors must educate employees in the proper use and maintenance of protective foot and leg gear.

10

Ergonomics

After reading this chapter, you will be able to:

- Describe various ergonomic problems and be alert for symptoms among your workers
- Understand the principles and goals of ergonomics and how physiology, anthropometrics, and biomechanics are used to solve work-stress problems
- Know the ergonomic guidelines for safe manual materials handling
- Understand how the workspace can be engineered to match workers' body characteristics
- Understand how to reduce ergonomic problems related to hand work, hand tools, whole-body vibration, video display terminals, and other work-related stresses

As a supervisor, you may be asked to understand and use some of the principles of ergonomics. Ergonomics is the study of how people interact with their work—in other words, how to create a good match between the employee and the workstation and tools with which he or she works. The goal of an ergonomics program is to minimize accidents and illness due to chronic physical and psychological stresses, while maximizing productivity and efficiency. Many corporations have found that effective ergonomics programs have reduced operating costs and increased worker morale.

You should be familiar with ways to recognize ergonomic problems, evaluate them, and suggest methods to control the risks. Many types of controls can be applied, from eliminating the problem, to changing work practices, to designing new tools, to training workers in how to use tools and equipment more effectively.

This chapter covers causes and symptoms of common ergonomic problems, understanding the field of ergonomics, and applying basic principles to eliminate or reduce work-stress problems.

WHAT ARE ERGONOMIC PROBLEMS?

Physical stress may arise when workstations, equipment, or tools do not fit the worker well. These stresses can cause immediate or long-term damage to muscles, nerves, and joints. Most illnesses due to ergonomic causes occur because of forceful or repetitive work activities, or because the workers are required to assume awkward postures over a period of time. While it takes a physician to diagnose a worker's illness, you as supervisor should be alert to any employee complaints of pain, tingling, numbness, swelling, or other discomforts. These symptoms may indicate that a work-related problem is developing.

Back, shoulder, and neck strains and sprains can all be the result of exposure to ergonomic stresses. These as well as hand, wrist, and other arm symptoms are often referred to as cumulative trauma disorders (CTDs). The following are some of the more common types of CTDs. (You may also use the term "repetitive strain injuries.")

Tenosynovitis. In this disorder the tendons in the wrist become sore and inflamed because of repetitive motion or awkward postures. Symptoms include pain, swelling, cracking sounds, tenderness, and loss of function. It is caused by poor workstation design, or by sudden changes in work habits.

Tendinitis. An inflammation of the tendon. Tendinitis causes symptoms similar to those of tenosynovitis. This disorder typically occurs in the shoulder, wrist, hands, or elbow.

Thoracic outlet syndrome. A disorder of the shoulder that affects the nerves in the upper arm, thoracic outlet syndrome is usually caused by doing tasks overhead for extended periods of time. There can be loss of feeling on the little finger side of the hand and arm, with pain, weakness, and deep, dull aching.

Ganglion cysts. These cysts are associated with cumulative trauma or repetitive motion. They appear as bumps on the wrist and can be surgically removed.

Carpal tunnel syndrome. This condition is caused by excessive flexing or twisting of the wrists, especially where force is used. It affects the median nerve, which runs through a bony channel in the wrist called the carpal tunnel. Symptoms include burning, itching, prickling or tingling feelings in the wrist or thumb and first three fingers. When the condition is severe, some of the thumb muscles may become much less strong, and there may be overall weakness in the hand.

DeQuervain's disease. This is an inflammation of the sheath of the tendons to the thumb. Scarring in the sheaths restricts the thumb's movement.

Trigger finger syndrome. This is another form of tendinitis caused by repetitive flexing of the fingers against vibrating resistance. Eventually the tendons in the fingers become inflamed, causing pain, swelling, and a loss of dexterity. Trigger finger may result from using a power tool with too large a handle.

Epicondylitis. Also known as tennis elbow, epicondylitis is the inflammation of tissues on the inner (thumb) side of the elbow. It may be caused by violent or highly repetitive action, such as use of a screwdriver, or other action that requires rotation of the forearm and wrist downward.

UNDERSTANDING ERGONOMICS

Many different fields of study have contributed to industry's understanding of ergonomic problems and solutions. These include:

- Anatomy and physiology of the human body
- Anthropometrics: the study of differences in the size of bodies and body parts among different groups of humans
- Biomechanics: the study of the way work activities produce forces on muscles, nerves, and bones
- Psychology: how people recognize and respond to signals in the environment
- Industrial design and engineering: the design of workplaces, tools, and processes

Ergonomists are people who have had training in these fields and who can develop practical solutions for workplace ergonomic problems.

In approaching an ergonomic problem, an organization may bring together specialists in these areas to analyze the way people interact with machines, tools, work methods, and workspaces. Safety professionals, industrial hygienists, and occupational physicians and nurses may add their expertise to help create solutions. Management and employees also have important roles in providing information, helping to evaluate work situations, and giving feedback on changes.

One way you as the supervisor can help in the effort is by keeping track of possible indicators of ergonomic problems. These include:

- Apparent trends in injuries and accidents
- Incidence of commutative trauma disorders
- Excessive use of sick days, high turnover rate
- Employees changing their workstation, e.g., adding padding to hand tools or equipment, using makeshift platforms to stand on, making changes in the work flow
- Poor product quality

Some of the causative factors you should look for include:

- Reliance on incentive pay systems that motivate people to work faster than is safely possible
- High overtime and increased work rate
- Manual material handling and other repetitive tasks
- Work requiring awkward postures
- High amount of hand force required for tasks
- Mechanical stress, e.g., hands or forearms resting on sharp table edges
- Raised elbows, bent wrists or hands
- Grasping or pinching objects
- Exposure to temperature extremes
- Use of vibrating tools

As supervisor, you should assist in the review

of any new manufacturing processes, designs, and major equipment purchases. In this way you will be able to point out potential difficulties in terms of how workers may actually use equipment in the workplace.

Physiology

Work tasks need to be designed to match the employee's physical capacities for the job. A person's ability to do manual work at a given level is usually determined by the ability of his or her respiratory and cardiovascular systems to deliver oxygen to the working muscles and to make use of energy derived from food.

One measure of the upper limit of a person's ability to do manual work is maximum oxygen uptake. In industry, maximum effort is usually required only for brief periods, such as when an employee must lift heavy loads onto a hand truck. Over an 8-hour shift, however, the amount of energy used is usually well below the maximum capacity of the worker.

The amount of work done can be measured in heat produced by the body; the measurement unit is kilocalories per hour (kcal/hr). For the fit young man, the heaviest work that can be sustained is 500 kcal/hr; for the general population the figure is closer to 400 kcal/hr. These figures are equivalent to about 40 percent of the maximum of which people are capable.

It is obviously important to match human capabilities with the demands of the job. If job demands equal or exceed the workers' capacities, they will be under strain and will not be able to perform the task. Several methods can be used to assess a person's ability to tolerate the demands of a task. However, many of these methods, such as oxygen uptake and heat output measurements, require sophisticated equipment not usually available in workplaces. On the other hand, one simple way to measure the task demands is by heart rate. For instance, light work is generally associated with a heart rate of 90 or fewer beats/minute; medium work, 100 beats/minute; and heavy work, about 120 beats/minute. Strenuous work produces a heart rate of 140–160 beats/minutes or more and cannot be sustained without rest or cessation of effort. At strenuous work loads, waste products of energy consumption build up in the blood and muscles, producing fatigue or a sensation of soreness.

A person's capacity for manual work is also limited by muscle strength; by flexibility, or lack of it, in the joints; and by the strength of the spinal column. The science of biomechanics allows us to measure the amount of force put on these parts when employees work in different positions and helps to determine which positions make best use of muscular strength.

The muscle's ability to perform is also affected by the way it is used. Activity can be either static or dynamic. A dynamic activity, like walking, is created by the rhythmic contraction and relaxation of the muscles. Dynamic work allows muscles to rest during the relaxation phase, so that the blood can supply oxygen and remove waste products. In contrast, static work is caused by the worker holding one object or body part in one position for any extended period. (Standing still is a static posture, for example.) The muscle is in a fixed and locked position, blood vessels are compressed, and blood flow is reduced; the muscle tires rapidly. A static muscle consumes more energy; it can also place more stress on tendons and joints, which may result in back, shoulder, and neck pain or inflammation of the tendons.

It is important to remember as well that heart rate (and the worker's ability to do the job) may also be affected by temperature, humidity, and the worker's age and fitness level. While engineers and work physiologists usually are responsible for assessing the physiological demands of a job, you need to understand the basics of how work demands affect body functioning.

Anthropometrics

Anthropometric studies provide information that can be used to design workspaces to match body dimensions. The idea that there is an "average"-sized person or worker is obsolete; even two people who are of "average" height, say 5 ft 8 in., may have very different body dimensions. Also, a person who has an average arm length might have long legs, or one with average leg length might have long arms.

Studies of military and civilian personnel have been used as the basis for tables of information on body sizes. These have defined the extremes of body and body part sizes in a particular population. This information is described by percentiles. For example, regarding height, a 5th percentile female is taller than 4 percent of the female population and shorter than 95 percent of the females in the group. A 95th percentile male is taller than 94 percent of the male population and shorter than the tallest 5 percent.

These percentiles are useful to engineers who try to design a workspace to fit as many in the population as possible. Usually they try to make the workstation adjustable so that it can fit anyone from the 5th percentile female to the 95th percentile male. (The number of people outside these extremes is relatively small; therefore, it is often not considered economically feasible to design for these groups.)

Biomechanics

The science of biomechanics explains characteristics of the human body in mechanical terms. Using these models allows us to describe the amount of force produced when a muscle works to move an object of a specific weight. The models may describe the arm as a level, for example. They are often illustrated as simple line drawings which show the different forces acting on the muscle and on the object. While these seem oversimplified, they can be very useful in helping to identify stressful postures. The calculations are derived from the laws of physics and can be quite complex, because often more than one muscle is involved.

Looking at the Workplace

Many operations in a workplace have potential for causing injury or illness, each in perhaps a different way. The amount of physical stress a job produces is often determined by the following:

- Posture the worker must assume to do his or her job
- Force needed to perform the task
- Number of repetitions of a stressful task

The next sections will discuss these principles of ergonomics and ways to apply them in the workplace.

MATERIALS MOVEMENT

When materials are moved, the appropriate equipment should be used. This can include mechanical systems such as conveyors, hand trucks, and carts to reduce manual lifting and carrying. Wherever possible, people should be engineered out of the flow of materials. Chapter 13, Materials Handling and Storage, discusses this topic in more detail.

Manual Materials Handling

Manual materials handling refers to the human activity necessary to move an object. This often means lifting and carrying but can also describe pushing, pulling, and shoving. Look for the following situations that are likely to cause injury:

- Lifting from the floor or while twisting
- Lifting heavy weights or bulky objects
- Repetitive lifts

- Lifting above shoulder height or while seated
- Pushing or pulling loads
- Bending while moving objects

Guidelines for Safe Materials Handling

Making manual materials handling easier may mean changing one or more of the following factors:

- Where possible, modify the object to make movement easier. Handholds should be provided to allow use of a power grip that will make carrying more efficient. In a power grip, the object is clasped between the flexed fingers; this technique allows less muscle force to be used than does a pinch grip, in which the hands cannot be placed entirely around the object. Cardboard cartons, for example, should have openings that can accommodate large gloved hands. The surface of the opening or handle should be smooth to avoid mechanical irritation of the skin (Figure 10-1). The center of gravity should be below the handholds and close to the body.
- Avoid lifting or carrying large, unwieldy, or heavy objects. Mechanical lifting and transporting means should be used wherever possible to reduce the amount of lifting necessary. These means can include conveyors, automatic or gravity-feed devices, load-leveling pallets, hoists, and lift trucks.
- Where mechanical means cannot be used, seek assistance from another person in lifting heavy or unwieldy objects.
- Where feasible, split loads in containers into smaller loads. Containers can be made of lighter materials, or the shape can be changed to allow the objects to be handled closer to the body.

The weight workers can lift is influenced by many factors. Various calculations have been developed to help determine acceptable and maximum loads for each worker. These calculations are usually based on *Work Practices Guide for Manual Lifting*, published by the National Institute for Safety and Health in 1981. The guidelines are modeled on lifting an object less than 30 in. wide with two hands, in front of the body, with good handholds and footing, and no twisting required. The NIOSH formula takes into account the distance of the load from the body, the location of the object before and at the end of the lift, and the frequency of lifting. (The *Guide* is undergoing review and possible revision; the most recent version should be consulted.)

Many industrial lifts do not meet these criteria. In general, the following guidelines apply to all lifts:

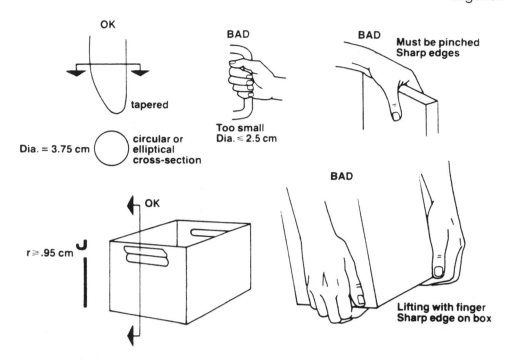

Figure 10-1. Handles should be smooth and shaped to permit a power grip.

- Lifts should be kept between knuckle and shoulder height.
- The lift should begin as close to the body as possible (Figure 10-2). Sometimes objects are bulky or have to be placed over or in containers or on shelving far back on counters. Job redesign should be considered in these circumstances, especially if a heavy object is involved. The longer the arm reach required to grasp the object, the less weight can be lifted. Lifting with arms outstretched puts far more stress on the back than lifting with the object close to the body.
- The height of the object above the floor at the start of the lift should be between shoulder and knuckle (mid-thigh). Placement of the object on the floor or other location that requires stooping should be minimized. Rearrange the workplace so the starting and ending points of the lift are close together. Avoid stacking items above shoulder height, store loads so they will be easy to retrieve, and avoid deep shelves. Use spring-loaded bottoms for bins or gravity-feed bins to bring the load into easy reach.
- Avoid or minimize repetitive lifting tasks. Design tasks so that twisting while lifting or carrying is not required.
- The location in which the task is performed should be set up for optimal materials handling. Footing should be comfortable and stable and should not be slippery or sticky. The space should have adequate room to work, but be small enough to limit the need to reach, stretch, or carry. Make sure workers have good lighting and clear sight lines.
- Eliminate the need for pushing or pulling by using conveyors, slides, or chutes. When objects must be pushed or pulled, push whenever possible.
- Carts should have large coasters to make the job easier. Limit ramp slope to 10 degrees or less.
- The worker must be considered as part of the system. Training in ergonomically correct methods for the job can sometimes help to reduce the risk of injury. However, training will probably not be the entire answer, especially where jobs with moderate or severe risk of ergonomic injury are involved. For example, many programs attempt to train employees in safe lifting techniques. Few of these programs have demonstrated success in reducing back injuries for the following reasons:

1. Where the job is physically stressful, modifying worker behavior will not change the inherent hazard.
2. People may revert to previous habits and customs if reinforcement on the job and retraining where necessary is not provided.

Figure 10-2. Pull an object close to the body before beginning the lift.

3. In emergency or unusual situations, sudden quick movement may strain the body as workers respond to the problem.

It is therefore generally better to design a job for safe lifting than to try to train people to work safely in an unsafe job.

However, some experts believe that lifting training is appropriate, although there is still controversy over what constitutes proper lifting techniques. The older rule was to lift with a straight back from a bent-knee position. However, some workers' leg muscles (and knee joints) may not be strong enough to lift this way repetitively.

Some experts advocate other styles of lifting, such as the "kinetic lift" or the "free-style lift," but it is not clear that these will work for all people. Another approach includes improving workers' physical fitness through exercises and warm-ups.

Therefore, while there are no comprehensive rules for safe lifting, simple commonsense rules described in the previous section may apply. Additional guidelines are summarized below:

- Eliminate manual lifting and lowering from the task, where feasible.
- Be in good physical shape. Workers who are not used to lifting and vigorous exercise should not attempt difficult lifting tasks.
- Think before acting. Place material in a convenient place and within easy reach. Use the handling aids available.
- Get a good grip on the load. Test the weight before trying to lift it. If it is too heavy or bulky, get an assistant, mechanical lifting aid, or both.
- Get the load close to the body with feet close to the load. Stand in a stable position with feet pointing in the direction of movement. Lift mostly by straightening the legs.
- Don't lift with sudden or jerky movements.
- Don't twist the back or bend sideways.
- Don't lift or lower with arms extended.
- Don't heave a heavy object.
- Don't lift a load over an obstacle.

WORKSPACE AND BODY CHARACTERISTICS

Because each person has unique body characteristics, the workspace must be flexible to allow for these differences. Ideally, all work platforms, tables, and charts should be adjustable. If that is not possible, the workspace should allow for the tallest person, with adjustable chairs, foot rests, or adjustable platforms used to accommodate shorter workers.

Equipment controls should be placed between shoulder and waist level. The normal work surface should be just below the elbow; the surface should be higher for precise work, lower for heavy work.

As discussed previously, workstation surfaces and leg room need to be designed for the largest potential user; adjustments can then be made to accommodate smaller individuals. However, reaching distances in the workstation need to be designed for the smaller worker. For example, at a seated assembly operation that involves reaching into bins for parts, the chairs and work-surface adjustments should be provided so that short individuals with short arms will not have to overreach, putting stress on the shoulders and arm muscles. Infrequent reaching behind or to the sides is acceptable but should not be required on a routine basis.

The workspace for the hands should be between hip and chest height in front of the body. Most work should be done just below elbow height. In heavy work, the hands should be lower, while in fine assembly work that involves precise visual inspection, the worker will need a higher surface, perhaps with rests for forearms. Work-surface height should allow the arms to be kept low and close to the body.

Work objects should be located close to the front edge of the work surface to prevent the employee from having to bend over and lean across the surface to grasp items. Sufficient leg room must be allowed for seated operators.

Any visual symbols or displays should be placed in front of the body and below the eye level, between 10 degrees and 40 degrees below the horizontal line of sight. The most important information for the task should be placed in the worker's viewing area.

Posture

The workstation should be designed to reduce static effort, and to keep the body in what is called the neutral posture:

- Body is relaxed, with arms hanging loosely at sides
- Wrists are straight
- Shoulders are relaxed
- Elbows are close to body

Working surfaces and seats should be designed to eliminate the need to work with a bent spine. In looking for ergonomic problems, the supervisor should investigate workstations where workers have to:

- Bend necks forward more than 15 degrees
- Lean forward or sideways
- Crouch over their work
- Work with arms above head or out, away from the body
- Bend their wrists, especially if work is repetitive or requires forceful movements

Standing Work

Many jobs can be performed with less effort when the worker is standing. However, prolonged standing can create stresses on the legs and lower back. Workers who stand for a long time on hard concrete surfaces should be supplied with padded antifatigue mats. The height of the work surface should usually be 2 to 6 in. below the level of the worker's elbow when the arm is hanging in a relaxed posture (Figure 10-3).

Elevating one foot while standing can help reduce low-back stress. Provide adjustable footrests where possible. When workers alternate between sitting and standing, a stool that leans with the body is sometimes useful.

Seating

Industrial seating is often poorly designed or uncomfortable. On the assembly line, providing a chair or stool is frequently an afterthought. As a result, workers tend to rig their own stools, or salvage discarded chairs, cushions, or car seats to use at their workstations.

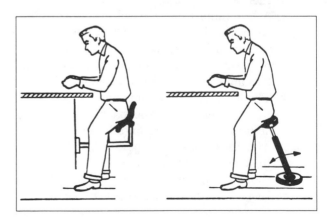

Figure 10-3. "Stand-seats" help relieve leg and lower back stress. (Courtesy *American Industrial Hygiene Association Journal.*)

The same care should be given to providing comfortable seating on the shop floor as to providing ergonomically correct chairs and equipment to office workers. Obviously, not all jobs will allow a worker to sit, but for those who do, the following guidelines should be considered:

1. The weight-bearing seat area should be large enough to accommodate the person without putting pressure on the buttocks or thighs (at least 18.2 in. wide). Seat surface should be comfortably but firmly upholstered to distribute pressure evenly. Preferably, upholstery should be a woven fabric to reduce sweating in hot months. The seat should allow the worker to change his or her seated posture. The front of the seat should be rounded: it should not produce pressure on the back of the thighs or calves, which can occur when the seat is too deep. To relieve subsequent discomfort, workers may sit too far forward, causing them to lean forward in a static posture, or they may slouch, which will put stress on the back. Generally, seats should be between 15 and 17 in. deep.

2. The seat height should be adjustable within a range of approximately 15 to 20 in. if the work surface is at a standard desk height. Adjustments of seat height will have to be made upward where the work surface is higher. For example, on an assembly line where the work surface is 32 in. high, the seat should be adjustable from 20 to 26 in. off the floor. Where higher seats are used, foot rests must be provided to prevent legs from dangling, which can cause circulatory problems. Foot rests must also be used for shorter workers at any seat height where their legs may dangle.

3. A minimum 8-in. vertical clearance should be provided under the work surface, with 26 in. forward clearance for feet and legs. Foot rests should be between 1 and 9 in. high, and be slanted, with a depth of about 12 in.

4. The back should be supported in the lumbar region. Preferably, the backrest should be adjustable both horizontally and vertically. The backrest should be at least 6 to 9 in. high and 12 to 14 in. wide. The support should maintain the back in its natural S-shaped position. Too small a backrest will provide inadequate support and too large a backrest may prevent workers from moving about if required by the work cycle (Figure 10-4).

Displays and Controls

A worker's interface with the job may be through a display console. On these jobs, workers are required to receive information through continual sensory signals, to process the information mentally, and to make a decision or take action. The display must be arranged so that the communication provided is clear, concise, and easily understood. The controls must also be arranged to minimize error.

Basic ergonomic principles of control and display design help ensure that this critical process is performed in the best manner. Controls should be within easy reach of the operator while he or she views displays or the field of operation. They should also be easily located by touch, with adequate spacing between controls.

Good sensory input (primarily involving sight and hearing) is a key principle, and requires proper lighting, visual stimuli, sound levels, and sound patterns. Urgent information is best conveyed by auditory signals, such as a bell or a buzzer. Complicated information can be displayed visually. A combination of visual and auditory signals often improves the effectivnesss of alerting the operator. For example, a buzzer could alert an operator to an emergency situation while a visual display would give details of the emergency.

Because a person's capacity for perceiving information is limited, it is important to group and simplify displays and controls. The following design concepts can help an operator to interpret and make decisions:

- Displays (gages, meters, etc.) should be grouped so that normal ranges are all in one area and pointers are in the same direction. The most frequently used displays and controls should be in the center of the visual field. A deviation from normal is then easy to detect because it stands out. Controls should be clearly labeled and color coded, if appropriate to distinguish machines, operation, and department.
- Symbols should be bold and simple. Use broad, rotating pointers on dials. The pointer should touch the gradations on the dial but should not obscure the numbers. The simpler and fewer the dial gradations, the better.
- Controls should be grouped with the related display and should follow the worker's usual expectations. In the United States, for instance, switches are usually moved *up* to activate electrical equipment and *down* to turn them off. Use distinctive shapes to aid in identification by touch.

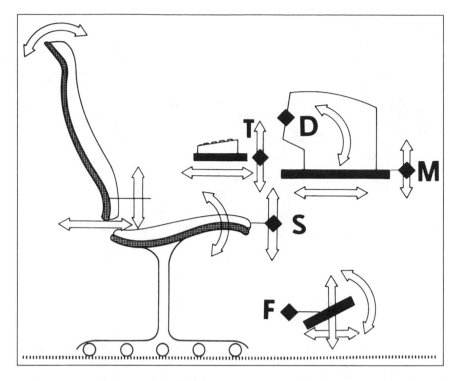

Figure 10-4. Adjustment features of a VDT office workstation. Key: S = seat height; T = table; F = footrest; D = monitor; M = support. (Reprinted with permission from Kroemer, 1985.)

- Controls should have adequate space between them.
- Controls must be arranged so workers will be able to shift their bodily positions over the course of the workday.
- Switches must be guarded with cowls to prevent accidental operation.
- Hand controls should be used where precision and speed are required; foot controls may be used where greater effort is needed.
- Workers should have time to make decisions about their work. For more complicated jobs, special training in decision making may be needed. Give operators immediate feedback on their decisions and actions.

HAND WORK AND USE OF TOOLS

Most jobs require some manipulation of objects with the hand. Although the human hand is a highly evolved grasping tool, it cannot always withstand the stresses of occupational tasks. Cumulative trauma problems are especially likely to occur where there are less than 30 seconds per cycle, or where more than 1,000 parts per shift are handled or processed.

Principles of Reducing Job Stress

The following principles should be considered when determining how much stress the job may produce:

- Activities should be performed in what ergonomists call the "natural range of movement." For the upper limbs, a work posture that allows the arms and wrists to stay in generally the same posture as if they were hanging at the sides is ideal. This means avoiding bent wrists and keeping elbows at right angles when using the arms to apply force. It also means avoiding extreme inward or outward rotation of the arms and wrists; instead, keep the wrist and hands in the "handshake" position (Figure 10-5).
- Repetitive motion and the amount of force needed to perform work should also be minimized. Jobs that require these two conditions put the worker at greatly increased risk of carpal tunnel syndrome and/or tendonitis.

Ways to reduce the effect of task repetitiveness and forcefulness include:

- Reducing the weight of the object (Figure 10-6), picking up fewer objects at a time, lifting

Figure 10-5. This worker can keep her hands in a natural, "handshake" position while working.

the object with two hands instead of one, or changing the size and shape of the objects.

- Allowing rotation to jobs where the demands on the muscles and nervous system are different.
- Alternating use of limbs.
- Allowing workers to pace themselves, taking short breaks as needed.
- Enlarging the job to include a variety of tasks.
- Using mechanical aids that perform in one motion the many tasks a worker might perform. A pneumatically operated lug wrench at a tire installation operation on an auto assembly line, for example, can reduce repetitive forceful wrist motion for the operator.
- Using jigs and fixtures to reduce the amount of holding an assembly line worker must do.
- Modifying the work surface to angle the object so that it is easier to work on.
- Using foot pedals that do not require the worker to continually depress them during an assembly operation.
- Avoiding impulse or impact forces on parts of the body; for example, the worker should use a tool rather than hammer small parts together with his or her palms.

Good Tool Design and Use

Most workers use hand tools for part of their jobs. As the time spent using tools increases, good tool design becomes more important. Improper design can result in inefficiency and in cumulative trauma disorders. Tools that require effort to hold or operate, have awkward handles that cause the wrist to be bent, or produce excessive pressure or vibration may injure the user.

Static loading of the muscles will occur where the tools or arms have to be held without support. This can lead to fatigue and sometimes injury in susceptible individuals. At times, bending a tool, such as a straight soldering iron, will reduce the need to keep the arm lifted. To reduce forces on the fingers, the tool should be activated by a bar or grip switch instead of a single finger trigger.

Minimize awkward postures, especially bent wrists. Because grip strength is greatest when the wrist is straight, tools designed with bent handles may allow the worker to keep the wrist in a more natural posture. However, each job should be analyzed to ensure that the right tool is being used.

Tool handles should be long enough to extend past the palm of the user's hand. The surfaces should be broad enough to distribute pressure evenly and should be padded or slip-resistant.

Figure 10-6. This reaming tool is supported above the work to relieve stress on hands, arms, and shoulders.

Tools such as hammers and screwdrivers should have handles about 1.5 in. in diameter. Those for fine work should be about half an inch in diameter. Handle spans, on two-handled tools, should be in the 2.5 to 3.5 in. range. They should be long enough to avoid putting pressure on the flesh of the palm or at the base of the thumb. Textured, but not highly grooved surfaces are preferred for hand tools.

Purchase tools that allow the worker to use a power rather than a pinch grip. Avoid using tools that expose workers to mechanical pinch points. Also, avoid tools with form-fitting handles that may pinch a larger worker's flesh and may reduce a smaller worker's efficiency. Hand-tool handles should be insulated against electricity, heat, and cold. Avoid repetitive use of hand tools that put pressure on the back of the hand; for example, conventional scissors have been redesigned to remove the oval handle that creates pressure on the backs of the fingers.

Whenever possible, use power tools that are balanced with a neutral buoyancy tool balancer. These items are better than a balancer that automatically reels the tool upward because they reduce the amount of reaching for and pulling on the tool.

All surfaces with which the arms, elbows, and other body parts are in contact should be rounded or padded to avoid mechanical contact stresses on the nerves.

Use of power tools that vibrate excessively can produce a disease known as Raynaud's syndrome or vibration white finger, which affects the ability of the blood to circulate in the hands, leading to numbness, stiffness, pain, and loss of strength. The affected fingers appear white. If the worker is not removed from exposure, severe tissue damage can occur.

Hand-arm vibration at frequencies below 1,000 hertz (cycles per second) should be avoided. Where workers use hand tools that vibrate, hazards may be reduced by decreasing exposure time or by dumping or isolating equipment vibration.

Driving a fastener into material with a power tool transfers torque to the worker's hand when the fastener bottoms out. The worker experiences this as a snapping action, which is a potential cause of stress. This stress can be reduced by using slip clutches or torque limiters, by keeping the torque setting low, by mounting the tool on an articulating arm, or by providing an extra handle so the worker can use two hands to help counter the torque effort.

The grip of a tool can be dampened by making it out of flexible materials. Vibration-damping gloves may be used, although if they are too large or bulky, the worker will need to use extra force to hold onto the tool.

Other tool-handling guidelines include:

- Reducing tool weight so that the worker does not need to use excessive grip strength to hold onto the tool
- Counter-balancing tools so their weight will not twist the tools out of the workers' hands
- Using tools that can be held in one hand and guided by the other hand
- Balancing and replacing worn-out parts
- Using vibrating tools less frequently or rotating personnel to other jobs

WHOLE-BODY VIBRATION

Whole-body vibration can also create physical stress. Driving a truck, bus, or car, or operating large vibrating production machinery can produce discomfort and health stresses. When in contact with an object that vibrates at very low frequencies, the body, or body parts, will resonate (pick up the vibration). Workers exposed to very low frequency (1-20 Hz) vibrations may experience difficulty breathing, pains in the chest and abdomen, backache, headache, muscular tension, and other problems. Workers exposed to chronic excessive vibration may suffer spinal problems and intestinal complaints. Strong vibration may impair visual perception, mental processing of information, and ability to complete skilled movements (Grandjean, p. 299).

Heavy vibrating tools such as jack hammers can cause damage to bones, joints, and tendons, as well as Raynaud's syndrome. The vibration from large equipment can be reduced by:

- Mounting it on springs or compression pads.
- Using structural materials that produce less vibration. These will also possibly reduce noise generated by the equipment.

The person can also be isolated from the vibrating source. For instance, for a truck driver or other seated task, you can provide cushioning or springs to isolate the seat from the vibrating surface. For standing tasks at large industrial equipment, make sure workers have a vibration-absorbing (rubber or vinyl) floor mat.

VIDEO DISPLAY TERMINALS

Video display terminals (VDTs) are widely used in offices and are becoming more common in plant and other industrial situations. They provide an example of the need for good displays and controls (Figure 10-7).

The VDT screen conveys the information for the worker. It is important that the screen be free from glare. Shielding windows with drapes or blinds can help to reduce glare. Reflected glare from shiny objects can be decreased by using dark and/or matte finishes. Glare from lighting can be reduced by avoiding overhead lighting, lowering the light level where possible, and adding task lighting on documents as needed.

- Screens should tilt back 10 to 15 degrees. The top of the screen should not be higher than eye level.
- The screen should be adjustable in placement on the desk. A comfortable viewing distance is usually between 6 and 20 in., but this varies with each individual.
- Keyboard height and placement should be adjustable, as should seat and desk height.

LIGHTING, NOISE, AND HEAT

Lighting must be appropriate to the task. For example, low light may be acceptable in storage areas, while very high lighting may be needed where workers do fine visual tasks. The amount of light needed will depend on the job, the operator's age and eyesight, and other factors. A full discussion of the technical aspects of lighting levels is beyond the scope of this book. However, the same types of guidelines apply to the general workplace as to the VDT operations.

- Provide adequate but not excessive light; excessive light can create eye fatigue.
- Provide adjustable blinds at windows.
- Avoid direct or reflected light sources in the worker's field of vision. The angle from the worker's line of sight to the source should be greater than 30 degrees.
- Shade or diffuse all light sources. Light should be reflected down from the ceiling, not up from the floor.
- Use task lighting where extra illumination is required.
- Ensure sufficient contrast for written material (i.e., use a black ink rather than pencil).
- Increase the size of small critical details.

Figure 10-7. The monitor, keyboard, and seating should be adjustable features of a VDT workstation. Note that the screen is about eye height and the keyboard is about elbow height.

- Put the task material perpendicular to the operator's line of sight (Kodak, vol. 1, p. 225).

Noise and heat stress are discussed in Chapter 8, Industrial Hygiene. When considering the ergonomics of the workplace, remember that too little heat can be as detrimental as too much. In the cold, workers' muscles stiffen, making them less able to make precise movements, and making muscles and nerves more vulnerable to injury.

SUMMARY OF KEY POINTS

This chapter covered the following key points:

- Ergonomics is the study of how people interact with their work, or how to create a good match between the employee and the workstation and tools with which the person works. The goal of ergonomics is to minimize accidents and illness due to chronic physical and psychological stresses on the job and maximize productivity and efficiency. Supervisors must be familiar with ways to recognize ergonomic problems, evaluate them, and suggest methods to control the risks.
- Ergonomic problems may arise when workstations, equipment, or tools do not fit the workers well. These stresses can cause im-

mediate or long-term damage to muscles, nerves, and joints. Most injuries are caused by forceful or repetitive motion activities or because workers are required to assume awkward postures over a period of time. These disorders are referred to as cumulative trauma disorders and include tenosynovitis and tendinitis (an inflammation of the tendons in shoulder, arm, elbow, wrist, or hands).

- Ergonomics draws on anatomy and physiology, anthropometrics, biomechanics, and psychology to solve work-related problems. Anthropometrics provides information used to design workspaces to match body dimensions. Workspaces are generally designed for the tallest or shortest worker and adjusted for individual differences. Biomechanics explains characteristics of the human body in mechanical terms. These models help to identify stressful postures and to develop countermeasures to relieve stress. In looking at the workspace, supervisors should be aware that the amount of physical stress a job produces is often determined by the posture a worker must assume to do the work, the force needed to perform the task, and the number of repetitions of a stressful task.
- To help companies reduce or eliminate ergonomic problems, supervisors can provide key information by keeping track of trends in

injuries and accidents, incidence of cumulative trauma disorders, sick leave and turnover rates, employee adjustments to workstations and equipment, and product quality levels. Causes of ergonomic problems include high overtime and increased work rate, manual materials handling, repetitive tasks, mechanical stress on the body, temperature extremes, and vibrating tools.

- According to ergonomic principles, work tasks should be designed to match the employee's physical capacity for the job. This capacity can be measured by maximum oxygen uptake or by heat produced by the body. Job demands that exceed an employee's maximum capacity; that put strain on tendons, muscles, or joints; or that disregard environmental factors (heat, humidity, etc.) can cause ergonomic injuries.

- Ergonomic principles can be applied to manual materials handling to eliminate or reduce stress-related injuries. Workers should be taught the correct ways to lift, push, or pull objects; to use mechanical means wherever feasible; to ask for assistance; and to modify workloads to make moving objects safer.

- Because each person has unique body characteristics, the workspace must be flexible to allow for these differences. Ergonomic principles applied to workplace conditions include designing furnishings and equipment to keep the body in a neutral posture as much as possible; providing relief when employees must stand to do their work; designing proper seating to prevent lower back and leg problems; and providing display consoles that allow easy viewing, maximize information obtained, and permit rapid, nonstressful operation.

- Ergonomic principles can also be applied to reduce job stress in hand work and when using hand tools. Activities should be performed as much as possible in the "natural range of motion," avoiding awkward postures and bent wrists and keeping elbows at right angles. Repeated motions should be minimized, and the amount of force required to do the work reduced as much as possible. Good tool design can minimize cumulative trauma disorders, and tools should always be kept in good condition.

- Whole-body vibration, such as riding a truck or operating large vibrating tools, can create physical stress. Ergonomic principles can be used to reduce this stress by dampening the vibration and/or isolating the person from the vibrating source.

- In addition, video display terminals must be free of glare and set at the proper height and angle to prevent neck and shoulder problems. Keyboard height and placement should be adjustable, as should seat and desk height.

- Lighting, noise, and heat must all be controlled and appropriate to the work environment or workers will suffer from work-related problems. Proper lighting is particularly important to prevent fatigue and chronic eye disorders.

11

Machine Safeguarding

After reading this chapter, you will be able to:

- **Understand the principles and benefits of safeguarding machines and equipment on the job**
- **Describe the basic requirements for effective safeguard design**
- **Understand the primary hazards of machinery and equipment mechanisms and the safeguards needed for each hazard**
- **Explain the safeguards needed for automated machines and equipment**
- **Establish a maintenance and repair program for all safeguards in your department**

The first step in any operation is to engineer the hazards out of a job as much as possible. This step helps to ensure continued production, employee safety and health, a good profit margin, and reduced equipment damage. However, at times redesign, replacement, or finding a better way is not feasible—or perhaps the technical knowledge is not sufficiently advanced. As supervisor, you must then use safeguarding to protect employees, equipment, and material from injury and damage. If such precautions are not taken, the consequences can be serious:

- When equipment must be shut down for any unscheduled reason, production capacity falls and cost of operations rises.
- When an employee is injured while operating machines or equipment, production capacity and cost of operations are again affected. Poorly designed, improperly safeguarded, or unguarded machinery or equipment is an ongoing threat to production capacity and to the well-being of employees.

An injury can take operators away from their jobs for a long time, sometimes permanently. They must be replaced if the department is to keep going. Even if a qualified person is transferred from an-other department, eventually a new person must be hired. Finding, hiring, and training the right person takes time and costs money.

Safeguards are a critical factor in controlling hazards and in preventing accidents. No matter how much training or experience a worker may have, no one can keep his or her mind focused on work every minute. Also, some machines look so deceptively easy to run that the supervisor might allow semi-skilled or even unskilled persons to operate them. Often the results can be disastrous in terms of personal injury and machine or material damage.

As a supervisor, you must have a complete understanding of the principles of machine safety and communicate these concerns to your workers. This chapter discusses guarding points of operation and power-transmitting parts of machinery or equipment, safeguarding mechanisms, use of safeguards in automation, and the maintenance of various safeguards.

PRINCIPLES OF GUARDING

Guarding is frequently thought of as being concerned only with the points of operation or with the means of power transmission (Figure 11-1). Although guarding against these hazards is required,

Figure 11-1. The cam and cam shaft of this turret lathe are completely enclosed, but a door (in open position here) is provided for maintenance.

this step also can prevent injuries from other causes, both on and around machines and from equipment and damaged material.

Guards or barriers can prevent injuries from the following sources:

1. Direct contact with exposed moving parts of a machine—either points of operation on production machines (power presses, machine tools, or woodworking equipment), or power-transmitting parts of mechanisms (gears, pulleys and sheaves, slides, or couplings), as shown in Figure 11-2.
2. Work in process, for example, pieces of wood that kick back from a power ripsaw, or metal chips that fly from tools or from abrasive wheels.
3. Machine failure, which usually results from lack of preventive maintenance, overloading, metal fatigue, or abuse.
4. Electrical failure, which may cause malfunctioning of the machine or electrical shocks or burns.
5. Operator error or human failure caused by lack of knowledge or skill, also by emotional distractions, misunderstandings, laziness, unsafe operation, illness, fatigue, and so on.

The experience of more than six decades of organized accident prevention proves that it is unwise to rely entirely on education and training of operators or their cooperation to avoid mishaps. Many factors can affect a person's attitude, judgment, and ability to concentrate. For example, skilled workers with emotional or physical problems cannot pay strict attention to their production responsibilities and cannot give their best effort.

Figure 11-2. A barrier guard covers the spindle of this turret lathe to prevent injuries from direct contact with moving parts.

Safeguarding Hazards Before Accidents

Guarding the hazard is a fundamental principle of accident prevention and is not limited to machinery. When you survey your department solely from the point of view of safeguarding against all hazards, you are likely to list a good many potential sources of injury that should be protected by barricades, rails, toeboards, enclosures, or other means. As you list electric switches and equipment, motors, engines, fixed ladders, stairs, platforms, and pits, you might well ask yourself, "Can an accident occur here, or here . . . or here?" See Chapter 1, Safety Management, for details on how to eliminate accidents before they happen.

Benefits of Safeguarding

The primary benefits of proper safeguarding, in the eyes of the supervisor, are that it reduces the possibility of injury and it may improve production. When operators are afraid of their machines or are afraid of getting close to moving parts, they obviously cannot pay strict attention to their production responsibilities. Once their fears have been removed, workers can concentrate on the operation at hand and often are more productive. In addition, whenever moving parts are left exposed, other objects may fall into them or get caught and damage the machine or material in process. Such damage can cause expensive shutdowns and loss of raw materials or finished products.

Well-designed and carefully maintained safeguards assure workers that management means what it says about preventing accidents and ensuring safe production. When employees realize this fact, they are more inclined to contribute to the safety effort.

An added benefit of a proper safeguarding program is that it can raise employee morale, especially when the workers who will have to use the safeguards are consulted before the safeguards are made or bought. Employees often have good ideas that contribute to both safety and economy. Even if they have no suggestions, workers feel flattered at being consulted about their jobs. As a result, they take a personal interest in the project, and are, therefore, less likely to remove the guards or barriers or fail to replace them. Finally, a proper safeguarding program will help to reduce litigations resulting from accidents and help to ensure worker compliance with state and federal regulations.

Types of Safeguards

A great many standardized devices, as well as improvised barriers, enclosures, and tools, have been developed to protect machine operators, particularly their hands, at point-of-operation areas. Table 11-1 presents concise descriptions of the nature, action, advantages, and limitations of a variety of such safeguards and devices.

Sheet metal, perforated metal, expanded metal, heavy wire mesh or stock may be used for types of guards. The best practice is to follow the requirements of the OSHA and ANSI standards when material for new guards or barriers is selected. Figures 11-3 through 11-5 (pages 155-56) show examples.

If moving parts must be visible, transparent impact plastic or safety glass can be used where the strength of metal is not required. Guards or barriers may be made of aluminum or other soft metals where resistance to rust is essential or where iron or steel guards can cause damage to the machinery. Wood, plastic, and glass fiber barriers have the advantage of usually being inexpensive, but they are low in strength compared with steel. However, these materials do resist the effects of splashes, vapors, and fumes from corrosive substances that would

react with iron or steel and reduce its strength and effectiveness.

SAFEGUARD DESIGN

It is easier to establish effective methods for safeguarding power transmissions than it is for guarding points of operation, because power transmissions are more standardized. *Safety Code for Mechanical Power-Transmission Apparatus*, ANSI/ASME B15.1, applies to all moving parts of mechanical equipment. Point-of-operation guarding can be found in ANSI standards for specific machinery, such as the following safety requirements:

Bakery Equipment, Z50.1
Construction, Care, and Use of Drilling, Milling, and Boring Machines, B11.8
Construction, Care, and Use of Grinding Machines, B11.9
Construction, Care, and Use of Lathes, B11.6
Construction, Care, and Use of Mechanical Power Presses, B11.1
Construction, Care, and Use of Metal Sawing Machines, B11.10
Construction, Care, and Use of Power Press Brakes, B11.3
Safety Standard for Stationary and Fixed Electric Tools, ANSI/UL 987
Safety Standard for Mechanical Power Transmission Apparatus, ANSI/ASME B15.1

Most states and OSHA regulate the safeguarding of mechanical equipment. *Code of Federal Regulations*, Part 1910, Subpart "O," covers safeguarding of machinery and equipment. It is, of course, a violation of law to ignore these regulations.

For safeguards to be effective, whether built by the user or purchased from manufacturers, they should meet certain performance or design standards. If a company does not have a safety professional, it should be able to get the advice of an insurance company safety engineer, state factory inspector, consulting engineer, or local safety council engineer as to what design is most effective in safeguarding a specific operation. The National Safety Council offers guidance to members; the Council's *Accident Prevention Manual for Industrial Operations, Engineering and Technology* volume, *Guards Illustrated*, and *Power Press Safety Manual* give many ideas in addition to those presented here.

To be generally acceptable, a safeguard should:
(text continues on page 154)

TABLE 11-1. Point-of-Operation Protection

Type of Guarding Methods	Action of Guard	Advantages	Limitations	Typical Machines on Which Used
		Enclosures or Barriers		
Complete, simple fixed enclosure	Barrier or enclosure which admits the stock but which will not admit hands into danger zone because of feed opening size, remote location, or unusual shape	Provides complete enclosure if kept in place Both hands free Generally permits increased production Easy to install Ideal for blanking on power presses Can be combined with automatic or semiautomatic feeds	Limited to specific operations May require special tools to remove jammed stock May interfere with visibility	Bread slicers Embossing presses Meat grinders Metal square shears Nip points of inrunning rubber, paper, and textile rolls Paper corner cutters Power presses
Warning enclosures (usually adjustable to stock being fed)	Barrier or enclosure admits the operator's hand but warns him before danger zone is reached	Makes "hard to guard" machines safer Generally does not interfere with production Easy to install Admits varying sizes of stock	Hands may enter danger zone— enclosure not complete at all times Danger of operator not using guard Often requires frequent adjustment and careful maintenance	Band saws Circular saws Cloth cutters Dough brakes Ice crushers Jointers Leather strippers Rock crushers Wood shapers
Barrier with electric contact or mechanical stop activating mechanical or electric brake	Barrier quickly stops machine or prevents application of injurious pressure when any part of the operator's body contacts it or approaches danger zone	Makes "hard to guard" machines safer Does not interfere with production	Requires careful adjustment and maintenance Possibility of minor injury before guard operates Operator can make guard inoperative	Dough brakes Flat roll ironers Paper box corner stayers Paper box enders Power presses Paper calenders Rubber mills
Enclosure with electrical or mechanical interlock	Enclosure or barrier shuts off or disengages power and prevents starting of machine when guard is open; prevents opening of guard while machine is under power or coasting. (Interlocks should not prevent manual operation or "inching" by remote control.)	Does not interfere with production Hands are free; operation of guard is automatic Provides complete and positive enclosure	Requires careful adjustment and maintenance Operator may be able to make guard inoperative Does not protect in event of mechanical repeat	Dough brakes and mixers Foundry tumblers Laundry extractors, driers, and tumblers Power presses Tanning drums Textile pickers, cards

(continued)

TABLE 11-1. Point-of-Operation Protection (continued)

Type of Guarding Methods	Action of Guard	Advantages	Limitations	Typical Machines on Which Used
Automatic or Semiautomatic Feed				
Nonmanual or partly manual loading of feed mechanism, with point of operation enclosed	Stock fed by chutes, hoppers, conveyors, movable dies, dial feed, rolls, etc. Enclosure will not admit any part of body	Generally increases production Operator cannot place hands in danger zone	Excessive installation cost for short run Requires skilled maintenance Not adaptable to variations in stock	Baking and candy machines Circular saws Power presses Textile pickers Wood planers Wood shapers
Hand-Removal Devices				
Hand restraints	A fixed bar and cord or strap with head attachments which, when worn and adjusted, do not permit an operator to reach into the point of operation	Operator cannot place hands in danger zone Permits maximum hand feeding; can be used on higher-speed machines No obstruction to feeding a variety of stock Easy to install	Requires frequent inspection, mainte-nance, and adjust-ment to each operator Limits movement of operator May obstruct work space around operator Does not permit blanking from hand-fed strip stock	Embossing presses Power presses
Hand pull-away device	A cable-operated attachment on slide, connected to the operator's hands or arms to pull the hands back only if they remain in the danger zone; otherwise it does not interfere with normal operation	Acts even in event of repeat Permits maximum hand feeding; can be used on higher-speed machines No obstruction to feeding a variety of stock Easy to install	Requires unusually good maintenance and adjustment to each operator Frequent inspection necessary Limits movement of operator May obstruct work space around operator Does not permit blanking from hand-fed strip stock	Embossing presses Power presses
Two-Hand Trip				
Electric	Simultaneous pressure of two hands on switch buttons in series actuates machine	Can be adapted to multiple operation Operator's hands away from danger zone No obstructions to hand feeding Does not require adjustment Can be equipped with continuous pressure remote controls to permit "inching" Generally easy to install	Operator may try to reach into danger zone after tripping machine Does not protect against mechanical repeat unless blocks or stops are used Some trips can be rendered unsafe by holding with the arm, blocking or tying down one control, thereby permitting one-hand operation Not used for some blanking operations	Dough mixers Embossing presses Paper cutters Pressing machines Power presses Washing tumblers
Mechanical	Simultanous pressure of two hands on air control valves, mechanical levers, controls interlocked with foot control, or the removal of solid blocks or stops permits normal operation of machine			

(continued)

TABLE 11-1. Point-of-Operation Protection (continued)

Type of Guarding Methods	Action of Guard	Advantages	Limitations	Typical Machines on Which Used
		Miscellanous		
Limited slide travel	Slide travel limited to ¼ in. or less; fingers cannot enter between pressure points	Provides positive protection Requires no maintenance or adjustment	Small opening limits size of stock	Foot power (kick) presses Power presses
Electric eye	Electric eye beam and brake quickly stop machine or prevent its starting if the hands are in the danger zone	Does not interfere with normal feeding or production No obstruction on machine or around operator	Expensive to install Does not protect against mechanical repeat Generally limited to use on slow-speed machines with friction clutches or other means to stop the machine during the operating cycle Can be circumvented	Embossing presses Power presses Rubber mills Squaring shears Press brakes
Special tools or handles on dies	Long-handled tongs, vacuum lifters, or hand die holders which avoid need for operator's putting his hand in the danger zone	Inexpensive and adaptable to different types of stock Sometimes increases protection of other guards	Operator must keep his hands out of danger zone Requires good employee training, close supervision	Dough brakes Leather die cutters Power presses Forging hammers
Special jigs or feeding devices	Hand-operated feeding devices of metal or wood which keep the operator's hands at a safe distance from the danger zone	May speed production as well as safeguard machine Generally economical for long jobs	Machine itself not guarded; safe operation depends upon correct use of device Requires good employee training, close supervision Suitable for limited types of work	Circular saws Dough brakes Jointers Meat grinders Paper cutters Power presses Drill presses

1. Conform to or exceed applicable ANSI standards and the requirements of OSHA and/or the state inspection department having jurisdiction.
2. Be considered a permanent part of the machine or equipment.
3. Afford maximum protection, not only for the operator and those performing lubrication, but also for passersby.
4. Prevent access to the danger zone or point of operation during operation.
5. Not weaken the structure of the machine.
6. Be convenient. Guards must not interfere with efficient operation of the machine, cause discomfort to the operator, or complicate light maintenance and/or cleaning the area around the machine.
7. Be designed for the specific job and specific machine. (Provisions must be made for oiling, inspecting, adjusting, and repairing of the machine parts.)
8. Be resistant to fire and corrosion. Be easy to repair.
9. Be strong enough to resist normal wear and shock, and durable enough to serve over a long period with minimum maintenance.
10. Not be a source of additional hazards, such as splinters, pinch points, sharp corners, rough edges, or other injury sources. If possible, a safeguard covering rotating parts

Figure 11-3. A sheet metal guard with transparent plastic viewing window guards this lathe operation.

Figure 11-4. A sheet metal hinged guard (in open position here) encloses the drive belt of this lathe.

should be interlocked with the machine itself so that the machine cannot be operated unless the safeguard is in place. On some equipment, this may be required by OSHA.

A Safeguarding Program

When installations of new or additional machinery are contemplated, persons in the organization who have a specific interest in them should be consulted. Management should seek out the opinions of machine operators, supervisors, setup and maintenance personnel, millwrights, oilers, and electricians. Also, check with manufacturers; frequently these companies have equipment safeguards available superior to any others that might be installed (see the discussion in this chapter under Built-in Safeguards). Remember to consult manufacturers if a proposed guard or barrier changes the original design of the equipment.

Pertinent regulatory codes should be studied before undertaking installation. Every safeguard, whether installed when machinery is purchased or constructed afterward, must meet or exceed state, local, and OSHA federal regulations. Since state and OSHA regulations at best provide only minimum requirements, the supervisor or the person responsible for buying, constructing, or maintaining guards should be familiar with the applicable ANSI standards and *CFR*, 1910, Subpart "O" (or other applicable portion).

A safeguarding program should include all

Figure 11-5. The transparent plastic splash guard on this index screw machine protects workers from operating oils.

equipment in the establishment. It is clear that additional protection may be required under the following conditions:

- Where the protection is clearly below that recommended or required by an ANSI standard or the state or federal authority having jurisdiction
- Where the protective devices are in need of repair or replacement

A good safeguarding program is a strong selling point when promoting observance of other preventive measures. If hands and arms are protected by guards or barriers, it makes more sense to shield eyes, feet, ears, and other parts of the body. Like a pair of safety glasses on a person's face, a safeguard is a warning signal—a constant reminder to workers that they are near a potential hazard.

Maximum-Sized Openings

An important factor to consider in the design of safeguards is the maximum size of openings. If a barrier is to provide complete protection, the openings in it must be small enough to prevent a person or object from getting into the danger zone. If, for operational purposes, an opening must be larger than that specified by the standard (Figure 11-6), further protection must be provided.

Built-in Safeguards

The best guard or barrier is the one provided by the manufacturer of a machine. For many years, most standard machinery manufacturers have designed first-class, serviceable safeguards for their equipment, available when specified on a purchase order. The manufacturer's safeguards are usually designed to be an integral part of the machine and are therefore superior in appearance, placement, and function to those made onsite.

Companies may fail to implement the principles of correct machine safeguarding for several reasons. They may believe that partial or makeshift barriers can do the job well enough, or they may be reluctant to spend the additional money required for built-in safeguards. Neither reason is valid.

The disadvantages of makeshift safeguards are

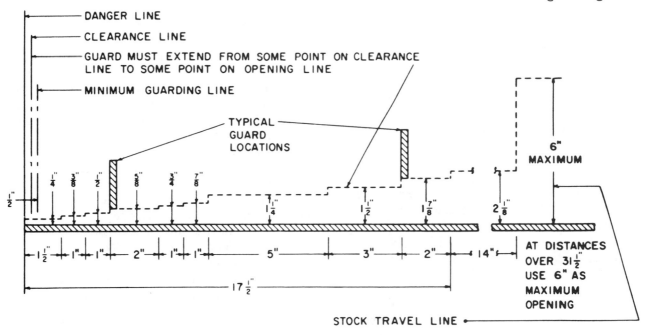

Figure 11-6. These point-of-operation guard locations show where barriers should be placed to prevent hand or forearm from contacting a hazard. The danger line is the point of operation. The clearance line marks the distance required to prevent contact between the guard and the moving parts. Minimum guarding line is ½ in. from the danger line. (From OSHA requirements 29 *CFR*, 1910.217.)

obvious. They give a false sense of security and may, therefore, be more harmful than no safeguard at all. They require the operators to be constantly alert in order to make up for the guard's inadequacy and thus give little protection against human failure. Makeshift guards are often flimsy and are certain to become damaged and ineffective, sometimes shortly after installation. At times they may be damaged on purpose by workers tired of using them.

Usually, it is just as easy to install an effective barrier as it is to use one with limited protection. A completely effective safeguard is one that eliminates the hazard completely and permanently and can withstand handling and normal wear and tear.

Too often, even today, machinery is purchased and installed without the necessary safeguards to protect the operator and nearby workers. Frequently, the excuse given is that purchase of a machine as a capital expense is closely budgeted, while construction of a "homemade" barrier is categorized as an operating cost and is more acceptable. As a result, a large number of companies buy stripped-down machinery and make safeguards after the machinery has been installed.

This tactic is poor economy. Machine safeguards can be provided by the manufacturer more cheaply because the costs are spread over a large number of machines. Aside from cost savings, of course, the hazards presented by an unguarded ma-

chine make a powerful argument in favor of specifying safeguards from the manufacturer at the time a machine is ordered.

In summary, the advantages of securing as much built-in protection as possible from the manufacturer are as follows:

1. The additional cost of safeguards designed and installed by the manufacturer is usually lower than the cost of installing safeguards after a machine has been purchased.
2. Built-in safeguards conform more closely to the contours of the machine.
3. A built-in safeguard can strengthen a machine, act as an exhaust duct or oil retainer, or serve some other functional purpose, thereby simplifying the design and reducing the cost of the machine.
4. Built-in safeguards generally improve production and prevent damage to equipment or material in process.

Substitution as a Safeguard

Like the installation of safeguards, substituting one type of machine for another sometimes can mean elimination or reduction of machine hazards. For example, substitution of direct-drive machines or individual motors for overhead line-shaft transmission decreases the hazards inherent in transmission equipment (assuming that some older line-shaft installations still remain). Speed reducers can replace multicone pulleys. Remote-controlled automatic lubrication can eliminate the need for employees to get dangerously close to moving parts.

Matching Machine or Equipment to Operator

Thus far in this chapter, the discussion has emphasized safeguarding transmission parts or points of operation. Safe operation of machinery, however, involves more than eliminating or covering hazardous moving parts. The overall accident potential of the machine operation must be considered. These basic questions should be asked:

- Is there a materials handling hazard?
- Are the limitations of a person's manual effort—lifting, pushing, and pulling—recognized?
- Is the design of existing or proposed safeguards based on physiological factors and human body dimensions?

All physical or design features of a production machine and the workplace should be evaluated as though the machine were an extension of a person's body and can only do what that person wants it to do. To match the machine or equipment to the operator, consider the following factors:

- **The workplace.** Machines and equipment should be arranged so that the operator does a minimum amount of lifting and traveling. Conveyors and skids to feed raw stock, and chutes or gravity feeds to remove finished stock should be considered.
- **The work height.** The workstation should be of optimal height in relation to stand-up or sit-down methods of operation, whichever is used. The proper height and type of chair or stool must be determined. Elbow height, actually, is the determining factor in minimizing worker fatigue. In general, an effective work level is 41 in. (1 m) from floor to work surface, with a chair height from 25 to 31 in. (0.6 to 0.8 m).
- **Controls.** Machine speed and ON-OFF controls should be readily accessible. Position and design of machine controls—such as dials, push buttons, and levels—are important. Controls should be standardized on similar machines; operators then can be switched back and forth, as necessary, without having to use different controls.

- **Materials handling aids.** These aids should be provided to minimize manual handling of raw materials and in-process or finished parts, both to and from machines. Overhead chain hoists, belts or roller conveyors, and work positioners are typical examples.
- **Operator fatigue.** Workers become fatigued at a machine station usually as the result of combined physical and mental activities, not simply from expending energy. Excessive speed-up, boredom from monotonous operations, and awkward work motion or operator position also contribute to fatigue.
- **Adequate lighting and other environmental considerations.**
- **Excessive noise.** More than just an annoyance, excessive noise can be a real hazard because it can cause permanent hearing damage. Methods of protecting workers are given in Chapters 8 and 9 and are covered by OSHA regulations.

SAFEGUARDING MECHANISMS

Machines have certain basic mechanisms which, if exposed, always need safeguarding. These mechanisms, which incorporate the primary hazards involved in machinery, can be grouped under the following headings:

- Rotating mechanisms
- Cutting or shearing mechanisms
- In-running nip points
- Screw or worm mechanisms
- Forming or bending mechanisms
- Impact mechanisms

A piece of equipment may involve more than one type of hazardous exposure. For instance, a belt-and-sheave drive is a hazardous rotating mechanism and also has hazardous in-running nip points.

Rotating Mechanisms

A rotating part (Figure 11-7) is dangerous unless it is safeguarded. Mechanical power transmission apparatus represents the large percentage of this type of hazardous mechanism. Although relatively few injuries are caused by such apparatus, the injuries often are permanently disabling. Transmission equipment should, therefore, be safeguarded as effectively as possible.

Small burrs or projections on a shaft can easily catch hair or clothing, or grab hold of a cleaning rag

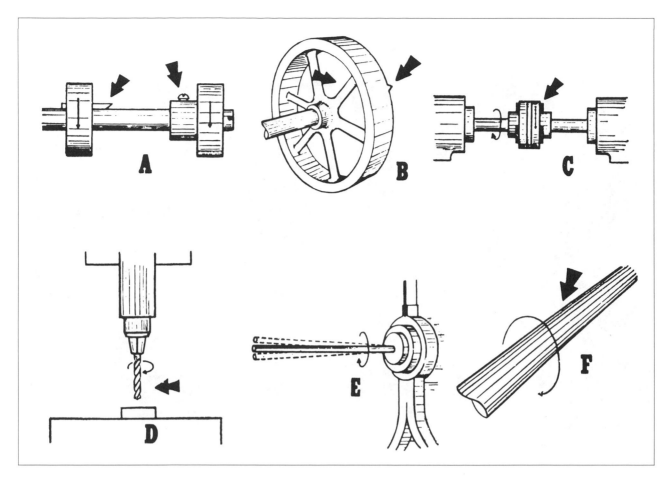

Figure 11-7. Rotating mechanisms can seize and wind up loose clothing, belts, hair, etc. They should, therefore, be guarded. Left to right, they are (A) projecting key and set screw, (B) spokes and burrs, (C) coupling bolts, (D) bit and chuck, (E) turning bar stock, and (F) rotating shaft.

or apron and drag a person against and around the shafting. Vertical or horizontal transmission shafts, rod or bar stock projecting from lathes, set screws, flywheels and their cross members, drills, couplings, and clutches are common hazardous rotating machine parts.

Shafting, flywheels, pulleys, gears, belts, clutches, prime movers, and other types of power transmission apparatus usually seem safe by virtue of their location. However, many accidents happen in places where such apparatus is located—places "where no one ever goes." The supervisor, the oiler, and maintenance people go into these seldom-entered places. For their safety, the hazards in such areas should also be safeguarded (Figure 11-8).

When a flywheel, shaft, or coupling is located in such a way that any part of it is less than 8 ft (2.4 m) (ANSI/ASME B15.1-1984) above a floor or work platform, it must be safeguarded in one of several methods set forth in the federal standards. For example, exposed pulleys, belts, and shafting with rotating parts within 8 ft of the floor should have substantial safeguards designed to protect employees from all possible contact. The underside of belts running over passageways or work areas should be protected by screening so that they can cause no harm if they break.

Enclosures should be removable, or provided with hinged panels, and interlocked to facilitate inspection, oiling, or repairing of the parts of the mechanism. However, enclosures should be taken off only when the machine is stopped and the power disconnect is locked in the OFF position, and then preferably only by the supervisor's special order. Machines should never be operated until the enclosures have been properly replaced and secured.

Ends of shafting projecting into passageways or into work areas should be cut off or protected by nonrotating caps or safety sleeves. Hangers should be securely fastened and well lubricated. Setscrews

Figure 11-8. This hinged screw machine gearbox guard (shown in open position) is required even if the gears seem to be protected by location.

with slotted or hollow heads should be used instead of the projecting type. Couplings should be protected to prevent contact not only with coupling, but with the exposed rotating shafts also.

Exposed gears should be entirely enclosed by substantial barriers or protected in some other equally effective way. Sheet metal is preferable. The in-running side of the gearing should be treated with special care. Removable barriers should be equipped with interlocks and the power disconnect locked in the OFF position for inspection and maintenance. Wood barriers can be used where excessive exposure to water or chemicals might cause rapid deterioration of metal barriers. Since wood barriers can also deteriorate, they should be monitored regularly.

It is essential to enclose the gearing in power presses because the large forces involved may cause overloading, which, in turn, can cause fatigue cracks to start in the shaft or gear. The fact that gears have split in half, and have fallen, indicates the need for strong enclosures.

Clutches, cutoff couplings, and clutch pulleys with projecting parts 8 ft or less from the floor should also be enclosed by stationary safeguards.

Some clutches within the machine may be considered "guarded by location," since they are out of normal reach. But if any possibility of contact exists, a complete enclosure should be provided.

Cutting or Shearing Mechanisms

The hazards of cutting or shearing mechanisms (Figure 11-9) lie at the points where the work is being done, and where the movable parts of the machine approach or cross the fixed parts of the piece or machine. Guillotine cutters, shear presses, band and circular saws, milling machines, lathes, shapers, and abrasive wheels are typical of machines that present cutting or shearing hazards.

Saws. A circular saw must be safeguarded by a guard that covers the blade at all times to at least the depth of the teeth. The hood must adjust itself automatically to the thickness of the material being cut in order to remain in contact with it. The hood should be constructed in such a way that it protects the operator from flying splinters or broken saw teeth. A table saw should be equipped with a spreader or splitter. It should also have an antikick-

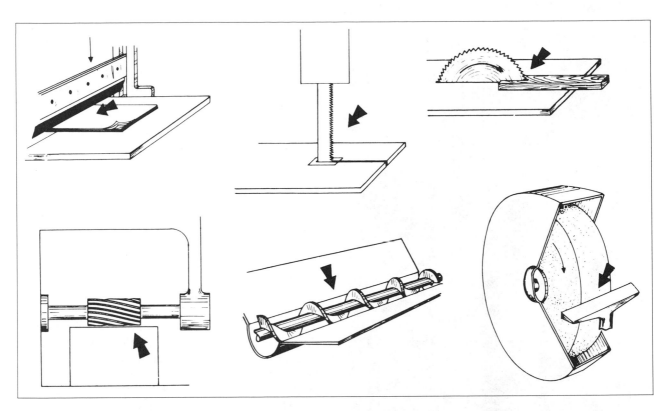

Figure 11-9. Protection should be provided for all variations of common cutting or shearing mechanisms.

back device to prevent materials from being thrown back at the operator.

All portions of a band saw blade must be enclosed or otherwise safeguarded except the working portion of the blade between the bottom of the guide rolls and the table. The enclosure for the portion of the blade between the sliding guide and the upper saw wheel guard should be self-adjusting, if possible. Band saw and band knife wheels should be completely enclosed, and the barrier should be constructed of heavy material, preferably metal (Figure 11-10).

A swing cutoff saw should be equipped with a complete enclosure for the upper half of the saw. The lower half must be designed so it will ride over the fence and drop down on the table or on the work being cut. Also, a counterweight or other device should automatically return the saw to the back of the table when the saw is released at any point in its travel. Limit chains or other equally effective devices should be provided to prevent the saw from swinging past the table toward the operator. Saw blades should be kept sharp at all times to prevent force feeding and the development of cracks.

Cutters. On a milling machine, cutters should be shielded by an enclosure that provides positive protection for the operator's hands. A rotary cutter

or slitter should have a barrier that completely encloses the cutting disks or knives. This makes it impossible for the operator to come into contact with the cutting edges while the machine is in motion.

Shears. The knife head on both hand- and power-operated shears should be equipped with a barrier to keep the operator's fingers away from the cutting edge. The barrier should extend across the full width of the table and in front of the hold-down. This barrier may be fixed or automatically adjusted to the thickness of the material to be cut.

If material narrower than the width of the knife blade is being cut, adjustable finger barriers should be installed. These will protect the open area at the sides of the material being sheared so that the operators cannot get their hands caught at the sides under the blade. The barrier may be slotted or perforated to allow the operator to watch the knife, but the openings should not exceed ¼ in. (6 mm), according to ANSI B11.1, and should preferably be slotted vertically to provide maximum visibility. A cover should be provided over the entire length of the treadle on the shears, leaving only enough room between the cover and the treadle for the operator's foot. This device prevents accidental tripping of the machine if something should fall on the treadle.

Figure 11-10. A transparent plastic shield has been formed into a guard for this band saw.

Grinding wheels. Since the abrasive grinding wheel is a common power tool and is often used by untrained workers, it is the source of many injuries. Stands for grinding wheels should be heavy and rigid enough to prevent vibration and should be securely mounted on a substantial foundation. The wheels should neither be forced on the spindle nor should they be too loose.

A person trained in correct and safe procedures should be in charge of wheel installations. Before a wheel is mounted, it should be carefully examined for cracks or other imperfections that might cause it to disintegrate. The work rest should be rigid and set no farther away than 1/8 in. (3 mm) from the face of the wheel so that the material cannot be caught between the wheel and the rest. Wheels should be kept true and in balance.

Each wheel should be enclosed with a substantial hood made of steelplate to protect the operator in case the wheel breaks. The threaded ends of the spindle should be covered so that clothing cannot get caught in them. Hoods should be connected to effective exhaust systems to remove chips and prevent harmful dust from entering the grinding operation.

The manufacturer's recommended wheel speed should be adhered to at all times. Eye and face protection should be provided and properly fitted. Instruct workers that respirators should be used if considerable dust is generated, or if a poisonous or dangerous material is used and the air contaminants cannot be properly exhausted. Respirators should be frequently cleaned and sterilized. Respirators should not be issued for use by more than one employee.

It is important that grinding wheels be properly stored and handled to prevent cracking or other damage. They should be stored in a dry place where they will not absorb moisture and should be kept in racks, preferably in a vertical position. The manufacturers of the wheels can supply additional information on proper storage and handling methods.

Buffing and polishing wheels. This equipment should be equipped with exhaust hoods to catch particles thrown off by the wheels. Not only do hoods protect the operator—they also prevent accumulation of particles on the floor and in the area. Eye or face protection should be provided for operators. They should not wear gloves and loose clothing. Protruding nuts or the ends of spindles, which might catch on the operator's hands or clothing, should be covered by caps or sleeves.

Wire brush wheels. The same machine setup and conditions that apply to polishing and buffing wheels apply to brushes. The speed recommended by the manufacturer should be followed strictly.

The hood on scratch wheels should enclose the wheel as completely as the nature of the work allows and should be adjustable so the protection is maintained as the diameter of the wheel decreases. The hood should cover the spindle end, nut, and flange protection. Personal protective equipment is especially important in the operation of scratch wheels because wires can break off and fly into the operator's face.

The materials should be held at the horizontal center of the brush. The wire tips of the brush should do the work. Forcing the work into the brush only results in (a) merely wiping or dragging the wires across the material, with no increase in cutting action, (b) increasing wire breakage, and (c) snagging the work. Small pieces should be held in a jig or fixture. More details about abrasive wheels, buffers, and scratch brushes are given in Chapter 12, Hand Tools and Portable Power Tools.

In-running Nip Points

Whenever two or more parallel shafts that are close together rotate in opposite directions (Figure 11-11), an in-running nip point is formed. Objects or parts of the body may be drawn into this nip point and be crushed or mangled. Typical examples of nip points are found on rolling mills, calenders, chains and sprockets, conveyors, belts and sheaves, racks and pinions, and at points of contact between any moving body and a corresponding stationary body.

Nip points should be made inaccessible by fixed barriers (Figure 11-12), or should be protected by instantaneous body-contact cutoff switches with automatic braking devices. The in-running side of rolls such as those used for corrugating, crimping, embossing, printing, or metal graining should be protected. Arrange a barrier so that operators can feed material to the machine without catching their fingers between the rolls or between the barrier and the rolls.

Protect calenders and similar rolls with a device that the operator can use to shut off the rolls immediately at the feed point by means of a lever, rod, or treadle. Otherwise, the nip should be guarded by an automatic electronic device that will stop the rolls if anything but stock approaches the intake points. Enclosures are usually the most satisfactory way to protect chains and sprockets, racks and pinions, belts and pulleys (or sheaves), and drive mechanisms for conveyors.

Screw or Worm Mechanisms

The hazards involved in the operation of screw or worm mechanisms are the shearing action set up between the moving screw and the fixed parts of the housing. Screw or worm mechanisms are generally used for conveying, mixing, or grinding materials. Examples are food mixers, meat grinders, screw conveyors, dry material mixers, and grinders of various types.

Screw conveyor covers should not be used as a walkway. If they must be walked on, additional protection should be provided. Corrosion and/or ab-

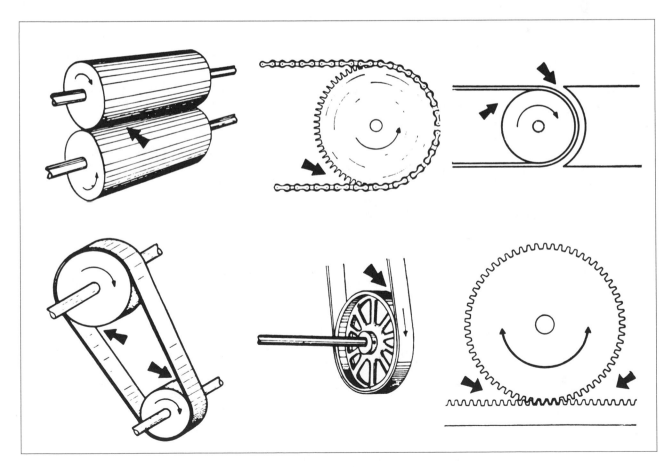

Figure 11-11. These typical in-running nip points require guarding.

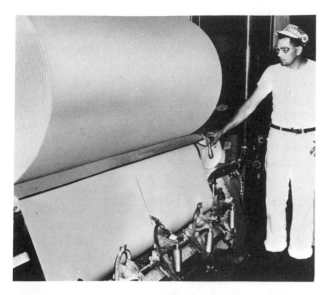

Figure 11-12. A horizontal guard protects the operator from the in-running nip point created by rollers on a paper-rewind machine.

rasive material in this type of conveyor can erode the metal from the underside of the conveyor cover so that a person might, without thinking, step through it. When screw conveyors are constantly fed while in motion and the use of an interlocked cover is impractical, install a heavy screen or mesh guard or grid. In this way, when the cover is removed for product inspection, the mesh will prevent anyone from falling through the opening.

Covers should also be provided for all mixers. The cover should be hinged to prevent removal and should have an interlock that will cut off the power source to stop action immediately when the cover is raised. Power switches for such machines should be locked out during maintenance or cleaning operations.

Rigid grids should cover the openings on grinder hoppers. Such coverings should be large enough to permit materials to be fed to the grinder, but small enough to prevent any part of the operator's body from touching the cutting knives or the worm. Removable hoppers should be interlocked so that the grinders cannot be operated when the hoppers are removed.

Other safeguards for screw or worm mechanisms may include mechanical and electrical devices that require the operator to use both hands on the controls. Operating controls should be located so that they cannot be activated while any part of the body is in a position to be caught in the machine.

The points of operation in many instances can be guarded by regulating the size, shape, and location of the feed opening.

Forming or Bending Mechanisms

The use of power, foot, and hand presses for stamping and forming pieces of metal and other materials has grown rapidly. Hand and finger injuries on these presses, as they are commonly operated, have become so frequent that misuse or abuse of these machines constitutes one of the most serious sources of mechanical problems in accident prevention.

Factors that make the problem difficult are variations in operations and operating conditions—the size, speed, and type of press; the size, thickness, and kind of pieces to be worked; construction of dies; degree of accuracy required in the finished work; and length of run. It is unwise to simply depend on the skill of the operator for protection.

If you want to make certain that operating methods are safe, insist that the die setter not only set the dies for a new run and test the machine for proper operation, but also set and adjust all other safeguards. Then ask the operator to run the machine a few strokes to make sure the safeguards are in place, and the adjustments are correct. See to it that operators know they must complete each of these steps before starting a production run. (This procedure applies equally to shearing mechanisms: for example, guillotine cutters, veneer chippers, paper-box corner cutters, leather dinking machines. As a group, these and similar machines present the worst kind of point-of-operation hazards, yet supervisors often believe that safeguarding them would slow down production. The right safeguards, properly designed and utilized, will not only cover the hazard, but will usually increase production.)

A procedure that omits any of these steps provides no guarantee that the dies are in alignment, that the kickout is working properly, that the clutch is in proper condition, or that the guard (if it is removable) is in place and operating. Supervisors with long experience in hazard control customarily require the machine setup crew to fill out a tag as written evidence that the procedure was thorough and complete for each setting. (*CFR* 1910.217 (e)(1) states, "Inspection, maintenance, and modification of presses—inspection and maintenance records. It shall be the responsibility of the employer to establish and follow a program of periodic and regular inspections of his power presses to insure that all their parts, auxiliary equipment, and safeguards are in a safe operating condition and adjustment. The employer shall maintain records of these inspections and the maintenance work performed.")

Each job poses questions. Should fixed guards, pull-away devices, two-hand devices, or electric-eye trips be used? The answer depends upon such factors as the kind of stock (piece, strip, roll); type of feed (hand, slide, dial, automatic); and type of knockout or ejection (mechanical or pneumatic) used to complete the operation.

Power presses sometimes repeat a cycle unexpectedly (usually referred to as a double trip) due to machine failure of one kind or another. The guard, therefore, must be designed to prevent injuries that could result from such occurrences. Workers lost fingers most often when they reach impulsively for misplaced stock after the press has been tripped. Details of how to prevent these accidents are given later in this chapter under the heading Automatic Protection Devices. Even the best training, experience, and supervision cannot substitute for well-designed guards and the constant observance of safety practices.

Primary and Secondary Operations

Power-press operations consist of primary and secondary operations. In primary operations, workers are not required to place their hands between the punch and the die. Examples are blanking, piercing, corner cutting, and other operations on long-strip stock or in processing a large part. Primary operations should have die enclosures or fixed-barrier guards (Figure 11-13).

In secondary operations, a preshaped part is further processed by being placed in a nest under the upper die. These operations include coining,

Figure 11-13. This die-enclosure guard made of pre-slotted material can be easily made in the shop.

drawing, and forming. Secondary operations, which account for most power-press accidents, should have the most effective protection within the limitations of the die—if possible, automatic or semiautomatic feeds and ejection and some form of protective guard.

Wherever possible, high-production or long-run dies should be guarded individually. This procedure saves the time involved in setup, and a permanent guard will not be detached or lost.

If die enclosures or fixed barriers cannot be used on secondary operations, use an adjustable-barrier device, a gate or movable barrier device, or a device that either prevents normal operation of the press until the hands are removed from the point of operation or the device itself removes the hands from the point of operation. Unless the press has automatic or semiautomatic feeding and ejection with a die-enclosure or a fixed-barrier guard, hand tools should be used. However, a hand-feeding tool is not a point-of-operation guard or protection device, and should never be used instead of it.

These safety devices must be maintained and their use supervised strictly in accordance with the manufacturer's instructions. Otherwise they may fail, resulting in injuries and loss of workers' confidence. Usually an insurance company safety professional or the state or local OSHA representative is willing to assist in the design and installation of the best guard for a specific operation.

Feeding Methods

Automatic or semiautomatic feeds, if they can be used, are generally successful safeguards. In most cases, they either increase production or reduce costs or both. With these feeds, operators do not need to place their hands under the slide during ordinary feeding. They may, however, be tempted to do so if a piece sticks, or they may do so inadvertently. Therefore, it is necessary to provide a wood or soft metal stick or pick with which the operators can remove the material, if necessary, and an enclosure to prevent their putting a hand under the slide. Use of an air-ejection jet can help in removing parts from the die. The enclosure should be interlocked with the clutch brake mechanism.

The choice of feed will depend on the design of the die, the shape and quantity of parts being processed, and the type of equipment available. Feed types include gravity or chute, push, follow, magazine, automatic magazine, dial, roll, reciprocating, hitch, and transfer. Feeds can be automatic, semiautomatic, or manual.

Among the advantages gained with automatic feeding are (a) the operator does not have to reach into the point of operation to feed the press; (b) the feeding method usually makes it possible to enclose the die completely; and (c) the operator can load the feed mechanism, start the press, and then leave the vicinity of the press for a considerable number of strokes. Sometimes, the operator may be able to run several presses at once. Production volume must be sufficient to justify the expense of automatic feeding.

With semiautomatic feeds, the operator does not have to reach into the point of operation, but usually must manually load the feed mechanism repeatedly or at frequent intervals. The feeding method usually makes it possible to enclose the die completely. Semiautomatic feeding is not adaptable for certain blanking operations or for nesting off-shaped pieces (Figure 11-14).

When manual feeding is required, provision should be made to eliminate the need for operators to place their hands or fingers within the point of operation. If this cannot be done, some method must be employed to protect the employee should the ram descend. Special tools have been developed and used successfully on operations where automatic feeds or enclosure guards are impractical. Such tools include

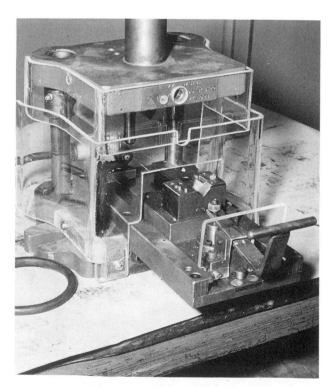

Figure 11-14. The slide feed allows loading of the die outside the danger zone. The permanent plastic barrier guard permits full visibility of the operation. The guard is separated at the top to allow die maintenance. The overlap of the guard at the separation eliminates shear hazard during travel of the slide. (Courtesy Allis-Chalmers Mfg. Co.)

pushers, pickers, pliers, tweezers, forks, magnets, and suction disks and are usually made of soft metal to protect the die. Strict discipline is necessary to force operators to use them consistently. Such hand tools are not substitutes for guards, but should be used in conjunction with guard devices.

Bear in mind that few kinds of press guards provide complete protection. An automatic or semiautomatic feed may make it unnecessary for operators to place their hands in the danger zone, but may not prevent their doing so. If the operator attempts to straighten a part just before it passes under the slide, this method of safeguarding provides no real protection. Therefore, good practice combines such guards with a two-hand trip (for actuating the clutch) that requires constant pressure or control during the downward stroke of the press.

Automatic or semiautomatic methods of feeding usually can be installed on jobs that are fed manually. Automatic feeds should be supplemented by a substantial enclosure at the point of operation, especially on slow-moving equipment, for complete protection. If possible, machine parts should be adjusted to reduce the hazard. For instance, the stroke on a press may be limited so that the fingers cannot enter between the dies. More details are given later in this chapter.

Ejecting Material

The safe removal of material is as necessary as safe feeding. Since the way in which the finished piece is removed from the machine may influence the choice of feeding method, the removal method must be considered at the time an automatic or semiautomatic feed is selected.

Various methods of ejection may be used: compressed air, punch, knockouts, strippers, and gravity. Operators should not be required to remove the finished parts or scrap material from the die manually because of the hazards involved. Air blowoff systems, crankshaft-operated scrap cutters, and other devices may be used.

Controls

Power presses should have actuating devices that prevent the operator's hands from getting under the slide or ram when the press is operated. Such devices include two-hand switches or levers, treadle bars, pedals, and switches, located away from the point of operation. If two-hand switches or levers are used, relays or interlocks should be installed so that one switch or lever cannot easily be made inoperable and permit the press to be controlled with one hand, thus defeating the safeguard. Pedals, foot switches, and pedal bars should be used only when it is absolutely necessary for the operator to have both hands away from the point of operation when the press is operated.

Press brakes are the source of many accidents because the gaging stops may be too low, and the piece being processed may slip beyond them. Because the motion of the ram is slow, some operators reach through the area between the ram and the die to adjust the work and, in doing so, are caught. The proper type of starting device will make it impossible for the operator to reach under the ram after the press has begun to operate.

Foot controls should be covered by stirrup-type covers that extend over the entire length of the treadle arm (an inverted U-shaped metal shield above the control) to prevent accidental tripping (Figure 11-15). When two or more operators run a press, foot controls or hand controls should be connected in series so that each person is in the clear before the press can be operated.

On die-casting machines, two-hand tripping devices have been widely used. However, many consider it safer to install a sliding door that covers the die area. As the door closes, protecting the hazardous zone, it activates switches that set the machine in motion. Such a door virtually eliminates burns from splashing material.

It is a prerequisite for a safe die casting that no operator be allowed to place the other hand or arm between the dies at any time. Long-handled pliers or tongs or similar tools should be used to place and remove stock; mechanical feeds and ejectors are even safer. Effective safeguards for these machines include two-hand tripping devices, sliding doors, treadle bars, and electrical or mechanical interlocking devices. To eliminate the dangers involved in machine operation, enclosures may either be built and installed over the hazardous areas or the equipment may be redesigned to eliminate such exposed parts.

The modern lathe is a good example of machinery made safe through improved design. Its motor drive and gear box are enclosed so that line shafts, pulleys, and belts are dispensed with. The modern power press, in which all the working parts with the exception of the slide (ram) are enclosed, is another good example.

Safeguards used to make machinery safe include the fixed guard or enclosure, the interlocking guard or barrier, and the automatic protection device. Automatic or semiautomatic feeding and ejection methods are also ways of safeguarding machine operations.

Fixed Guards or Enclosures

The fixed guard or enclosure (Figure 11-16) is considered preferable to all other types of protection and should be used in every case, unless it has been definitely determined that this type is not at all practical. The principal advantage of the fixed guard is that it prevents access to the dangerous parts of the machine at all times. (If a fixed barrier is to provide complete protection, the openings in it must be small enough to prevent a person from getting into the danger zone. See Figure 11-6 earlier in this chapter.) Another advantage to this type of guard is that when the production job is finished, the guard remains with the die until it is needed for the next run.

Fixed safeguards may be adjustable to accommodate different sets of tools or various kinds of work. However, once they have been adjusted, they should remain fixed; under no circumstances should they be detached or moved.

Typical examples of the application of fixed safeguards are found on power presses, sheet-leveling or flattening machines, milling machines, gear trains, drilling machines, and guillotine cutters. Some fixed barriers are installed at a distance from the danger point in association with remote feeding arrangements that make it unnecessary for the operator to approach the danger point.

Figure 11-15. Foot pedals must be covered to prevent unintentional tripping of the press by the operator or by falling objects.

Figure 11-16. These expanded metal enclosure guards protect workers from protruding moving parts and power transmissions.

Interlocking Guards or Barriers

Where a fixed safeguard cannot be used, an interlocking guard or barrier should be fitted onto the machine as the first alternative. Interlocking may be mechanical, electrical, pneumatic, or a combination of types. The purpose of the interlock is to prevent operation of the control that sets the ma-chine in motion until the guard or barrier is moved into position. Operators subsequently cannot reach the point of operation, the point of danger.

When the safeguard is open, permitting access to dangerous parts, the starting mechanism is locked to prevent accidental starting, a locking pin or other safety device is used to prevent the basic mechanism from operating, for example, to prevent the main shaft from turning. When the machine is in motion, the enclosure cannot be opened. It can be opened only when the machine has stopped or has reached a fixed position in its travel.

To be effective, an interlocking safeguard must satisfy three requirements:

1. Guard the dangerous part before the machine can be operated.
2. Stay closed until the dangerous part is at rest.
3. Prevent operation of the machine, if the interlocking device fails.

Two-hand tripping devices are incorporated in many types of interlocking controls. These devices require simultaneous and sustained pressure of both hands on switch buttons, air-control valves, mechanical levers, or controls interlocked with a foot control, to name just a few. Two-hand operating attachments should be connected so that it is impossible to block, tie down, or hold down one button, handle, or lever, and still operate the machine.

When gate devices or hinged barriers are used with interlocks, they should be arranged so that they completely enclose the pinch point or point of operation before the operating clutch can become engaged.

Interlocking controls are often installed on bakery machinery, guillotine cutters, power presses, dough mixers, some kinds of pressure vessels, centrifugal extractors, tumblers, and other machines on which covers or barricades must be in place before the starting control can be operated.

Automatic Protection Devices

An automatic protection device may be used, subject to certain restrictions, when neither a fixed barrier nor an interlocking safeguard is practicable. Such a device must prevent the operator from coming in contact with the dangerous part of the machine while it is in motion, or must be able to stop the machine in case of danger.

An automatic device functions independently of the operator, and its action is repeated as long as the machine is in motion. The advantage of this type of device is that tripping can occur only after the operator's hands, arms, and body have been removed from the danger zone.

An automatic protection device is usually operated by the machine itself through a system of linkage, through levers, or by electronic means, and there are many variations. It can also be a hand-restraint or similar device, or a photoelectric relay.

Pull-away or hand-restraint devices are attached to the operator's hands or arms and connected to the slide or ram plunger, or outer side of the press in such a way that the operator's hands or fingers will be withdrawn from the danger zone as the slide or ram plunger, or outer slide descends. These devices should be readjusted at the start of every shift so they will properly pull the operator's hands clear of the danger zone.

All electronic safety devices for power presses are made to perform the same function when energized—they interrupt the electric current to the power press and shut off the machine (just as if the STOP button had been pushed). Electronic safety devices are effective only on power presses having air, hydraulic, or friction clutches. Such devices are not effective on power presses with positive clutches. This is because once the operating cycle of this type of power press starts, nothing can prevent completion of the cycle.

To be effective, the electronic device should be operated from a closed electric circuit so that interruption of the current will automatically prevent the press from tripping. As the supervisor, it is your responsibility to make sure these devices are properly adjusted and maintained in peak operating condition. Many injuries have occurred because of improper adjustments and because parts have been allowed to become worn or need repair.

One advantage claimed for electronic devices is the absence of a mechanism in front of the operator. This is particularly advantageous on large presses. The electric-eye device should be installed far enough away from the danger zone so that it will stop the slide or ram before the operator's hand can get underneath. Sufficient light beams should cover the bad-break area with a curtain of light.

Indexing is another press-shop term used to describe a mechanical method of feeding stock into press dies. One method is by a dial feeder. The dial feed is constructed, as its name implies, in the form of a dial having multiple stations that progress into the die by the indexing motion of the dial. The indexing of the dial should take place in conjunction with the up-stroke of the ram. When hand feeding, the index circuit should be controlled by dual-run buttons. Release of either button, during indexing, should stop the index cycle by releasing a safety clutch in the table. Safety guards should be connected to switches that stop the crank motion of the press whenever they are bumped. Both of these safety precautions will prevent accidents should the operator be tempted to reach into the die area to correct an improperly positioned part on the dial.

AUTOMATION

Automation, a somewhat misused term, is defined in this text as the mechanization of processes by the use of automatically controlled conveying equipment. Automation has minimized the hazards associated with manually moving stock in and out of machines and transferring it from one machine to another. It has also minimized exposure to the causes of hernias, back injuries, and foot injuries.

However, experience shows that those who work with automated equipment must have a thorough knowledge of its hazards and must be well trained in safe work practices to avoid accidents. Most finger and hand injuries result from operator exposure to the closing or working parts of a machine in the process of loading and unloading. The use of indexing fingers, sliding dies, and tongs or similar hand tools reduces the hazard from such exposure. Nevertheless, as supervisor, you must make sure that these devices are used consistently and correctly.

Like most innovations, automation has brought not only benefits, but also hazards. Because each automatic operation is dependent on others, machine breakdown or failure must be corrected quickly. Also, because speed is highly important, maintenance employees may expose themselves, inadvertently perhaps, to working parts of the equipment. It is therefore imperative to have a mandatory policy that equipment must be completely de-energized and locked out at the power source before servicing the machine. A well-defined lockout procedure should be written down and followed carefully.

Automated handling also has increased the use of stiles or crossovers. These should be constructed and installed in accordance with ANSI A12.1, *Safety Requirements for Floor and Well Openings, Railings, and Toeboards.* Automation eliminates or greatly reduces exposure to mechanical and handling hazards. In the single-operation process, however, the basic principles of safeguarding of equipment must still be applied. These principles are:

- Engineer the hazard out of the job as far as possible.
- Guard against the remaining hazards.
- Educate and test the performance of workers.
- Insist on the use of safeguards provided.

Robotics

Robots—machines specifically designed and programmed to perform certain operations—are rapidly becoming a part of the work environment. You should know that these machines can and have caused accidents to unsuspecting people working in the vicinity.

As early as possible, consult the engineer who designed or worked on the robot to learn about its capabilities, features, hazards, and operation. Allow for proper clearances when the robot is working with peripherals, such as machine tools, presses, transfer lines, palletizers, gaging stations, and so on. Even the robot with the best conceptual design may be efficient only if it has these clearances. Providing clearance for personnel, however, is an even more important factor. Although robots perform repetitive tasks for long periods of time, they are still machines and require periodic preventive maintenance. Eventually, the robots will need repairs to keep their devices, modules, and/or tooling working properly. Clear spaces and safe areas must be incorporated in the design stage (Figure 11-17).

A major hazard of robots involves their computer programming. When an operation is being performed, it may be virtually impossible to shut it off or reverse the cycle because of the automatic programming feature. Therefore, workers should be warned not to get near these machines. In some cases, it may be possible to totally enclose the robot, but in many instances, this may be impractical.

Control systems for robots usually work with a "priority interrupt" scheme. For example, higher priorities are interrupted by special hardware devices to signal that personnel are entering the work area or that tooling needs to be changed or repaired. As the supervisor, you should follow these basic rules:

1. Get as much information as possible on the operation of the robot.
2. Provide proper clearances and barrier guards around the robot (Figure 11-17).
3. Discuss all facets of the operation, including safety, with your people. All personnel should be trained.
4. Do not allow workers near robots in operation; instead, wait for the cycle to be completed and the machine de-energized before approaching.

Some jobs may require work to be done with power "on" (such as alignment and repair of servo systems). This must be realized in the design stage by the user's engineers. It cannot be assumed that work always will be performed in a power-off state. The simplest guarantee that equipment is safe to work on is to insist that repair personnel and the maintenance crew place their company-issued padlocks on the power source. (Lockout procedures are discussed later in this chapter.)

Safe Practices

Safeguards are of primary importance in eliminating machine accidents, but they are not enough. Employees who work around mechanical equipment or operate a piece of machinery must have a healthy respect for safeguards.

Figure 11-17. This robot is guarded by a chain-link fence with gates. Clear spaces and areas for personnel who must maintain and repair the robot must be provided for in the design phase.

Before being permitted to run a piece of equipment, operators should be instructed in all the practices required for safe operation of the machine. Even experienced operators should be given refresher training, unless you are certain that they know the hazards and the necessary precautions to be taken. In addition, employees who do not themselves operate machinery, but who work in machine areas, should also receive instruction in basic safety practices.

Positive safety procedures should be established to prevent employees from misunderstanding instructions. You should enforce the following rules:

1. No guard, barrier, or enclosure should be adjusted or removed for any reason by anyone unless that person has permission from you, has been specifically trained to do this work, and machine adjustment is considered a normal part of the job.
2. Before safeguards or other guarding devices are removed so that repair or adjustments can be made or equipment can be lubricated or otherwise serviced, the power for the equipment must be turned off and the main switch locked out and tagged.
3. No machine should be started unless the safeguards are in place and in good condition.
4. Defective or missing safeguards should be reported to you immediately.
5. Employees should not work on or around mechanical equipment while wearing neckties, loose clothing, watches, rings, or other jewelry.

MAINTENANCE OF SAFEGUARDS

As supervisor you are responsible for scheduling inspection of machine safeguards as a regular part of machine inspection and maintenance. Such inspections are necessary, because employees tend to operate their machines without safeguards if they are not functioning properly, if they have been removed for repairs, or if they interfere in any manner with their operations. A guard or enclosure that is difficult to remove or to replace may never be replaced once it has been taken off. An inspection checklist can be developed for each machine to simplify the job and to provide a convenient record for follow-up.

Maintenance and Repairs

Machines are subject to wear and deterioration and can become unsafe to operate. Wear cannot be prevented, but it can be reduced to a minimum by controlling loads through proper manufacturing methods, alert supervision, attention of employees, and good maintenance.

Lubrication is a basic maintenance function. Centralized lubrication will reduce the hazards to which the oiler is subjected when climbing ladders to reach fairly inaccessible points. Perhaps changes can be made so that most lubrication can be done at floor level. When oilers or repair crews must get to the tops of presses to lubricate flywheel bearings, motors, and other parts, good work practices suggest that permanent ladders with sturdy enclosures (cages or wells) be installed.

Regardless of the type of lubrication necessary or the method used, it is your responsibility to ensure that the machinery—including driving mechanisms, gears, motors, shafting hangers, and other parts—is being lubricated properly. Where automatic lubrication is not possible or feasible, extension grease or oil pipes should be attached to machines so that the oiler can avoid coming in contact with moving parts. However, automatic lubrication should not be used in cases where the oil or grease can congeal in the pipes.

Lockout/Tagout Procedures

Lockout or tagout procedures are designed to isolate or shut off machines and equipment from their power sources before employees perform any servicing or maintenance work. Employees must be trained in these procedures and instructed to replace safeguards after the work is completed. You must use all of your authority to enforce safe work practices.

Maintenance personnel and oilers should be provided with and use padlocks to lockout the power-driven apparatus on machinery or equipment they are to clean or oil. This is especially important when the working place is some distance from machine controls or is hidden from view. Each worker should have his or her own lock and key, and no two locks or keys should be alike. Only locks bought by the company should be used. They can be painted different colors, according to craft, shift, or department, to facilitate identification. The employee's name or clock number should be stamped on the lock, or a metal tag bearing the owner's name should be attached.

Where a lockout system is to be set up, equipment must have built-in locking devices. They must be designed for the insertion of padlocks or have attachments on which locks can be placed. Methods include special tongues that hold several locks or sliding rods that can be extended and then locked to prevent operation of control handles (Figure 11-18). The lockout procedure, however, will be effective only if the supervisor trains employees to follow it and then watches constantly for deviations. A typical lockout routine for maintenance and repair workers is given here:

1. Notify the operator that repair work is to be done on a machine or piece of equipment.
2. Make sure the machine cannot be set in motion without your permission.
3. Place your padlock on the power disconnect, even though another worker's lock is already on it and blocks the mechanism. Another person's lock will not protect you.
4. Place a MACHINE UNDER REPAIR sign at the control and block the mechanism. Make sure that neither the sign nor the blocking can be easily removed.
5. When the job is finished or the shift has ended, remove your own padlock and blocking. Never let another person remove it for you. Make sure first that you will not expose others to danger by removing your block or sign.
6. If the key to a lock is lost, it must be reported to you at once and you should issue a new set.

Figure 11-18. This supervisor checks the lockout/tagout of electrical circuitry. Note that there is space for several locks.

If two or more people are to work on a machine or piece of equipment where they might become separated, they should agree upon a foolproof set of signals. Establish a rule that when a machine or group of machines is stopped, only a specific, authorized person may start them again. In any case, the safe procedure is to give a warning and make sure that everyone is in the clear before power is turned on again.

It is also good practice for you to check machines back into operation after repairs have been made and before new operations are undertaken. Every worker should be warned not to start the machine unless safeguards are in place. This is your responsibility—not only to teach employees this practice, but to make sure they always comply with it. More details can be found in ANSI Z244.1–1982, *Safety Requirements for the Lock Out/Tag Out of Energy Sources.*

OSHA has put into effect a new rule for installation of locks to guarantee that power to machinery is cut off during maintenance and repair. The rule went into effect October 31, 1989. It also requires training and annual retraining for all workers, plus annual retraining for all workers exposed to moving machinery (*Federal Register*, volume 54, no. 169, Friday, September 1, 1989, pages 36644 through 36696). Compliance was required by January 2, 1990. This rule is in 24 *CFR*, 1910.147, Subpart J, "Control of Hazardous Energy Service (Lockout/Tagout)." The original section 1910.147 has been redesignated as section 1910.150 to allow for the new section.

Replacements

If is generally up to you whether to alert management whenever replacement of machinery and/or its safeguards is advisable. This step may seem like a production problem, but it is actually related to accident prevention. When machinery, safeguards, and safety devices become so worn that repairs cannot restore them to original operating efficiency, they constitute hazards, often, unfortunately, concealed hazards. For example, if a nonrepeat device on a press becomes so worn that it does not operate at all or fails to operate properly, workers can lose a hand or arm in the machines. These hazards can be eliminated only by replacing the machine part.

Several methods are used to determine when a machine needs maintenance or an overhaul to prevent a breakdown or when replacements are in order. Some supervisors rely on reports of the machine operators; some have occasional inspections made by oilers, machine setters, or similar mechanics; others set up inspections on a regular schedule and do it themselves, or have it done, routinely. The best plan is a system of frequent, regularly spaced inspections.

SUMMARY OF KEY POINTS

This chapter covered the following key points:

- The basic principles of safeguarding are (1) engineer the hazard out of the job as much as possible, (2) guard against remaining hazards, (3) educate and test safety performance of workers, and (4) use safeguards provided. If such precautions are not taken, the consequences can be serious in terms of injury and property loss.
- Safeguarding is often thought of as being concerned only with the point of operation or with the means of power transmission. However, safeguards can prevent accidents from direct contact with exposed moving parts, work in process, machine failure, electrical failure, and operator error or human failure. The primary benefits of safeguarding are reducing the possibility of injury and improving production.
- A great many standardized devices, as well as improvised barriers, enclosures, and tools, have been developed to protect machine operators on power and hand-operated equipment. OSHA and ANSI standards present detailed guidelines for safeguarding such equipment. Establishing effective methods of safeguarding power transmissions is easier than guarding points of operation because power transmissions are more standardized.
- An acceptable safeguard should conform to or exceed standards, be a permanent part of the machine, afford maximum protection, prevent access to the danger zone, be convenient, be designed for the specific machine and job, resist fire and corrosion, resist wear and shock, be long lasting, and not be the source of additional hazards. Built-in safeguards are generally more effective and less costly in the long run than "homemade" barriers. Supervisors should be alert to employees who build their own makeshift safeguards or who try to get around the safeguards built into the equipment.
- Safeguarding programs should adhere to pertinent regulatory codes and be instituted before machinery or equipment is installed. At times, substituting a less hazardous machine can solve a safeguarding problem. Regardless, supervisors should make every effort to match equipment to operators in terms of workplace, work height, controls, materials handling aids, operator fatigue, adequate lighting, noise levels, and other environmental factors.
- Machines have certain basic mechanisms which, if exposed, always need safeguarding. These include rotating mechanisms, cutting or shearing mechanisms, in-running nip points, screw or worm mechanisms, forming or vending mechanisms, and impact mechanisms. Each of these parts must have special safeguarding covers, barriers, or other equipment to shield operators and nearby workers from injury and to prevent equipment or product damage and loss.
- Power presses consist of primary and secondary operations in which workers are exposed to potential injury. Wherever possible, high-production or long-run dies should be guarded individually. These devices should shut off the presses or prevent normal operation if workers' hands, clothing, or tools become caught in the press.
- Automatic or semiautomatic feeds, if they can be used, are generally successful safeguards. They can increase production, reduce costs, or both. The main advantages of these safeguards are that the operator does not have to reach into the point of operation to feed the press, the die can be enclosed completely, and the operator can leave the press untended for short periods of time.
- The safe removal, or ejection, of material is as necessary as safe feeding. Various methods of ejection include compressed air, punch, knockouts, strippers, and gravity. Operators should not be required to remove the finished parts or scrap material manually.
- Controls should be designed so that the machinery can be operated without endangering the workers. Foot controls, for example, should be covered by stirrup-type covers to prevent accidental tripping. Dangers involved in machine operation can be eliminated either by enclosing the hazardous parts of the machine, or by redesigning it to eliminate exposed parts.
- Safeguards used to ensure safe production include the fixed guard or enclosure, the interlocking guard or barrier, and the automatic protection device. Fixed guards or enclosures are preferred to all other types because they cannot be removed by workers. When this safeguard is not practicable, interlocking guards or barriers can be fitted onto the ma-

chine to prevent operation of the controls until the guard or barrier is moved into position.

When neither the fixed nor interlocking safeguard can be used, an automatic protection device can be applied. This safeguard prevents the operator from coming into contact with the dangerous part of the machine while it is in motion, or it can stop the machine in an emergency.

- Automation has minimized many of the hazards associated with manually moving stock in and out of machines but has brought its own hazards. Automated machinery must be completely de-energized and locked out at the power source before anyone is allowed to service or reset the machines. To apply effective safeguards on robots, supervisors must learn as much about the operation of the robot as possible, provide proper clearance and barriers around the robot, discuss all facets of operation and safety issues with workers, and do not allow workers near robots in operation until the cycle is completed and the machine is de-energized.

- In addition to the use of effective safeguards, supervisors must instruct workers in safe practices to prevent accidents and injuries. All workers, before being permitted to operate machinery, should be thoroughly trained in how to use the safeguards on equipment. Positive safety procedures should be established and vigorously enforced.

- Supervisors must set up a schedule for inspecting and maintaining machine safeguards as a regular part of their operations. Wear and tear on these parts can be reduced by proper machine-handling methods, alert supervision, attention of employees, and regular maintenance. Before allowing workers to service machines—for example, lubricating parts—supervisors should establish precise lockout/tagout procedures to isolate or shut off machines from their power source. Worn out, corroded, or damaged machine parts or safeguards should be replaced or repaired immediately.

12

Hand Tools and Portable Power Tools

After reading this chapter, you will be able to:

- Establish a program to control accidents based on four safe work practices: selecting the right tools for the job, using tools correctly and safely, keeping them in good condition, and storing them properly
- Describe the major hazards and safeguards for a wide variety of hand and portable power tools
- Understand the need for a centralized tool supply room or other method of controlling access to company tools
- Establish procedures for regular inspection, repair, and replacement of tools in your department

One of your major responsibilities as a supervisor is to train your people in the safe handling of hand and power tools. This chapter discusses ways in which hand and portable power tool injuries can be avoided. To pinpoint where major problems lie, you should review the department or company accident records first. Try to identify specific hazards, then check to see if any employees have had an unusual number of injuries. These workers should have additional training and closer supervision. Make up written procedures for tools used rarely in any operation to remind workers how to use the equipment properly and safely.

Training and close supervision are also needed when new tools, processes, and equipment are introduced in the operation or changes are made to an existing job. A job safety analysis of each task will serve to clarify tool needs and safe work practices.

SAFE WORK PRACTICES

To be effective, a program to reduce tool injuries must include training in four basic safe practices. Your supervisory program to control accidents should include the following activities:

1. Train employees to select the right tools for each job. Good job safety analysis and job instruction training helps here. (See Chapter 4, Employee Safety Training.)
2. Train and supervise employees in the correct use of tools.
3. Set up regular tool inspection procedures (including inspection of employee-owned tools), and provide good repair facilities as insurance that tools are being maintained in safe condition.

4. Set up a procedure for control of company tools. A check-out system at tool cribs is ideal. Provide proper storage facilities in the tool room and on the job (Figure 12-1).

Select the Right Tool

While many tools can be used to do a job, proper selection means choosing the right tool that can be most safely used by the employee.

Using improper tools can result in errors, accidents, and damage to the product. For example, an adjustable wrench can tighten a nut, but it would be better if the employee used a box end or a socket wrench of the proper size. Ergonomically designed tools, of course, provide the best fit for the employee and the job.

Tool selection is also related to other factors, such as job setup, workspace available, work height, and so forth, each bearing on the most efficient tool selection. Cost and quality of tools are additional

Figure 12-1. Accident prevention begins with proper inspection and storage of tools. They should be inspected when checked out and again when checked into the tool room. Only properly adjusted tools in good condition should be issued.

items to consider. Employers generally are expected to provide the correct tool(s) to do the job safely.

Use Tools Correctly and Safely

As the supervisor, you show employees how to use tools safely and correctly. Far too often workers select tools that are neither safe nor necessary for the task, such as substituting a screwdriver for a crowbar or using pliers instead of the proper wrench. In many situations the employee should study a tool's manual, which may suggest the use of proper personal protective equipment. The work area needs to be clean, dry, well lighted, and ventilated. Supervisors should ensure that the workers understand the correct use of their tools and are aware of other workers in the vicinity. Guards on tools must be checked to be sure they operate effectively.

Keep Tools in Good Condition

Every tool should be examined before use. Employees in the tool crib and workers using tools on the job should be taught how to check equipment properly so that worn or damaged items can be repaired or discarded. Tools must be kept in good repair through a planned tool maintenance program.

Tools that have deteriorated should not be used until repaired to meet manufacturers' specifications—examples: wrenches with cracked or worn jaws, screwdrivers with broken bits (points) or broken handles, hammers with loose heads, dull saws, deteriorated extension cords or power cords on electric tools or broken plugs, and improper or removed grounding systems. Cutting edges must be kept sharp.

All tools need to be replaced from time to time. However, employees are usually not expected to do tool maintenance work. If they are, they should be trained and supervised. Good housekeeping is closely related to good, clean tools.

Store Tools in a Safe Place

Tools need to be stored properly either in or on the work area or in a common tool crib. If employees own their own tools, safe storage is of equal importance.

Also make a complete check of operations to determine the need for special tools to do the work more safely. Special tools may require unusual handling and storage. For example, some tools (like powder-actuated hand tools) should be kept under lock and key.

USE OF HAND TOOLS

The misuse of common hand tools is a major source of injury to industrial workers. In many instances, injury results because it is assumed that "anybody knows how to use" common hand tools. Observation and the records of injuries show that this is not the case. As the supervisor, you should study each job and train new workers and retrain old employees on correct procedures for using tools. Where employees have the privilege of selecting or providing their own tools, you should advise them on the hazards of the job and insist on their using the safest equipment for each task. This training is so important that considerable attention is given in this chapter to safe work practices.

Be sure to enforce all safety rules. You should check the condition of tools frequently to see they are maintained and sharpened correctly and that guards are not altered or removed. Also make routine checks regarding the use and the condition of personal protective equipment.

Adopt specific rules for using hand tools in each operation. For example, personal protective equipment, such as gloves, is often required when using hand tools. The hazards identified on a job safety analysis can be useful in setting up personal protective equipment needs.

Cutting Tools

All edged cutting tools should be used in such a way that, if a slip occurs, the direction of force will be away from the body. For efficient and safe work, edged tools should be kept sharp and ground to the proper angle. A dull tool does a poor job and may stick or bind. A sudden release may throw the user off balance or cause his or her hand to strike an obstruction. Thus, all cuts should be made along the grain when possible.

Metal chisels. Factors determining the selection of a cold chisel are (a) the materials to be cut, (b) the size and shape of the tool, and (c) the depth of the cut to be made. The chisel should be heavy enough so that it does not buckle or spring when struck. For best results and for maximum safety, flat and cape chisels should be ground in such a way that the faces form an angle of 70 degrees for working on cast iron, 60 degrees for steel, 50 degrees for brass, and about 40 degrees for babbitt and other soft metals.

A chisel only large enough for the job should be selected so that the blade rather than the point or corner is used. Also, a hammer heavy enough to do the job should be used. Some workers prefer to hold the chisel lightly in the hollow of their hands with the palms up, supporting the chisel by the thumb and first and second fingers. If the hammer glances from the chisel, it will strike the soft palm rather than the knuckles. Other workers think that a grip with the fist holds the chisel steadier and minimizes the chances of glancing blows. Moreover, in some positions this is the only grip that is natural or even possible.

Hand protection can consist of a rubber pad, forced down over the chisel to provide a hand cushion. Workers should chip in a direction away from the body. When shearing with a cold chisel, workers should hold the tool at the vertical angle that permits one bevel of the cutting edge to be flat against the shearing plane. Employees should wear safety goggles when using chisels and should set up a shield or screen to prevent injury to other workers from flying chips. If a shield does not afford enough protection to all exposed employees, then these workers should wear safety glasses with side protection.

Bull chisels held by one person and struck by another require the use of tongs or a chisel holder to guide the tool so the worker is not exposed to injury. Both workers should wear safety goggles, hats, and shoes. The one who swings the sledge should not wear gloves.

Dress all heads at the first sign of mushrooming because mushroomed heads often produce flying chips that can be very dangerous to the eyes. It is suggested that when using a 1/2-in. steel chisel, a hammer with 1 1/2-in. diameter face be used.

Tap and die work. This type of work requires certain built-in safeguards. The work should be firmly mounted in a vise. Only a T-handle wrench or adjustable tap wrench should be used. Steady downward pressure should be applied on the taper tap. Excessive pressure causes the tap to enter the hole at an angle or bind the tap, causing it to break. A properly sized hole must be made for the tap, and the tap should be lubricated as necessary.

Keep hands away from broken tap ends. Broken taps should be removed with a tap extractor. If a broken tap is removed by using a prick punch or a chisel and hammer, the worker should wear safety goggles. When threads are being cut with a hand die, the hands and arms should be kept clear of the sharp threads coming through the die, and metal cuttings should be cleared away with a brush.

Hack saws. These tools should be adjusted in the frame to prevent buckling and breaking, but should not be so tight that the pins that support the blade break. Install blades with teeth pointing forward.

The preferred blade to be used is shown in Table 12-1. A general rule is that at least two teeth should be in the cutting piece. It is advisable to loosen

TABLE 12-1. Selector for Hack Saw Blades

Pitch of Blade (Teeth per Inch)	Stock to Be Cut	Explanation

14	Machine Steel
	Cold Rolled
	    Steel
	Structural Steel
		The course
		    pitch makes
		    saw free and
		    fast cutting

18	Aluminum
	Babbitt
	Tool Steel
	High-Speed
	    Steel
	Cast Iron
		Recommended
		    for general
		    use

24	Tubing
	Tin
	Brass
	Copper
	Channel Iron
	Sheet Metal of
	    18 gage or over
		Thin stock will
		    tear and strip
		    teeth on a
		    blade of
		    coarser pitch

32	Sheet Metal of
	    less than 18 gage
	    (18 gage = 1.27 in.)	

blades before storing hack saws and retighten the blades before starting work.

Files. Selection of the right kind of file will prevent injuries, lengthen the life of the file, and increase production. A file-cleaning card or brush should be used to keep the file in peak condition. Files should not be hammered or used as a prying tool. Such abuse frequently results in the file's chipping or breaking, causing an injury to the user. A file should not be made into a center punch, chisel, or any other type of tool because the hardened steel could fracture. Use a vise, whenever possible, to hold the object being filed.

The correct way to hold a file for light work is to grasp the handle firmly in one hand and use the thumb and forefinger of the other to guide the point, using smooth file strokes. This technique gives good control, and thus produces better and safer work.

A file should never be used without a smooth, crack-free handle; otherwise, if the file binds, the tang may puncture the palm of the hand, the wrist, or other part of the body. Under some conditions, a clamp-on, raised offset handle can provide extra clearance for the hands. Files should not be used on lathe stock turning at high speeds (faster than three turns per file stroke). The end of the file may strike the chuck, dog, or face plate and throw the file (or metal chip) back at the operator and inflict serious injury. To avoid contact with the turning parts of the lathe, the operator should always cross file.

Tin snips. This tool should be heavy enough to cut the material easily so that the worker needs only one hand on the snips and can use the other to hold the material. The material should be well supported before the last cut is made so the cut edges do not press against the hands. When cutting long sheet-metal pieces, push down the sharp ends next to the hand holding the snips.

The jaws of snips should be kept tight and well lubricated. When not in use, tin snips should be hung up or laid on a shelf. Burrs on the cut usually indicate a need for adjusting or sharpening the snips.

Workers should always wear gloves and use safety goggles when trimming corners or slivers of sheet metal. Small particles can fly with considerable force into a worker's face.

Cutters used on wire, reinforcing rods, or bolts should have ample capacity for the stock; otherwise, the jaws may spring or spread. Also, a chip may fly from the cutting edge and injure the user.

Cutters are designed to cut at right angles only. They should not be "rocked," hammered, or pushed against the floor to facilitate the cut because they are not designed to take the resulting strain. This practice will nick the cutting edges and result in future problems. A good rule is, "If it doesn't cut with ease, use a larger cutter or use a cutting torch." Cutters require frequent lubrication. To keep cutting edges from becoming nicked or chipped, cutters should not be used as nail pullers or pry bars.

Cutter jaws should have the hardness specified by the manufacturer for the particular kind of material to be cut. By adjusting the bumper stop behind the jaws, workers can set the cutting edges to have a clearance of 0.003 in. (0.076 mm) when closed. Always store cutters in a safe place.

Punches. Punches are used like chisels. They should be held firmly and securely and struck squarely. The tip should be kept shaped as the manufacturer has specified. Punches should be held at right angles to the work.

Wood chisels. All employees should be instructed in the proper method of holding and using wood chisels. While molded plastic and metal handles are often available, wood handles are also found on wood chisels. If the handle is wood, it should be free of splinters and cracks. The wood handle of a chisel struck by a mallet should be protected by a metal or leather cap to prevent its splitting.

The work to be cut should be free of nails to avoid damage to the blade or to prevent a chip from flying into the user's face or eye. The steel in a chisel is hard so that the cutting edge will hold, and is, therefore, brittle enough to break if the chisel is used as a prying bar. When not in use, the chisel should be kept in a rack, on a workbench, or in a slotted section of the tool box so that the sharp edges will be out of the way. Sharp edges can be safeguarded, as shown in Figure 12-2.

Saws. These tools should be carefully selected for the work they are to do. For fast crosscut work on green wood, a coarse saw (4 to 5 points per in.) is best. A fine saw (over 10 points per in.) is better for smooth, accurate cutting in dry wood. Saws should be kept sharp and well set to prevent binding. When not in use, make sure they are placed in racks.

Saw set (the amount of angle or lean of a point from the blade) is needed to cut wood or other materials properly and cleanly. Inspect the material to be cut to avoid sawing into nails or other metal.

Sawing should be a one-hand operation. When starting a cut, workers should guide the saw with the thumb of their free hand held high on the saw. They should not place a thumb on material being cut. Begin with a short, light stroke toward the body. After the cut begins, increase pressure and increase stroke length. Use light, shorter strokes as the cut is completed.

Axes. To use an axe safely, workers must be taught to check the axe head and handle, clear the area for an unobstructed swing, and swing correctly and accurately. Accuracy is attained by practice and proper handhold—and good supervisory training. All other workers must be kept a safe distance away from the direct line of the axe swing.

A narrow axe with a thin blade should be used for hard wood, and a wide axe with a thick blade for soft wood. A sharp, well-honed axe gives better chopping speed and is much safer to use because it bites into the wood. A dull axe will often glance off the wood being cut and strike the user in the foot or leg.

The person using the axe should make sure that there is a clear circle in which to swing before starting to chop. Also, vines, brush, and shrubbery within the range, especially overhead vines that may catch or deflect the axe, should be removed.

Axe blades must be protected with a sheath or metal guard wherever possible. When the blade cannot be guarded, it is best to carry the axe at one's side. The blade on a single-edged axe should be pointed down.

Hatchets. Hatchets, which are used for many purposes, frequently cause injury. For example, when workers attempt to split a small piece of wood while holding it in their hands, they may strike their fingers, hand, or wrist. Hatchets are dangerous tools in the hands of an inexperienced worker. To start the cut, it is good practice to strike the wood lightly with the hatchet, then force the blade through by striking the wood against a solid block of wood.

Hatchets should not be used for striking hard metal surfaces. Flying chips from the tempered head may injure the user or others nearby.

Knives. These tools cause more disabling injuries than does any other hand tool. In the meat packing industry, for example, hand knives account for more than 15 percent of all disabling injuries. The principal hazard in using knives is that the hands may slip from the handle onto the blade or that the knife may strike the body or the free hand. A handle guard or a finger ring (and swivel) on the handle reduces these hazards (Figure 12-3).

The cutting stroke should be away from the body. If that is not possible, then the hands and body should be in the clear, a heavy leather apron or other protective clothing should be worn, and, where possible, a rack or holder should be used for the material to be cut. Jerky motions should be avoided to help maintain balance. Be sure employees are trained and supervised. Training in handling

Figure 12-2. Guard sharp edges of hand tools with metal, fiber, or heavy cardboard sleeves that fit over them. Check tool boxes regularly to assure that sharp edges of tools are covered.

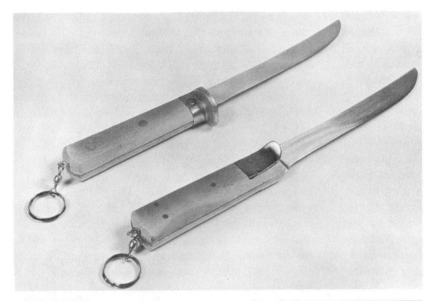

Figure 12-3. Knives equipped with ring-and-swivel guards and handle guards prevent the user's hand from sliding over the handle onto the blade in the event that the knife stabs against a solid object.

knives is very important in the food-service industry.

Workers who must carry knives with them on the job should keep them in sheaths or holders. Never carry a sheathed knife on the front part of a belt—always carry it over the right or left hip, toward the back. This will prevent severing a leg artery or vein in case of a fall.

Knives should never be left lying on benches or in other places where they could cause hand injuries. When not in use, they should be kept in racks with the edges guarded. Safe placement and storage are important to knife safety.

Ring knives—small, hooked knives attached to a finger ring—are used where string or twine must be cut frequently. Supervisors should make sure that the cutting edge is kept outside the hand, not pointed inside. A wall-mounted cutter or blunt-nose scissors would be safer.

Carton cutters are safer than hooked or pocket knives for opening cartons. They not only protect the user, but eliminate the deep cuts that could damage carton contents. Frequently, damage to contents of soft, plastic bottles may not be detected immediately; subsequent leakage may cause chemical burns, damage other products, or start a fire.

To cut corrugated paper, a hooked linoleum knife permits good control of pressure on the cutting edge and eliminates the danger of the blade suddenly collapsing as pocket knives can do. Be sure hooked knives are carried in a pouch or heavy leather (or plastic) holder. The sharp tip must not stick out.

Make certain that employees who handle knives have ample room in which to work so they are not in danger of being bumped by trucks, the product, overhead equipment, or other employees. For instance, a left-handed worker should not stand close to a right-handed person; the left-handed person might be placed at the end of the bench or otherwise given more room. Workers should be trained to cut away from or out of line with their bodies.

Be particularly careful about employees leaving knives hidden under the product, scrap paper, or wiping rags, or among other tools in work boxes or drawers. Knives must be kept separate from tools to protect the cutting edges and the employee.

Work tables should be smooth and free of slivers. Floors and working platforms should have slip-resistant surfaces and be kept unobstructed and clean. If sanitary requirements permit mats or wooden duck boards, they should be in good repair, so workers do not trip or stumble. Conditions that cause slippery floors should be controlled as much as possible by good housekeeping and frequent cleaning.

Careful job and accident analysis may suggest some changes in the operating procedure that will make knives safer to use. For instance, on some jobs, special jigs, racks, or holders may be provided so it is not necessary for the operator to stand close to the piece being cut.

The practice of wiping a dirty or oily knife on the apron or clothing should be discouraged. The blade should be wiped with a towel or cloth with the sharp edge turned away from the wiping hand. Sharp knives should be washed separately from other utensils and in such a way that they will not be hidden under soapy wash water.

Horseplay should be prohibited around knife operations. Throwing, "fencing," trying to cut objects into smaller and smaller pieces, and similar practices are not only dangerous but reflect inadequate supervision.

As a supervisor, you should make sure that nothing is cut that requires excessive pressure on the knife, such as frozen meat. Food should be thawed before it is cut or else frozen food should be sawed. In addition, knives should not be used as substitutes for openers, screwdrivers, or ice picks.

Miscellaneous Cutting Tools

Planes, scrapers, bits, and draw knives should be used only by experienced personnel. Keep these tools sharp and in good condition. When not in use, they should be placed in a rack on the bench or in a tool box to protect the user and prevent damage to the cutting edge.

Torsion Tools

Many tools used to fasten or grip materials are available to facilitate work. Proper tool selection and usage are important if employees are to work safely.

Wrenches. All wrenches should be pulled, not pushed, in operation. Workers' footing should be secure and allow plenty of clearance for their fingers. Use a short, steady pull. If a nut does not loosen or tighten fully, a larger wrench may be needed. Damaged tools should be removed for service. Open-end and box wrenches (Figure 12-4) should be inspected to be sure that they fit properly. Wrenches should not be hammered or struck.

Socket wrenches give great flexibility. The use of special types should be encouraged on jobs where workers may be injured. Socket wrenches not only are safer to use than adjustable or open-end wrenches, but they also protect the bolt head or nut.

Wrench jaws that fit (and are not sprung or

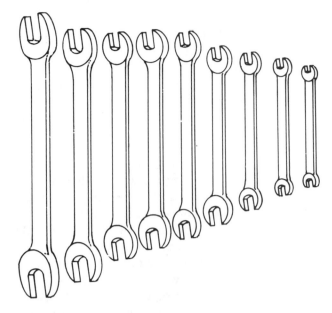

Figure 12-4. Open-end wrenches, solid nonadjustable wrenches with openings in one or both ends, come in both English and metric systems. Wrenches with openings greater than 1 in. are widely used on pipelines, in marine operations, and on heavy equipment. The smaller the opening, the shorter the length; this proportions the leverage to the size of the bolt or nut and helps reduce the force applied in order to keep the threads from stripping or the bolt from being twisted in two.

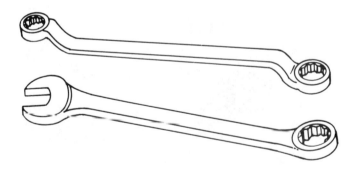

Figure 12-5. Box-end wrenches (also called box wrenches or 12-point wrenches) can be used in close quarters and are best for breaking loose tight nuts. If securely rooted, there is little chance of the box wrench slipping off the nut; also, the wrench opening cannot spread. The lower drawing shows a combination wrench.

cracked) prevent damage to the heads of nuts or bolts, and are not likely to slip and cause injury to the user (Figure 12-5).

Adjustable wrenches are used for many purposes. They are not intended, however, to take the place of open-end, box, or socket wrenches. They are used mainly for nuts and bolts that do not fit a standard wrench. Pressure is always applied to the fixed jaw and the wrench is pulled toward the body (Figure 12-6).

Pipe wrenches. Workers, especially those on overhead jobs, have been seriously injured when pipe wrenches slipped on pipes or fittings, causing them to lose their balance and fall. Pipe wrenches, both straight and chain tong, should have sharp jaws and be kept clean to prevent slipping.

The adjusting nut of the wrench should be inspected frequently. If it is cracked, the wrench should be taken out of service. A cracked nut may break under strain, causing complete failure of the wrench and possible injury to the user.

Using a wrench of the wrong length is another source of accidents. A wrench handle too small for the job does not give proper grip or leverage. An oversized wrench handle may strip the threads or

Figure 12-6. Adjustable (crescent) wrenches are probably the most often misused wrenches. They are made for use on odd-sized nuts and bolts that other wrenches might not fit. To use, the wrench should be placed on the nut so that when the handle is pulled, the moveable jaw is closer to the user's body. The pulling force will then be applied to the fixed jaw, which is the stronger of the two.

break the fitting or the pipe suddenly, causing a slip or fall.

A pipe wrench should never be used on nuts or bolts, the corners of which will break the teeth of the wrench, making it unsafe to use on pipe and fittings. Also, a pipe wrench, when used on nuts and bolts, can damage their heads. The wrench should not be used on valves, struck with a hammer, or used as a hammer unless, as with specialized types, it is specifically designed for such use.

Torque wrenches. These have a scale to indicate the amount of force to be applied to the nut or bolt (usually foot pounds). It is important that a torque wrench be well cared for to ensure accurate measurements. Cleanliness of the bolt or nut threads is also important.

Tongs. Although tongs are usually purchased, some companies make their own to perform specific jobs. To prevent pinching hands, the end of one handle should be upended toward the other handle, to act as a stop. It is also possible to braze, weld, or bolt bumpers on the handles a short distance behind the pivot point so that the handles cannot close against the fingers.

Pliers. Side-cutting pliers sometimes cause injuries when the short ends of wire are cut. A guard over the cutting edge and the use of safety glasses will help to prevent eye injuries.

The handles of electrician's pliers should be insulated. In addition, workers should wear insulated gloves if they are to work on energized lines. Because pliers do not hold the work securely, they should not be used as a substitute for a wrench. Vise-type pliers can be locked on the object, although they may pinch the fingers or hand.

Special cutters. These tools include those for cutting banding wire and strap. Claw hammers and pry bars should not be used to snap metal banding material. Only cutters designed for the work provide safe and effective results.

Nail band crimpers. These crimpers make it possible to keep the top band on kegs and wood barrels after nails or staples have been removed. Use of these tools eliminates injury caused by reaching into kegs or barrels that have projecting nails and staples.

Pipe tongs. Such tongs should be placed on the pipe only after the pipe has been lined up and is ready to be fashioned. A 3-in. or 4-in. (7.5- or 10-cm) block of wood should be placed near the end of the travel of the tong handle and be parallel to the pipe to prevent injury to the hands or feet in the event the tongs slip.

Workers should neither stand nor jump on the tongs nor place extensions on the handles to obtain more leverage. They should use larger tongs, if necessary, to do the job.

Screwdrivers. The screwdriver is probably the most commonly abused tool. Discourage workers from using screwdrivers for punches, wedges, pinch bars, or pries. If used in such a manner, they become unfit for the work they are intended to do. Furthermore, a broken handle, bent blade, or a dull or twisted tip may cause a screwdriver to slip out of the slot and injure a worker's hand (Figure 12-7).

The tip must be kept clean and sharp to permit a good grip on the head of the screw. A screwdriver tip should fit the screw snugly. A sharp, square-edged bit will not slip as easily as a dull, rounded one, and requires less pressure. The part to be worked upon should never be held in the hands; it should be laid on a bench or flat surface, or held in a vise. This practice will lessen the chance of injury to the hands if the screwdriver should slip from the work. In addition, workers should observe the following precautions:

- Never use any screwdriver for electrical work unless it is insulated.
- In wood and sheet metal, make a pilot hole for the screw.
- Don't carry screwdrivers in your pockets.
- Keep screwdriver handles clean.
- Keep screwdriver blades properly shaped.

Screwdrivers used in a shop are best stored in a rack. This layout allows easy selection of the right screwdriver.

Allen wrenches. These wrenches are used like screwdrivers. The key factor is choosing the correct wrench for the job. It is often easy to apply too much force on small Allen wrenches.

Vises. Vises in a variety of sizes, jaws, and attachments can be used to grip and hold objects being worked. Vises should be secured solidly to a bench or similar base. When sawing material held in a vise, make the cut as close to the jaw as possible. If clamping long pieces, support the other end of the piece. Lightly oil all moving parts of a vise.

Clamps. Many kinds of clamps are used in a variety of operations. Nearly all clamps can be used with pads to prevent marring the work. Over tightening a clamp can break it or damage the product. If there is a swivel, it must turn freely. Lightly oil moving parts like the threads. Clamps should be stored on a rack and not in a drawer.

Impact Tools

The term *impact tools* refers to a variety of hammers and related tools used to drive items such as nails into material by the use of manual or powered force. These tools are a common source of injuries, particularly to hands and arms. (See Chapter 10, Ergonomics, for more information on potential injuries.) It may be difficult to get employees to follow safe work practices because people often believe they know how to handle hammers. In your routine safety inspections, make sure employees are not being careless with these tools.

Hammers. All hammers should have a securely fitting handle suited to the type of head used. The handle, whether glass fiber, wood, or metal, should be smooth, free of oil, shaped to fit the hand, and of the specified size and length. Warn employees against using a steel hammer on hardened steel surfaces; instead, they should use a soft-head hammer, ballpeen hammer, or one with a plastic, wood, or rawhide head. Safety goggles should be worn to protect against flying chips, nails, or scale.

Hammers may chip or spall, depending on four factors:

1. The more square the corners of the hammer, the easier it chips.
2. The harder the hammer is swung, the more likely it is to chip.
3. The harder the object struck, the more chipping increases.
4. The greater the angles between the surface of the object and the hammer face, the greater the chances of chipping.

Selection of the proper hammer is important. One that is too light is as unsafe and inefficient as one that is too heavy.

To drive a nail, hold the hammer close to the end of the handle, use a light blow to start, and increase power after setting the nail. The fingers should hold the handle underneath and alongside or

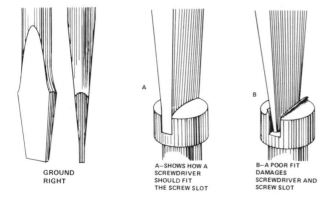

GROUND RIGHT

A—SHOWS HOW A SCREWDRIVER SHOULD FIT THE SCREW SLOT

B—A POOR FIT DAMAGES SCREWDRIVER AND SCREW SLOT

Figure 12-7. Screwdriver blades must be ground flat in that portion of the tip that enters the screw slot, and then gradually taper out to the diameter of the shank (left). When selecting a screwdriver, make sure the blade makes a good fit in the screw slot (right).

on top of the handle. Hold the hammer so that an angle of the face of the hammer and the surface of the object being hit will be parallel. The nail will drive straighter, and there will be less chance for damage. Placing the hammer on the nail before drawing it up to swing may increase the accuracy of the aim.

Sledge hammers. These tools have two common defects that present a hazard to workers: split handles and loose or chipped heads. Because these tools are used infrequently in some industries, a loose or chipped head may not be noticed. Some companies place a steel band around the head and bolt it to the handle to prevent the head from flying off. The heads should be dressed whenever they start to check or mushroom. A sledge hammer so light that it bounces off the work is hazardous; similarly, one that is too heavy is hard to control and may cause body strain.

Riveting hammers. Often used by sheet metal workers, riveting hammers should have the same kind of use and care as ballpeen hammers. Inspect them often for checked or chipped faces.

Carpenter's or claw hammers. These hammers are designed primarily for driving and drawing nails. When a nail is to be drawn from a piece of wood, a block of wood may be used under the hammer head to increase the leverage.

The striking faces should be kept well dressed at all times to reduce the hazard of flying nails while they are being hammered into a piece of wood. A checker-faced head is sometimes used to reduce this hazard.

Eye protection is advisable for all nailers and all employees working in the same area, such as in a shipping room, when blocking and bracing trucks or railway cars, or in carpenter shops and the like.

Spark-Resistant Tools

Spark-resistant tools of nonferrous materials generally should be used where flammable gases, highly volatile liquids, and explosive materials are stored or used. These tools will not produce sparks when striking hard surfaces, thus reducing the risk of fire.

PORTABLE POWER TOOLS

Portable power tools are divided into four primary groups according to the power source: electric, pneumatic, internal combustion, and explosive (powder actuated). Several types of tools, such as saws, drills, and grinders, are common to the first three groups, whereas explosive tools are used exclusively for penetration work and cutting.

A portable power tool presents similar hazards as a stationary machine of the same kind. Typical injuries caused by portable power tools are burns, cuts, and sprains. Sources of injury include electric shock, particles in the eyes, fires, falls, explosion of gases, and falling tools.

Because of the extreme mobility of power-driven tools, they can easily come in contact with the operator's body. At the same time, it is difficult to guard such equipment completely. There is also the possibility of breakage because the tool may be dropped or roughly handled. The source of power (electricity, compressed air, liquid fuel, or explosive cartridge) is close to the operator, thus creating additional hazards. When using powder-actuated tools (explosive cartridge equipment) for driving anchors into concrete, or when using air-driven hammers or jacks, it is imperative that hearing protection and eye and/or face protection be worn by operators, assistants, and adjacent personnel when the tool is in use.

All companies and manufacturers of portable power tools attach to each tool a set of rules for operating the equipment safely. These are meant to supplement the thorough training each power tool operator should have. (More details on using these tools are given below.)

Power-driven tools should be kept in safe places and not left in areas where they may be struck and activated accidentally by a passerby. The power cord should always be disconnected before accessories on a portable tool are changed, and guards should be replaced or put in correct adjustment before the tool is used again.

Selecting the Proper Tool

When you replace a hand tool with a power tool, you may be replacing a less serious hazard with a more serious one. Check with your safety professional to make sure the tools you select meet current safety standards. The tool manufacturer, too, can recommend the best tool to do the job. Describe the job you want accomplished, the material to be worked on, and the space available in the work area of your shop or plant. Tell the manufacturer if the operation is intermittent or continuous. Power tools are available in a variety of light to heavy-duty models.

Electric Tools

Electric shock is the chief hazard from electrically powered tools. Injury categories are electric flash burns, minor shock, and lethal shock. Serious electric shock is not entirely dependent on the voltage of the power input. Nearly all tools are powered at

110 or 220 volts. As explained in Chapter 14, Electrical Safety, the ratio of the voltage to the resistance determines the current. The current is regulated by the body's resistance to the ground and by environmental conditions. It is possible for a tool to operate with a defect or short in the wiring (Figure 12-8); however, the use of a ground wire protects the operator under most conditions.

Electric tools used in damp areas or in metal tanks expose the operator to conditions favorable to the flow of current through the body, particularly if the worker perspires. Most electric shocks from tools have been caused by the failure of insulation between the current-carrying parts and the metal frames of the tools. Only tools in good repair that are listed by an authorized testing laboratory should be used.

Insulating platforms, rubber mats, and rubber gloves provide an additional safety factor when tools are used in damp locations, such as in tanks, boilers, or on wet floors. Safe low voltage of 6, 12, or 24 volts, obtained through portable step-down transformers, will reduce the shock hazard in damp locations.

Double-insulated tools. Protection from electric shock while using portable power tools depends on third-wire protective grounding. However, "double-insulated" tools are available and provide reliable shock protection without third-wire grounding or a ground fault circuit interrupter (GFCI) (Fig-ure 12-9). The National Electrical Code permits double insulation for portable tools and appliances. Tools in this category are permanently marked by the words "double insulation" or "double insulated." Units designed in this category that have been tested and listed by Underwriters Laboratories also carry the UL mark:

Many manufacturers are using the symbol for a variety of tools to denote "double insulation." This double-insulated tool does not require separate ground connections; the third wire or ground wire is not needed and should not be used.

Failure of insulation is harder to detect than worn or broken external wiring, and points up the need for frequent inspection and thorough maintenance. Care in handling the tool and frequent cleaning will help prevent the wear and tear that cause defects. See Chapter 14, Electrical Safety, for more details.

Grounding. The most convenient way of safeguarding the operator is to ground portable electric tools. If any defect or short circuit inside the tool occurs, the current is drained from the metal frame

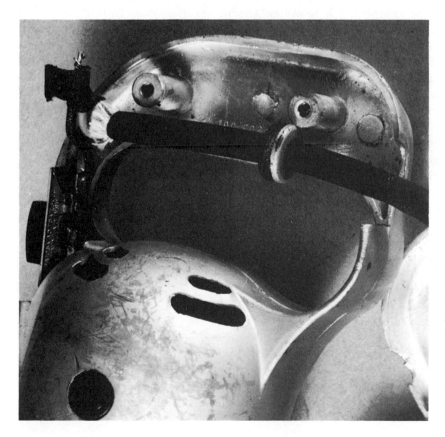

Figure 12-8. The exposed energized wire inside the handle of this tool can contact the metal shell. The tool will still run, but if the operator makes a good ground (such as by touching a water pipe), the flow of current through him or her could cause death. Many accidents primarily caused by electric shock are not reported as such and are charged against falls, tools dropped on feet, burns, and the like.

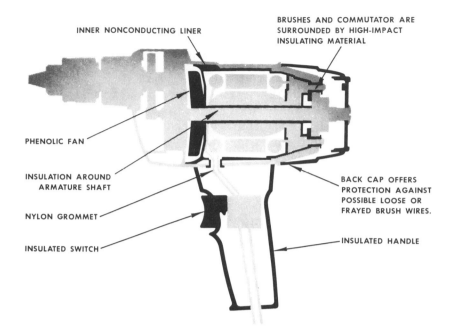

INNER NONCONDUCTING LINER

BRUSHES AND COMMUTATOR ARE SURROUNDED BY HIGH-IMPACT INSULATING MATERIAL

PHENOLIC FAN

INSULATION AROUND ARMATURE SHAFT

NYLON GROMMET

INSULATED SWITCH

BACK CAP OFFERS PROTECTION AGAINST POSSIBLE LOOSE OR FRAYED BRUSH WIRES.

INSULATED HANDLE

Figure 12-9. A double-insulated electric tool has an internal layer of protective insulation that isolates the electrical components from the outer housing.

through a ground wire and does not pass through the operator's body. All electric power tools should be effectively grounded unless they are double insulated or of the cordless type.

The noncurrent-carrying metal parts of portable and/or cord- and plug-connected equipment that require this safety measure may be grounded in two ways: first, by means of a grounding conductor or, second, by enclosing in metal the conductors feeding such equipment—provided an approved grounding-type attachment plug is used. The grounding conductor should be run with the power supply conductors in a cable assembly or flexible cord that is properly terminated in an approved grounding-type attachment plug having a fixed grounding contacting member. The grounding conductor may be noninsulated; if individually covered, however, it must be colored either in solid green or green with one or more yellow stripes.

A major safety concern is proper maintenance of the electric power system, especially the ground. Periodic tests of the electric system should be made by an electrician. A ground fault circuit interrupter, GFCI, is a device that protects the tool operator should a ground fault occur. The GFCI can be either part of the receptacle or located on an extension cord. GFCIs come in a variety of sizes and types and must be tested periodically. If a GFCI trips, it is often an indicator of a fault some place in the system. GFCIs are described in detail in Chapter 14, Electrical Safety.

Electric cords of power tools should be in-

spected frequently and kept in good condition. Have them checked by a qualified person to be sure they are the right wire size. Use heavy-duty plugs that clamp to the cord to prevent strain on the current-carrying parts if the cord is accidentally pulled. Employees should be trained not to pull on cords and to protect them from sharp objects, heat, and oil or solvents that might damage or soften the insulation.

Extension cords. Use only three-wire extension cords with three-prong, grounding-type plugs and three-pole receptacles that accept the tool's plug. Replace or repair damaged or worn cords immediately. An undersized cord will cause a drop in line voltage, resulting in a drop in power and overheating. (See discussion in Chapter 14.) Table 12-2 shows the proper gage wire to use for various lengths of extension cord.

Electric drills. These power tools cause injuries in several ways: a part of the drill may be

TABLE 12-2. Minimum Gage Wire Size for Extension Cords

Tool Nameplate Amperes	Cord Length In Feet			
	25	50	100	150
0–6	18	16	16	14
6–10	18	16	14	12
10–12	16	16	14	12
12–16	14	12		

Source: Power Tool Institute, Inc.

pushed into the hand, the leg, or other parts of the body; the drill may be dropped when the operator is not actually drilling; or the eyes may be hit either by material being drilled or by parts of a broken drill bit. Although no guards are available for drill bits, some protection is afforded if drill bits are carefully chosen for the work to be done.

Oversized bits should not be ground down to fit small electric drills; instead, an adapter should be used that will fit the large bit and provide extra power through a speed reduction gear. However, this condition again is an indication of improper drill size. When drills are used, the pieces of work should be clamped on or anchored to a sturdy base to prevent whipping.

Electric drills should always be of the proper size for the job. A ¼-in. drill means a maximum ¼-in. (diameter) drill bit can be used for wood and light metal, whereas a ½-in. drill bit would be needed to pierce steel and masonry. To operate, the chuck key should be attached to the cord, but removed from the chuck before starting the drill. If the drill has a side handle, it should be used. Make a punch mark to facilitate starting the drill and bit. Hold the drill firmly and at the proper angle for the job, then start slowly. Gradually increase speed as needed.

Routers. The widespread use of routers is based on their ability to perform an extensive range of smooth finishing and decorative cuts.

Safety in operating a router starts with an understanding that it operates at a very high speed, in the 20,000 RPM range, 25 times faster than a drill.

Always wear safety goggles or safety glasses with the side shields complying with the current national standard, and a full face shield when needed. Use a dust mask in dusty work conditions. Wear hearing protection during extended periods of operation.

Do not wear gloves, loose clothing, jewelry, or any dangling objects, including long hair, that may catch in rotating parts or accessories.

Install router bits securely, and according to the owner/operator's manual. Always use the wrenches provided with the tool.

Keep a firm grip with both hands on your router at all times. Failure to do so could result in loss of control, leading to possible serious injury. Read the operator's manual carefully regarding laminate trimmers and other small routers that are used one-handed. Always face the cutter blade opening away from the body.

When the router is equipped with carbide-tipped bits, start it beneath a work bench to protect operators from a possible flying cutter should the carbide be cracked.

Hold only those gripping surfaces of the router designated by the manufacturer. Check the owner/operator's manual. If a router is equipped with a chip shield, keep it properly installed. Keep your hands away from the bits or cutter area when the router is plugged in. Do not reach underneath the work while bits are rotating. Never attempt to remove debris while the router is operating.

Your desired cutting depth adjustments should be made only according to the tool manufacturer's recommended procedures for these adjustments. Tighten adjustment locks. Make certain that the cutter shaft is engaged in the collet at least ½ in. Check the owner/operator's manual carefully.

Be certain to secure clamping devices on the workpiece being used before operating the router. The switch should be in the OFF position before plugging into the power outlet. For greater control, always allow the motor to reach full speed before feeding the router into the work. Never force a router.

When removing a router from your workpiece, always be careful not to turn the base and bit toward the body. Unplug and store the router immediately after use.

Electric circular saws. These saws are well guarded by the manufacturer, but employees must be trained to use the guard as the manufacturer intended (Figure 12-10). The guard should be checked frequently to be sure that it operates freely and encloses the teeth completely when not cutting, and encloses the unused portion of the blade when it is cutting.

Circular saws should not be jammed or crowded into the work. The saw should be started and

Figure 12-10. The lower movable guard of this portable circular saw always returns to the guarded position. The operator should keep fingers away from the trigger when the saw is not being used. (Courtesy Black & Decker Manufacturing Company.)

stopped outside the work. At the beginning and end of the stroke, or when the teeth are exposed, the operator must take extra care to keep the body and the power cord away from the cutting line. All saws have a trigger switch to shut off power when pressure is released.

Injuries that occur when using portable circular saws are caused by contact with the blade; electric shock or burns; tripping over the electric cord or the extension cord, saws, or debris; losing balance; and kickbacks resulting from the blade being pinched in the cut.

Important requirements for safe operation of the portable circular saw are proper use, frequent inspection, and a rigid maintenance schedule. The manufacturer's recommendations for operation and maintenance must be followed faithfully and treated as standard procedure.

The following are specific safety "musts" when using any portable circular saw. Not doing so must be considered dangerous.

- Do not use a circular saw that is too heavy for workers to easily control.
- Be sure the switch actuates properly. It should turn the tool on and return to the OFF position after release.
- Use sharp blades. Dull blades cause binding, stalling and possible kickback, waste power, and reduce motor and switch life.
- Use the correct blade for the application. Check these points carefully: Does it have the proper size and shape arbor hole? Is the speed marked on the blade at least as high as the no-load RPM on the saw's nameplate?
- The workpiece must be securely clamped. For maximum control, use both hands to properly and safely guide the saw.

Check blades carefully before each use for proper alignment and possible defects. Be sure the blade washers (flanges) are correctly assembled on the shaft and that the blade is properly supported.

Is the blade guard working? Check for proper operation before each cut. Check often to be sure that guards return to their normal position quickly. If a guard seems slow to return or "hangs up," repair or adjust it immediately. Never defeat the guard to expose the blade—for example, tying back or removing the guard.

Before starting a circular saw, be sure the power cord and extension cord are out of the blade path and are sufficiently long to complete the cut. Stay constantly aware of the cord location. A sudden jerk or pulling on the cord can cause a worker to lose control of the saw and suffer a serious accident.

For maximum control, hold the saw firmly with both hands after securing or clamping the workpiece. Check frequently to be sure clamps remain secure. Never hold a workpiece in your hand or across your leg when sawing. Avoid cutting small pieces of material which can't be properly secured, and material on which the saw shoe cannot properly rest.

When making a "blind" cut (you can't see behind what is being cut), be sure that hidden electrical wiring, water pipes, or any mechanical hazards are not in the blade path. If wires are present, they must be avoided or disconnected at the power source by a qualified person. Contact with live wires could cause lethal shock or fire. Water pipes should be drained and capped.

Set blade depth to no more than $\frac{1}{8}$ in. to $\frac{1}{4}$ in. greater than the thickness of the material being cut. Always hold the tools by the insulated grasping surfaces. When you start your saw, allow the blade to reach full speed before the workpiece is contacted. Be alert to the possibility of the blade binding and kickback occurring. If a fence or guide board is used, be certain the blade is kept parallel with it. Don't overreach—ever!

When making a partial cut, or if power is interrupted, release the trigger immediately and don't remove the saw until the blade has come to a complete stop. Never reach under the saw or workpiece.

Portable circular saws are not designed for cutting logs or roots, trimming trees or shrubs. These are very hazardous practices.

Switch the tool off after a cut is completed, and keep the saw away from the body until the blade stops. Unplug, clean, and store the tool in a safe, dry place after use.

Kickback is a sudden reaction to a pinched blade, causing an uncontrolled portable tool to lift up and out of the workpiece toward the operator. Kickback is the result of tool misuse and/or incorrect operating procedures or conditions. Take specific precautions to help prevent kickback when using any type of circular saw:

- Keep saw blades sharp. A sharp blade will tend to cut its way out of a pinching condition.
- Make sure the blade has adequate set in the teeth. Tooth set provides clearance between the sides of the blade and the workpiece, thus minimizing the probability of binding. Some saw blades have hollow ground sides instead of tooth set to provide clearance.
- Keep saw blades clean. A buildup of pitch or sap on the surface of the saw blade increases the thickness of the blade and also increases

the friction on the blade surface. These conditions cause an increase in the likelihood of a kickback.

- Don't cut wet wood. It produces high friction against the blade. The blade will also tend to load up with wet sawdust and create a much greater probability of kickback.
- Be cautious of stock that is pitchy, knotty, or warped. These are most likely to create pinching conditions and possible kickback.
- Always hold the saw firmly with both hands.
- Release the switch immediately if the blade binds or the saw stalls.
- Support large panels so they will not pinch the blade. Use a straightedge as a guide for ripping.
- Never remove the saw from a cut while the blade is rotating.
- Never use a bent, broken, or warped saw blade. The probability of binding and resultant kickback is greatly increased by these conditions.

Overheating a saw blade can cause it to warp and result in a kickback. Buildup of sap on the blades, insufficient set, dullness, and unguided cuts can all cause an overheated blade. Never use more blade protrusion than is required to cut the workpiece—$\frac{1}{8}$ in. to $\frac{1}{4}$ in. greater than the thickness of the stock is sufficient. This minimizes the amount of saw blade surface exposed and reduces the probability of kickback and the severity if any kickback does occur. Minimize blade pinching by placing the saw shoe on the clamped, supported portion of the workpiece and allowing the cutoff piece to fall away freely.

When the saw is being used in a damp environment or outdoors, the operator should wear electrical insulating boots and gloves. GFCI protection or a double-insulated saw is necessary for outside use.

Cutoff wheels. Do not use any cutoff wheel beyond its rated speed. Check catalog RPM against safe wheel speed. Never try to cut through thick material in one try; make a series of shallow cuts that gradually deepen to the cut desired. Operators who are exposed to harmful or nuisance dusts should be required to wear approved respirators.

Electric reciprocating saws. The common saber saw and reciprocating saw are two basic types of electric saws. Because the blade is almost fully exposed when in use and storage, the tool must be handled with extreme care. The blade must be secured when not in use. The type of blade needed will be determined by the material being cut. Hold the shoe as securely as possible against the work. Turn off the tool when the job is completed and do not remove the saw from the material until the motor has fully stopped.

The versatility of the reciprocating saw in cutting metal, pipe, wood, and other materials has made it a widely used tool. By design, it is a simple tool to handle. Its few demands for safe use, however, are very important.

- Without exception, use the blade specifically recommended for the job being done. Check your owner/operator's manual carefully concerning this.
- Position yourself to maintain full control of the tool, and avoid cutting above shoulder height.
- To minimize blade flexing and provide a smooth cut, use the shortest blade that will do the job.
- The workpiece must be clamped securely, and the shoe of the saw held firmly against the work to prevent operator injury and blade breakage.

Abrasive wheels, buffers, and scratch brushes. These tools should be guarded as completely as possible. For portable grinding, the maximum angular exposure of the periphery and sides should not exceed 180 degrees, and the top portion of the wheel should always be enclosed. Guards should be adjustable so that operators will be inclined to make the correct adjustment rather than to remove the guard. However, the guard should be easily removable to facilitate replacement of the wheel. In addition to this mechanical guarding, the operator must wear safety goggles at all times to prevent eye injuries from broken wheels and spokes (Figure 12-11).

A portable grinding wheel is exposed to more abuse than is a stationary grinder. The wheel should be kept away from water and oil, which might affect its balance; protected against blows from other tools; and used carefully to avoid striking the sides of a wheel against objects or dropping the wheel. Cabinets or racks can help to protect the wheel against damage.

The speed and weight of a grinding wheel, particularly a large one, make it more difficult to handle than some other power tools. Since part of the wheel must necessarily be exposed, it is important that employees be trained in the correct way to hold and use the wheel so that it does not touch their clothes or body.

The wheels should be mounted by trained personnel only, with the wheels and safety guards conforming to ANSI standard B7.1, *Safety Requirements for the Use, Care, and Protection of Abrasive*

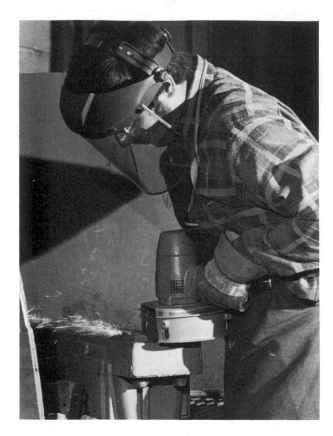

Figure 12-11. The operator of a portable abrasive wheel should wear proper protective equipment and be trained in the correct way to hold the tool being sharpened so that neither the tool nor the wheel contacts clothes or body.

Wheels. Grinders should be marked to show the maximum abrasive wheel size and speed. Abrasive wheels should be sound-tested (ring-tested) before being mounted; for details, see the standard.

Sanders. Whether the belt or disk type, sanders can cause serious skin burns when the rapidly moving abrasive touches the body. Because it is impossible to guard sanders completely, employees must be thoroughly trained to use them safely. The motion of the sander should be away from the body, and all clothing should be kept clear of the moving parts.

A vacuum system attached to the sander or an exhaust system must be used to remove the dust from the work area. Dust-type safety goggles or plastic face shields should be worn. If harmful dusts are created, a respirator certified for the exposure by the National Institute for Occupational Safety and Health (NIOSH) or the Mine Safety and Health Administration (MSHA) should be worn. (See Chapter 9, Personal Protective Equipment.)

Safety precautions for using sanders include the following:

- Do not wear loose clothing, jewelry, or any dangling objects, including long hair, that may catch in rotating parts or accessories.
- Stay constantly aware of cord location.
- Never lock a portable sander in the ON position when the nature of a job may require stopping the sander quickly, such as using a disc sander on an automobile's fender well, where the rotating disc could get jammed in the well and result in an accident.
- With portable sanders, be careful not to expose the tool to liquids, or to use in damp, wet locations.
- When adjusting the tracking of the belt on a portable sander, be certain that you have the sander supported and positioned to avoid accidental contact with yourself or adjacent objects.
- Your work area should be at least 3 ft to 4 ft larger than the length of stock you are sanding. On stationary sanders, maintain a 1/6 in. maximum clearance between the table and the sanding disc or belt—on all working sides.
- Always support your workpiece on a stationary sander with the table or backstop.
- Use jigs or fixtures to hold your workpiece whenever possible.

Sanders require especially careful cleaning because of the dusty nature of the work. If a sander is used steadily, it should be dismantled periodically, as well as thoroughly cleaned every day by being blown out with low-pressure air—less than 30 psig (208 kPa). If compressed air is used, the operator should wear safety goggles or work with a transparent chip guard between the body and the air blast.

Because wood dust presents a fire and explosion hazard, keep dust to a minimum. Sanders can be equipped with a dust collection or vacuum bag. Electrical equipment should be designed to minimize the explosion hazard. Fire extinguishers approved for Class C (electrical) fires should be available, and employees should be trained in what to do in case of fire.

Disc grinders. Portable grinders basically remove material by contact with an abrasive wheel or disc and buffing wheels. There are safety precautions that apply to all grinders and other specific recommendations that apply to specialized grinding operations.

Before operating a grinder, compare the data on the nameplate with the voltage source and be sure the voltage and frequency are compatible. Remove material or debris from the area that might be ignited by sparks. Be sure others are not in the path of the sparks or debris. Keep a properly charged fire extinguisher available.

Maintain proper footing and balance. Never attempt to grind in an awkward position. A portable grinder can kick and glance off the work if not properly controlled.

Always disconnect the tool from the power source before installing or changing discs. Use only those discs marked with a rated speed that is at least as high as or above the speed rating on the nameplate of the tool. Don't use an unmarked wheel. Handle or store discs carefully to prevent damage or cracking.

Be careful not to over tighten the spindle nut. Too much pressure will deform the flanges and stress the disc.

After mounting a wheel or brush and replacing the guard, stand to the side and allow a one minute run-up at no load to test the integrity of the wheel. Your grinder should come up to full speed each time before you contact the workpiece. Do not apply excessive pressure to the disc because that will stress the wheel, overheat the workpiece, and reduce your control.

Portable straight grinders should be used only with high-strength, bonded wheels. Position this type grinder away from the body and allow it to run for one minute before contacting work.

Tuck point grinders are a variation of straight grinders and are equipped with reinforced abrasive discs and the appropriate guard. Because using a tuck point grinder is very dusty work, a dust mask, face shield, and safety glasses with side shields are highly recommended. Maintain firm control of the tool. Never overreach. Carefully maintain balance. Do not allow the grinding wheel to bend, pinch, or twist in the cut because kickback may result.

Angle grinders are primarily used with reinforced abrasive discs or wire cup brushes for the removal of metal or masonry. Use of the proper wheel and guard combination is critical. The user must follow the manufacturer's recommendations contained in the owner/operator's manual.

Many angle grinders are equipped with guards that can be mounted with the opening in a variety of positions. Take care to position the guard to provide maximum protection.

Many angle grinders can be converted for use as sanders. When guards are removed for a sanding operation, it is essential that they be replaced before the tool is again used for grinding.

Soldering irons. Such tools are often the source of burns and of illness resulting from inhalation of fumes. Insulated, noncombustible holders will practically eliminate the fire hazard and the danger of burns from accidental contact. Ordinary metal covering on wood tables is not sufficient because the metal conducts heat and can ignite the wood.

Holders should be designed so that employees cannot accidentally touch the hot irons if they should reach for them without looking. The best holder completely encloses the heated surface and is inclined so that the weight of the iron prevents it from falling out. Such holders must be well ventilated to allow the heat to dissipate, otherwise the life of the tip will be reduced and the wiring or printed circuit may be damaged. Also see the National Safety Council Data Sheet 445-1986, *Soldering and Brazing*.

Local and federal regulations may require exhaust facilities if much lead soldering is done. Even if lead fumes are not present in harmful quantities, you should exhaust the nuisance fumes and smoke. Sample the air to verify that the amount of lead in the air is not harmful.

Lead solder particles should not be allowed to accumulate on the floor and on work tables. If the operation is such that the solder or flux may spatter, employees should wear face shields or do the work under a transparent shield.

Electrically heated glue guns. These tools are being used increasingly in industry. Although all guns have insulated handles, there is danger in the high temperature of the glue and tip of the gun. Proper holders and storage can minimize the exposure. Electric shock is also a risk and all tools should be properly grounded and insulated.

Percussion tools. This family of tools is primarily associated with masonry applications as varied as chipping, drilling, anchor setting, and breaking of pavement. They range from pistol grip types to large demolition hammers. Normal operating modes include hammering, hammering with rotary motion, and rotation or drilling only. Many models incorporate a varied combination of the above types.

Always wear safety goggles or safety glasses with side shields complying with the current national standard, and a full face shield when needed. Use a dust mask in dusty work conditions. Wear hearing protection during extended periods of operation. Do not wear gloves, loose clothing, jewelry, or dangling objects, including long hair, that may catch in rotating parts or accessories.

For maximum control, use the auxiliary handles provided with the tool. Do not tamper with clutches on those models that provide them. Have the clutch settings checked at the manufacturer's service facility at the intervals recommended in the owner/operator's manual.

Check for subsurface hazards such as electrical conductors or water lines before drilling or breaking blindly into a surface. If wires are present, they must be disconnected at the power source by a qualified person, or be certain they are avoided to prevent the possibility of lethal shock or fire. Water pipes must be drained and capped.

Always hold the tool by the insulated grasping surfaces. Do not force the tool. Percussion tools are designed to hit with predetermined force. Added pressure by the operator only causes operator fatigue, excessive bit wear, and reduced control.

When cutting, drilling, or driving into walls, floors, or wherever "live" electrical wires may be encountered, workers should never touch any metal parts of the tool. Instead, they should hold the tool only by the insulated gripping surfaces to prevent electric shock if contact is made with a "live" wire.

Air-Powered Tools

Operators of air tools should be instructed to:

- Keep hands and clothing away from the working end of the tool.
- Follow safety requirements applicable to the tool being used and the nature of the work being performed.
- Inspect and test the tool, air hose, and coupling before each use.

Most air-powered tools are difficult to guard. Therefore, the operator must be careful to prevent injury to hands, feet, or body if the machine slips or the tool breaks. Pistol or doughnut hand grips, or flanges in front of the hand grip, provide protection for the hands and should be used on all twist or percussion tools.

Air hose. An air hose presents the same tripping or stumbling hazard as do power cords on electric tools. Persons or material accidentally hitting the hose may throw the operator off balance or cause the tool to fall down. To prevent a tripping hazard, an air hose on the floor should be protected against trucks and pedestrians by placing two planks on either side or by building a runway over it.

Workers should be warned against disconnecting the air hose from the tool and using it for cleaning machines. Air hoses should not be used to remove dust from clothing.

Accidents sometimes occur when the air hose becomes disconnected and whips about. A short chain attached to the hose coupling and to the tool housing will keep the hose from whipping about if the coupling should break; other couplings in the air line should also be safely pinned or chained. Air should be cut off before attempting to disconnect the air hose from the air line. Air pressure inside the line should be released before disconnecting.

A safety check valve installed in the air line at the manifold will shut off the air supply automatically if a fracture occurs anywhere in the line. If kinking or excessive wear of the hose is a problem, it can be protected by a wrapping of strip metal or wire.

Air powered grinders. These tools require the same kind of guarding as electric grinders. Be sure the speed regulator or governor on these machines is carefully maintained to avoid overspeeding the wheel (runaway). Hold grinder properly. Regular inspection by qualified personnel at each wheel change is recommended.

Pneumatic impact (percussion) tools. Percussion tools, such as riveting guns and jackhammers, are essentially the same in that the tool is fitted into the gun and receives its impact from a rapidly moving reciprocating piston driven by compressed air at about 90 psig (625 kPa) pressure.

Two safety devices are needed to protect workers who operate these tools. First, the ON/OFF trigger should be located inside the handle where it is reasonably safe from accidental operation. The machine should operate only when the trigger is depressed. Second, make sure the tool has a device holding it in place so that it cannot be shot accidentally from the barrel (Figure 12-12). On small air hammers not designed to use this device, use a spring clip to prevent the tool and piston from falling from the hammer.

It is essential that employees be thoroughly trained in the proper use of pneumatic impact tools. Impress on all operators of small air hammers the following safety rule: *Do not squeeze the trigger until the tool is on the work.*

Air percussion tools either produce heavy blows or, due to rapid pulsating, vibrate strongly. Because tools such as rotary drills, saws, or grinders—especially those rotating at high speeds—produce excessive vibration, continued use of this equipment may cause damage to the body. Instruct workers to re-

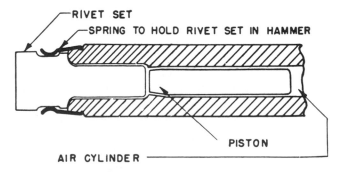

Figure 12-12. A spring clip, like the one shown here, should always be used to prevent a rivet set from falling from the air hammer.

duce transmission of this vibration by using rubber hand grips, air-cushion devices, and other vibration dampeners wherever possible. Prolonged use should be avoided. (See Chapter 10, Ergonomics, for more information on potential injuries.)

Jackhammers. Handling of heavy jackhammers causes fatigue and strain. Jackhammer handles should be provided with heavy rubber grips to reduce vibration and fatigue, and operators should wear safety shoes to reduce the possibility of injury should the hammer fall.

Two chippers should work away from each other, that is, back to back, to prevent face cuts from flying chips. Workers should not point a pneumatic hammer at anyone, nor should they stand in front of operators handling pneumatic hammers.

Many accidents are caused by the steel drill breaking because the operator loses balance and falls. Also, if the steel is too hard, a particle of metal may break off and strike the operator. The manufacturer's instructions for sharpening and tempering the steel should be followed.

Air-operated nailers and staplers. The principal hazard from these tools is the accidental discharge of the fastener. In such instances, the fastener can become a dangerous projectile and inflict serious injury at considerable distance. Operators should be trained in the use of these tools and must follow the manufacturer's operating instructions.

Personal protective equipment. As with all pneumatic impact tools, there is, of course, a hazard from flying chips. Operators should wear safety goggles; if other employees must be in the vicinity, they should be similarly protected. Where possible, screens should be set up to shield persons near chippers, riveting guns, or air drills in use. Also, operators should wear hearing protection when needed. The sound of pneumatic tools can be extremely loud—similar to firecrackers or gunfire.

Eye and hearing protection should also be provided when using powered saws, grinding wheels, screwdrivers, buffers, scratch brushes, and sanders. The noise levels from some of these tools can reach or exceed 90 dBA.

Special Power Tools

Flexible shaft tools. These tools require the same type of personal protective equipment as do direct power tools of the same type. Abrasive wheels should be installed and operated in conformance with ANSI B7.1, discussed earlier. The flexible shaft must be protected against denting and kinking, which may damage the inner core and shaft.

It is important that the power be shut off whenever the tool is not in use. When the motor is being started, the tool end should be held with a firm grip to prevent injury from sudden whipping. The abrasive wheel or buffer of the tool is difficult to guard. Because it is more exposed than the wheel or buffer on a stationary grinder, workers should use extra care to avoid damaging the equipment. When not in use, wheels should be placed on the machine or put on a rack, not on the floor.

Gasoline power tools. These tools are commonly used in logging, construction, and other heavy industry. The most well known is the chain saw. (OSHA regulations 1910.266(c)(5), *Chain saw operations,* and National Safety Council Data Sheet 320, *Portable Power Chain Saws.*)

Operators of gasoline power tools must be trained in their proper operation and follow the manufacturer's instructions. They must also be familiar with the fuel hazards (see the discussion in Chapter 13, Materials Handling and Storage).

Powder-actuated fastening tools. These are used for fastening fixtures and materials to metal, precast or prestressed concrete, masonry block, brick, stone, and wood surfaces, tightening rivets, and punching holes. Blank cartridges provide the energy and are ignited by means of a conventional percussion primer.

The hazards encountered in the use of these tools are similar to those encountered with firearms. The handling, storing, and control of explosive cartridges present additional hazards. Therefore, instructions for the use, handling, and storage of both tools and cartridges should be just as rigid as those governing blasting caps and firearms (see OSHA regulations 1910.243(d) and National Safety Council Data Sheet 236, *Powder-Actuated Hand Tools*).

Specific hazards associated with these tools are accidental discharge, ricochets, ignition of explosive or combustible atmospheres, projectiles penetrating the work, and flying dirt, scale, and other particles. In case of misfire, the operator should hold the tool in operating position for at least 30 seconds and follow the manufacturer's instructions for removing the load.

Powder-actuated tools can be used safely if workers are given special training and proper supervision. Manufacturers of the tools will aid in this training. Only properly qualified personnel should ever be permitted to operate or handle the tools. Workers can become qualified after a few hours of instruction (Figure 12-13).

Powder-assisted, hammer-driven tools are used for the same purposes as powder-actuated tools, and, generally, the same precautions should be followed. Workers should be trained to use powder charges of the correct size to drive studs into specific surfaces and should be made responsible for the safe handling and storage of the cartridges and the tools.

```
┌─────────────────────────────────────────────────┐
│  QUALIFIED OPERATOR OF POWDER-ACTUATED TOOLS      │
│                                                   │
│  Make(s)_____  Model(s)_____   │
│                                                   │
│  This certifies that _____   │
│                         (NAME OF OPERATOR)        │
│                                                   │
│  Card No._____  Soc. Sec. No._____   │
│                                                   │
│  Has received the prescribed training in the      │
│  operation of powder actuated tools               │
│  manufactured by                                  │
│  _____   │
│           (NAME OF MANUFACTURER)                  │
│                                                   │
│  Trained and issued by _____   │
│                    (SIGNATURE OF AUTHORIZED       │
│                          INSTRUCTOR)              │
│                                                   │
│  I have received instruction in the safe          │
│  operation and maintenance of powder actuated     │
│  fastening tools of the makes and models          │
│  specified and agree to conform to all rules      │
│  and regulations governing their use. Failure     │
│  to comply shall be cause for immediate           │
│  revocation of this card.                         │
│                                                   │
│  _____   _____  │
│      (SIGNATURE)                 (DATE)           │
└─────────────────────────────────────────────────┘
```

Figure 12-13. A wallet card for qualified powder-actuated tool operators is available from tool manufacturers. A list of instructors should be maintained by each manufacturer.

Powder-actuated tools should not be used on concrete less than three times the fastener shank penetration, or on very hard or brittle materials including cast iron, glazed tile, hardened steel, glass block, natural rock, hollow tile, or smooth brick. Fasteners should not be driven closer than 3 in. (7.5 cm) from an unsupported edge or corner.

Operators must wear adequate eye and face protection when firing the tool (Figure 12-14). Where the standard shield cannot be used for a particular operation, order special shields from the tool manufacturer. Hearing protection should also be worn to prevent damage from short, powerful bursts of sound. See Chapter 9, Personal Protective Equipment, for more details.

Propane torches. Commonly used in many industries, propane torches are dangerous because of their open flame. Workers should wear fire-retardant safety gear and make sure the work area is clear of flammable materials. Proper storage of propane cylinders is important to prevent leakage of this highly volatile flammable gas. Operating instructions are available from the manufacturer.

SUPERVISORY CONSIDERATIONS

As supervisor, you are responsible for controlling access to tools, for seeing that employee-owned tools meet company and government standards, and for ensuring that workers follow safe practices in carrying tools from place to place. You must be aware of these aspects of safety within your department, in addition to the hazards associated with particular tools.

Centralized Tool Control

Centralized tool control helps to ensure uniform inspection and maintenance of tools by a trained person. The toolroom attendant can promote tool safety by recommending or issuing the right type of tool, by encouraging employees to turn in defective or worn tools, and by stressing the safe use of tools. The attendant can recommend the correct protective equipment, such as welder's safety goggles or respirators, when the tool is issued. Centralized control and careful records of tool failure and other accidents will pinpoint hazardous conditions and unsafe practices.

Some companies issue each employee a set of numbered checks that are exchanged for tools from the toolroom. With this system, the attendant knows where each tool is and can recall it for inspection at regular intervals.

You may want to set up a procedure so that the toolroom attendant can send tools in need of repair to a department or to the manufacturer for a thorough reconditioning.

Companies performing work at scattered locations may find it impractical to maintain a central toolroom. In such cases, the job supervisor should inspect all tools frequently and remove defective items from service. Many companies have supervisors check all tools weekly, using a preprinted checklist.

Carrying Tools

When climbing a ladder or any structure, workers should never carry tools in any way that might interfere with using both hands freely. A strong bag, bucket, or similar container should be used to hoist tools from the ground to the job (Figure 12-15). Tools should be returned in the same manner, not brought down by hand, carried in pockets, or dropped to the ground.

Mislaid and loose tools cause a substantial portion of hand-tool injuries. Tools are put down on scaffolds, on overhead piping, on top of stepladders, or in other locations from which they can fall on people below. Leaving tools overhead is especially hazardous in areas of heavy vibration or where people are moving about or walking.

Chisels, screwdrivers, and pointed tools should never be carried in a worker's pocket. Instead, they should be carried in a toolbox or cart, in a carrying belt (sharp or pointed end down) like that used by electricians and steel workers, in a pocket tool pouch, or in the hand with points and cutting edges held away from the body.

Tools should be handed from one worker to another, never thrown. Edged or pointed tools should be passed, preferably in their carrying cases, with the handles toward the receiver. Workers carrying tools on their shoulders should pay close attention to clearances when turning around and should handle the tools so that they will not strike other people.

Figure 12-14. This powder-actuated tool drives studs into concrete slab. The operator should wear eye, ear, and head protection. Note the holster (left) for carrying the tool. (Courtesy of Hilti, Inc.)

Personal Tools

In trades or operations where employees are required or prefer to have their own personal tools, as supervisor, you may encounter a serious problem regarding the safety of such items. Personal tools are usually well maintained, but some people may purchase cheap tools, make inadequate repairs, or attempt to use unsafe tools (such as hammers with broken and taped handles, or electrician's pliers with cracked insulation).

To ensure that personal tools meet company standards, the company should spell out the general requirements that tools must meet before they can be used. First, where applicable, they should meet the appropriate ANSI standard(s) and/or be listed by a nationally recognized testing firm. Second, you should arrange a thorough inventory and initial inspection of personal tools, inspecting and listing additional tools employees buy as though the items had been purchased by the company. Do not permit workers to use inferior tools.

MAINTENANCE AND REPAIR

The toolroom attendant or tool inspector should be qualified by training and experience to pass judgment on the condition of tools for further use. Dull or damaged tools should not be returned to stock. Keep enough tools of each kind on hand so that when a defective or worn tool is removed from service, you can replace it immediately.

Efficient tool control requires periodic inspections of all tool operations. These inspections should cover housekeeping in the tool supply room, tool maintenance, service, inventory, handling routine, and condition of the tools. Responsibility for such periodic inspections is usually given to you as the department supervisor. This job should not be delegated. However, all employees need to recognize signs indicating when a tool is damaged, dull, or unsafe so they can report it at once.

Hand tools receiving the heaviest wear—chisels, wrenches, sledges, drills and bits, cold cutters, and screwdrivers—will require regular maintenance. Proper maintenance and repair of tools require ad-

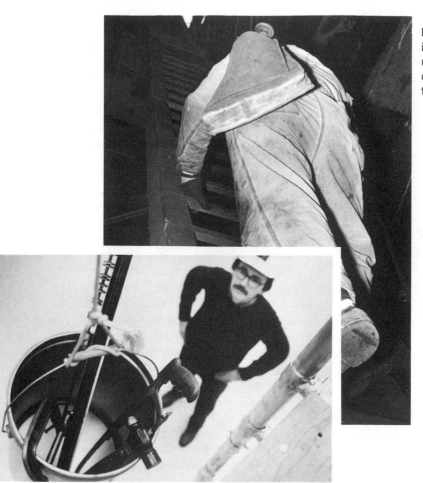

Figure 12-15. To avoid carrying tools in the hand when climbing a ladder, use a shoulder bag (top) or a bucket or similar container to hoist or lower tools.

equate facilities: workbenches, vises, a forge or furnace for hardening and tempering, safety goggles, repair tools, grinders, and good lighting.

Tempering Tools

Such tools as chisels, stamps, punches, cutters, hammers, sledges, and rock drills should be made of carefully selected steel and be heat-treated. This procedure makes them hard enough to withstand blows without excessive mushrooming, and yet not so hard that they chip or crack. Hardening and tempering of tools require special skills in manufacture.

Metal Fatigue

Metal fatigue is a common problem in or near the polar regions. Metals exposed to subzero temperatures undergo cold-soaking. A molecular change takes place that makes steel brittle and easily broken. This phenomenon shortens the useful life of such equipment as pneumatic impact tools, drill bits, and dies for threading conduit.

Ferrous metals, including carbon steel, alloy steel, and cast iron, are characterized by decreased toughness and corresponding brittleness at low tem-

peratures. Nonferrous metals, on the other hand, such as aluminum and aluminum alloys, nickel, copper and copper alloys, chromium, zinc and zinc alloys, magnesium alloys, and lead, are more resistant to low temperatures and are often used in areas of extreme cold.

Safe-Ending Tools

Such tools as chisels, rock drills, flatters, wedges, punches, cold cutters, and stamping dies should have their heads properly hardened by a qualified worker. Short sections of tight-fitting rubber hose can be set flush with the striking ends of chisels, hand drills, mauls, and blacksmith's tools to keep chips from flying, since they usually embed themselves in the rubber sleeve. Chisels, drift punches, cutters, and marking tools that manufacturers claim do not spall or mushroom are available. This feature is achieved by using a combination of alloys and scientific heat treatment to temper the metal.

Dressing Tools

Tools should have regular maintenance of their cutting edges or striking surfaces. In most cases, once the cutting or striking surfaces have been properly hardened and tempered, only an emery wheel, grindstone, or oilstone is needed to keep the tool in good condition. Be sure to grind in easy stages. Keep the tool as cool as possible with water or other cooling medium.

Tools that require a soft or medium-soft head should be dressed as soon as they begin to mushroom. A slight radius ground on the edge of the head, when it is dressed, will enable the tool to stand up better under pounding and will reduce the danger of flying chips. A file or oilstone, rather than an abrasive wheel, is recommended for sharpening pike poles and axes.

A wood-cutting tool, because of its fine cutting edge, can be initially dressed on a grinder with a wheel recommended by the manufacturer. An oilstone set securely in a wood block placed on a bench should be used to obtain a fine, sharp cutting edge. The oilstone should never be held in the hand; a slip off the face of the stone could cause a severe hand injury. Often a few finishing strokes on a leather strop will produce a keener edge.

Metal-cutting tools, because they generally have greater body, can be dressed or sharpened (or both) on an abrasive wheel. Do not use too much pressure against the wheel or the tool will overheat.

Make sure workers follow the manufacturer's recommendations for choosing the proper kind of abrasive wheel to sharpen wood- or metal-cutting tools. Each cutting edge should have the correct angle according to its use.

Handles

The wooden handles of hand tools should be of the best straight-grained material, preferably hickory, ash, or maple, and should be free from slivers. Make sure that they are properly attached; poorly fitted or loose handles are a hazard, can damage the work material, and make it difficult for the worker to control the tool.

No matter how tightly a handle may be wedged at the factory, both use and shrinkage will loosen it. Inspect tools regularly for damaged and loose handles. These should be removed from service and repaired, if feasible.

SUMMARY OF KEY POINTS

This chapter covered the following key points:

- One of a supervisor's major responsibilities is to train and supervise people in the safe handling of hand and portable power tools. Supervisors must pinpoint where major problems in tool handling lie, and devise a program to reduce tool injuries. Four basic safe work practices include (1) select the right tool for the job, (2) use tools properly and safely, (3) keep tools in good condition, and (4) store tools in a safe place.

- Supervisors must also establish procedures for regular inspection, repair, and replacement of tools and set up a central tool supply room or other method of controlling access to company tools. Special tools may need to be kept under lock and key.

- The misuse of common hand tools is a major source of injury to industrial workers. These tools include cutting, torsion, shock, and spark-resistant equipment. Supervisors must make sure that employees are trained in the proper selection and use of these tools and that they follow all safety guidelines and procedures. Personal protective equipment such as gloves, safety goggles and helmets, hearing protection, aprons, and safety shoes should be worn.

- Cutting tools include metal and wood equipment, such as chisels, tap and die work, hack saws, files, tin snips, punches, saws, axes, hatchets, knives, and miscellaneous tools. All edged tools should be used so that the direction of force is away from the body, and cuts should be made with the grain when possible. Edged cutting tools should be kept sharp and ground to the proper angle. Saws must be sharpened and oiled regularly or they may stick in the cut. Workers should wear personal protection equipment to shield them from flying chips or metal slivers and from cuts. Cutting tools should never be used frivolously in mock fights or contests of skill; they can inflict serious injuries in a moment of carelessness.

- Torsion tools are used to fasten or grip materials and include various types of wrenches, tongs, pliers, special cutters, nail band crimpers, screwdrivers, Allen wrenches, vises, and clamps. These tools must fit the job precisely to prevent slippage, the jaws or blades should be kept sharp and clean, and pressure should be applied away from the body. Workers must protect their hands, arms, and faces with special care when using these tools.

- Shock tools include regular hammers, sledge hammers, riveting hammers, and carpenter's or claw hammers. All hammers should have secure handles and properly dressed heads. Workers must be sure to select the proper weight and size hammer to perform work safely.

- Spark-resistant tools of nonferrous materials are used where flammable or explosive materials are stored. These tools will not create sparks when striking metal or other hard surfaces.

- Portable power tools are divided into four groups according to the power source: electric, pneumatic, internal combustion, and explosive. Typical injuries caused by portable power tools include burns, cuts, and sprains. These tools must be carefully handled and stored so that they cannot be activated accidently. Workers must be thoroughly trained in proper use. In addition to the hazards associated with the tool itself, the power source represents a risk to workers. All employees must wear hearing protection and eye and/or face protection when operating portable power tools or working near them.

- Electric shock is the chief hazard from electrically powered tools. Workers can suffer electric flash burns, minor shock, or lethal shock. Insulating platforms, rubber mats, rubber gloves, double-insulated tools, grounding wires, and grounded extension cords are all ways of protecting workers who use electric tools. Also, workers should wear appropriate personal protective gear to minimize risk of injury and hearing loss. All damaged or worn cords should be replaced immediately, and cords should be inspected frequently to be certain they can carry the required electric current safely.

- Important requirements for safe operation of electrically powered drills, routers, saws, wheels, sanders, soldering irons, percussion, and other portable power tools are proper use, frequent inspection, and a rigid maintenance schedule. The manufacturer's recommendations for operation and maintenance must be followed strictly.

- Air-powered tools such as grinders and pneumatic tools (jackhammers, nailers, and staplers) present serious risks to workers. Operators must be instructed to keep hands and clothing away from the working end of the tool, protect the air hose and check the safety value, follow safety requirements for the tool being used and work performed, and inspect and test the tool, air hose, and coupling before each use. Hands, feet, and torso should be protected in case the machine slips or the tool breaks, and hearing and eye protection should be used. Pistol or doughnut hand grips, or flanges in front of the hand grip, provide protection and should be used on all twist or percussion tools.

- Special power tools include flexible shaft tools, gasoline power tools, powder-actuated fastening tools, and propane torches. This equipment requires the same type of personal protection gear as do electric or air-powered tools. In addition, workers must be trained in fuel hazards and the dangers associated with explosive cartridges used in powder-actuated tools. Workers generally are given special instruction to operate these tools safely and must know the manufacturer's recommendations for using the tools under various conditions.

- Supervisors must establish routine procedures for controlling access to company tools. For example, centralized tool control helps to ensure uniform inspection and maintenance by a trained person. The attendant can recommend proper safety equipment and issue the right tool for the right job.

- When employees purchase their own tools, supervisors must ensure that such tools meet appropriate standards, are regularly inspected and repaired or replaced, and fit the work to be done. Also, supervisors should instruct employees on the proper way to carry tools. Workers should never carry tools while climbing ladders, should always carry pointed tools with the sharp end covered or held away from the body, and should always hand, never throw, tools to another worker.

- Supervisors must assume responsibility for inspection and maintenance and repair of tools in their department. Tools should be checked for quality construction and manufacture, for signs of metal fatigue in extremely cold environments, for properly hardened heads, for appropriately dressed cutting edges or striking surfaces, and for properly fitting handles.

13

Materials Handling and Storage

After reading this chapter, you will be able to:

- Describe common materials handling problems and precautions on the job
- Explain safe manual handling methods for lifting and carrying loads and handling specific container shapes
- Describe the hazards, safeguards, and operating guidelines for common materials handling equipment
- Establish a training program for your workers to ensure that they follow safe working practices while moving materials
- Establish a program of regular inspection, repair, and replacement of all materials handling equipment and tools
- Understand the principles and guidelines for safe storage of materials

Most workers, at one time or another, have had to handle materials on the job. Materials handling, whether done manually or with mechanical equipment, can be a major source of occupational injuries. The same hazards exist when handling materials off the job and can produce the same results—injuries and loss of productive time.

This chapter presents basic materials handling hazards and safeguards to help you prevent injuries and the destruction of property. The topics covered include materials handling problems; manual handling methods; equipment used for materials handling; inspection of ropes, chains, and slings; and safe storage of materials. As the supervisor, you must know how to instruct your workers to move materials safely.

MATERIALS HANDLING PROBLEMS

Manual handling of materials accounts for an estimated 25 percent of all occupational injuries. These injuries are not limited to the shipping department or the warehouse, but come from all operations, because it is impossible to run a business without moving or handling materials. Moreover, such injuries are not limited to back strains, but can affect legs or feet (by dropping whatever one is handling) or fingers, hands, and palms. Given the cost of injuries to workers and employees alike, the importance of proper materials handling techniques becomes glaringly apparent when all factors are considered.

Common injuries workers suffer include strains and sprains, fractures, and bruises. These are caused primarily by unsafe practices—improper lifting, carrying too heavy a load, incorrect gripping, failing to observe proper foot or hand clearances, and failing to use or wear proper equipment and/or personal protective equipment and clothing.

Another major cause of materials handling accidents can be traced to poor job design. Take a look at your operations and ask the following questions about your present operating practices:

- Can the job be engineered to eliminate manual handling of materials (Figure 13-1)?
- How do materials such as chemicals, dusts,

Figure 13-1. Manual handling can be engineered out of many operations by using the appropriate mechanical equipment.

and rough and sharp objects injure the people doing the handling?

- Can employees be given handling aids—properly sized boxes, adequate trucks, or hooks—that will make their jobs safer (Figure 13-2)?
- Will protective clothing, or other personal equipment, help to prevent injuries?

These are by no means the only questions that might be asked, but they serve as a start and an overall appraisal. It is beneficial for everyone to adhere to safe lifting and handling techniques. These methods are discussed in the next section.

MANUAL HANDLING METHODS

Since the largest number of injuries occurs to the fingers and hands, people need to be taught how to pick up and put down heavy, bulky, or long objects. Some general precautions are in order.

1. Inspect materials for slivers, jagged edges, burrs, rough or slippery surfaces.
2. Get a firm grip on the object.
3. Keep fingers away from pinch points, especially when putting materials down.
4. When handling lumber, pipe, or other long objects, keep hands away from the ends to prevent them from being pinched.
5. Wipe off greasy, wet, slippery, or dirty objects before trying to handle them.
6. Keep hands free of oil and grease.

In most cases, gloves, hand leathers, or other hand protectors can be worn to prevent hand injuries. In other cases, handles or holders can be attached to the objects themselves, for example, handles for moving auto batteries, tongs for feeding material to metal-forming machinery, or wicker baskets for carrying laboratory samples.

Feet and legs, primarily the toes, sustain a large share of materials handling injuries. One of the best ways to avoid such mishaps is to insist that people wear foot protection—safety shoes, instep protectors, and ankle guards.

The eyes, head, and trunk of the body are also vulnerable to injury. When opening a wire-bound or metal-bound bale or box, for example, make sure people wear eye protection as well as stout gloves. They should take special care to prevent the ends of the bindings from flying loose and striking the face or body. The same precaution applies to coils of wire, strapping, or cable. If a material is dusty or toxic, the person handling it should wear a respirator or other suitable personal protective equipment. Workers also need training in handling heavy

Figure 13-2. A drum hand truck minimizes strain in handling and moving drums. (Courtesy Liftomatic Material Handling, Inc.)

objects. See the directions below for lifting and carrying.

Lifting and Carrying

Obviously, the best means to reduce back injuries is to try to eliminate manual lifting. If this cannot be done, another way is to reduce exposure. This can be achieved by cutting weight loads, using mechanical aids, and rearranging the workplace. In spite of all these efforts, manual lifting cannot be entirely eliminated. The basic rules and instructions to be followed for manual lifting include:

1. Never let workers overexert themselves when lifting. If the load is thought to be more than one person can handle, assign another person to the job.
2. Lift gradually, without jerking, to minimize the effects of acceleration.
3. Keep the load close to the body.
4. Lift without twisting the body.
5. Follow the safe lifting procedures described below.

In reference to the safe lifting procedure, remember that some researchers working in the area of safe lifting now feel that it is better to let workers choose the lifting position most comfortable for them (see Figure 10-2).

Training for Safe Lifting Practices

Numerous attempts have been made to train materials handlers to do their work, particularly lifting, in a safe manner. Unfortunately, hopes for significant and lasting reductions of overexertion injuries through the use of training have been generally disappointing. There are several reasons for the disappointing results:

- If the job requirements are stressful, "doctoring the system" through behavioral modification will not eliminate the inherent risk. Designing a safe job is basically better than training people to behave safely in an unsafe job.
- People tend to revert to previous habits and customs if practices to replace previous ones are not reinforced and refreshed periodically.
- Emergency situations, the unusual case, the sudden quick movement, increased body weight, or impaired physical well-being may overly strain the body, since training does not include these conditions.

Thus, unfortunately, training for safe materials handling (which is not limited to lifting) should not be expected to really solve the problem. On the other hand, if properly applied and periodically reinforced, training should help to alleviate some aspects of the basic problem.

The idea of training workers in safe and proper manual materials handling techniques has been discussed for many years. Originally, the straight back/bent knees lifting posture was recommended. However, the frequency and intensity of back injuries was not reduced during the last 40 years while this lifting method has been taught. Biomechanical and physiological research has shown that leg muscles used in this lifting technique do not always have the needed strength. Also, awkward and stressful postures may be assumed if this technique is applied when the object is bulky, for example. Hence, the straight back/bent knees action evolved into the "kinetic" lift, in which the back is kept mostly straight and the knees are bent, but the positions of the feet, chin, arms, hands, and torso are prescribed. Another variant is the "free style" lift, which, however, may be better for male (but not female) workers than the straight back/bent knee technique. It appears that no single lifting method is best for all situations.

Training of proper lifting techniques is an unsettled issue. It is unclear what exactly should be taught, who should be taught, how and how often a technique should be taught. This uncertainty concerns both the objectives and methods, as well as the expected results. Claims about the effectiveness of one technique or another are frequent but are usually unsupported by convincing evidence.

A thorough review of the existing literature indicates that the issue of training for the prevention of back injuries in manual materials handling is confused at best. In fact, training may not be effective in injury prevention, or its effect may be so uncertain and inconsistent that the money and effort paid for training programs might be better spent on research and implementation of techniques for worker selection and ergonomic job design. Nevertheless, according to the National Institute for Occupational Safety and Health (NIOSH),

> The importance of training in manual materials handling in reducing hazard is generally accepted. The lacking ingredient is largely a definition of what the training should be and how this early experience can be given to a new worker without harm. The value of any training program is open to question as there appear to have been no controlled studies showing a consequent drop in the manual material handling (MMH) accident rate or the back injury rate. Yet so long as it is a legal duty for employers to provide such training or for as long as the employer is liable to a claim of negligence for

failing to train workers in safe methods of MMH, the practice is likely to continue despite the lack of evidence to support it. Meanwhile, it may be worth considering what improvements can be made to existing training techniques. (NIOSH, 1981, p. 99.)

Currently, it appears that two major training approaches are most likely to be successful. One involves training in awareness and attitude through information on the physics involved in manual materials handling and on the related biomechanical and physiological events going on in one's body. The other approach is the improvement of individual physical fitness through exercise and warm-ups. This approach, of course, also influences awareness and attitude, though indirectly.

Rules for lifting. There are no comprehensive and sure-fire rules for "safe" lifting. Manual materials handling is a very complex combination of moving body segments, changing joint angles, tightening muscles, and loading the spinal column. The following DOs and DO NOTs apply, however:

- DO design out manual lifting and lowering in the task and workplace. If it nevertheless must be done by a worker, perform it between knuckle and shoulder height.
- DO be in good physical shape. If you are not accustomed to lifting and vigorous exercise, do not attempt to do difficult lifting or lowering tasks.
- DO think before acting. Place material conveniently within reach. Have handling aids available. Make sure sufficient space is cleared.
- DO get a good grip on the load. Test the weight before trying to move it. If it is too bulky or heavy, get a mechanical lifting aid or somebody else to help, or both.
- DO get the load close to the body. Place the feet close to the load. Stand in a stable position with the feet pointing in the direction of movement. Lift mostly by straightening the legs.
- DO NOT twist the back or bend sideways.
- DO NOT lift or lower awkwardly.
- DO NOT hesitate to get mechanical help or help from another person.
- DO NOT lift or lower with the arms extended.
- DO NOT continue heaving when the load is too heavy.

Handling Specific Shapes

Boxes, cartons, and crates. These are best handled by grasping them at alternate top and bottom corners and drawing one corner between the legs.

Any box, carton, or crate that appears either too large or too heavy for one person to handle should either be handled by two persons or with mechanical handling equipment, if possible.

When two persons handle a crate, box, or carton, it is preferable that they be nearly the same size so both carry an equal portion of the load. The carriers must be able to see in front of the load so they don't run into pedestrians, plant trucks, machines, walls, or equipment.

Bags or sacks are grasped in the same manner as boxes. If a sack is to be raised to shoulder height, it should be raised to waist height first, and rested against the belly or hip before it is swung to the shoulder. It should rest on its side. Also, if two persons are to lift a large bag or sack, they should be nearly the same size. This makes the job easier and keeps the load balanced. The two should lift at the same time, on an agreed signal.

Barrels and drums. Workers need special training to handle barrels and drums safely. If two people are assigned to upend a full drum, they should use the following procedure:

- Stand on opposite sides of the drum and face one another.
- Grasp both chimes (rolled edges at both ends of the drum) near their high points. Lift one end; press down on the other.
- As the drum is upended and brought to balance on the bottom chime, release the grip on the bottom chime and straighten it up with the drum.

When two people are to overturn a full drum, they should use this procedure:

- Make sure that there is enough room. Cramped quarters can result in badly injured hands.
- Stand near one another, facing the drum. Grip the closest point of the top chime with both hands. Rest palms against the side of the drum, and push until the drum balances on the lower chime.
- The two workers step forward a short distance, and each person releases one hand from the top chime in order to grip the bottom chime. They ease the drum down to a horizontal position until it rests solidly on its side.

If one person is to overturn a drum, he or she should:

- Make sure there is enough room.
- Stand in front of the drum, reach over it, and grasp the far side of the top chime with both hands. (A short person can grasp the near side of the chime, if this is easier.) If the drum is tight against a wall or against other drums, pull on the chime with one hand and push against the wall (or other drum) with the other hand for additional control.
- Pull the top of the drum toward you, until it is balanced on the edge of the lower chime.
- Transfer both hands to the near side of the top chime. Keep the hands far enough apart to avoid their being pinched when the drum touches the floor.
- Lower the drum. Keep the back straight, inclined as necessary. Bend the legs so that the leg muscles take the strain.

If one person is to upend a drum, reverse this procedure. Actually, if one person must handle a drum, he or she should have a lifter bar that hooks over the chime and gives powerful leverage and excellent control. Barrel and drum lifters are commercially available. To roll a barrel or drum, push against the sides with the hands. To change direction of the roll, grip the chime. No one should kick the drum with the feet.

To lower a drum or barrel down a skid, turn it and slide it on end. Do not roll it. To raise a drum or barrel up a skid, two workers stand on opposite sides of the skid (outside the rails, not inside, and not below the object being raised). Then, roll the object up the incline.

Handling drums and barrels can be hazardous, even when using utmost care. Special handling equipment and tools (approved by the safety professional or safety department) should be available to make this job safer and easier (Figure 13-2).

Long objects, like ladders, lumber, or pipe, should be carried over the shoulder. The front end should be held as high as possible to prevent its striking other employees, especially when turning corners. When two or more people carry an object, they should place it on the same shoulder, respectively, and walk in step.

MATERIALS HANDLING EQUIPMENT

Hand Tools

Many hand tools are available for specific jobs and for specific materials. They should be used only for their designated purpose. Further, hand tools should always be kept in peak condition for best results and for safety reasons. For instance, if a chisel is used, the blade should be sharp and free of burrs; a mushroom-headed chisel should be repaired or discarded (see Chapter 12, Hand Tools and Portable Power Tools, for more details).

Crowbars are probably the most common hand tools used in materials handling. Select the proper kind and size for the job. Because the bar can slip, people should never work astride it. They should position themselves to avoid being pinched or crushed if the bar slips or the object moves suddenly. The bar should have a good "bite" and not be dull or broken. When not in use, crowbars should be hung on a rack or otherwise placed so they cannot fall on or trip someone.

Rollers are used for moving heavy or bulky objects. Workers should be careful not to crush fingers or toes between the rollers and the floor. They should use a sledge or bar, not their hands and feet, to change the direction of the object. They should avoid unnecessary turns by making sure the rollers are properly placed before moving. Compressed gas cylinders must never be used as rollers.

Handles of tongs and pliers should be offset so hands and fingers will not be pinched.

Hand hooks used for handling materials should be kept sharp so that they will not slip when applied to a box or other object. The handle should be strong, securely attached, and shaped to fit the hand. The handle and the point of long hooks should be bent on the same plane so that the bar will lie flat and will not constitute a tripping hazard. The hook point should be shielded when not in use.

Shovel edges should be kept trimmed and handles should be checked for splinters. Workers should wear safety toe shoes with sturdy soles. Their feet should be well separated to provide balance and spring in the knees. The leg muscles should take much of the load.

To reduce the chance of injury, the ball of the foot—not the arch—should be used to press the shovel into clay or other stiff material. If the instep is used and the foot slips off the shovel, the sharp corner of the shovel may cut through the worker's shoe and into the foot.

Dipping the shovel into a pail of water occasionally will help to keep it free from sticky material. Greasing or waxing the shovel blade will also prevent some kinds of material from sticking. When not in use, shovels should be placed upright against a wall or kept in racks or boxes.

Nonpowered Hand Trucks

Two-wheeled hand trucks. These vehicles look as though they would be easy to handle, but workers must follow several safe procedures:

1. Keep the center of gravity of the load as low as possible. Place heavy objects below lighter objects.
2. Place the load well forward so the weight will be carried by the axle, not by the handles.
3. Position the load so it will not slip, shift, or fall. Load only to a height that will allow a clear view ahead.
4. Let the truck carry the load. The operator should only balance and push.
5. Never walk backwards with a hand truck.
6. When going down an incline, keep the truck in front of you. When going up, keep the truck behind you.
7. Move the truck at a safe speed. Do not run. Keep the truck under control at all times.

A truck that is designed for a specific purpose should only be used for that purpose—a curved bed truck should be used for handling drums or other circular materials, and a horizontal platform truck should be used for handling acetylene or compressed gas cylinders. Foot brakes can be installed on wheels of two-wheeled trucks so that operators need not place their feet on the wheel or axle to hold the truck. Handles should have knuckle guards, unless they are narrower than the distance between the outside of the wheel axle or the chisel, whichever is greater (Figure 13-3).

Four-wheeled hand trucks. Operating rules for these vehicles are similar to those for two-wheeled trucks. Extra emphasis should be placed on

Figure 13-3. With this design, it is more natural for the operator to grasp the handle away from the outside framework.

proper loading, however. Four-wheeled trucks should be evenly loaded to prevent tipping. These trucks should be pushed rather than pulled, except for those with a fifth wheel and a handle for pulling.

Trucks should not be loaded so high that operators cannot see where they are going. If there are high racks on the truck, two persons should move the vehicle—one to guide the front end, the other to guide the back end. Handles should be placed at protected places on the racks or truck body so that passing traffic, walls, or other objects will not crush or scrape the operator's hands. Truck contents should be arranged so that they will not fall or be damaged if the truck or the load is bumped.

General precautions. Truckers should be warned of four major hazards: (a) running wheels off bridge plates or platforms; (b) colliding with other trucks or obstructions; (c) jamming hands between the truck and other objects; and (d) running wheels over feet.

When not in use, trucks should be parked in a designated area, not in the aisles or in other places where they constitute tripping hazards or traffic obstruction. Trucks with drawbar handles should be parked with handles up and out of the way. Two-wheeled trucks should be stored on the chisel with handles leaning against a wall or the next truck.

Powered Hand Trucks

The use of powered hand trucks, controlled by a walking operator, is increasing. The principal hazards are (a) catching the operator between the truck and another object, and (b) possible collisions with objects or people.

The truck should be equipped with a dead-man control, wheel guards, and an ignition key that can be taken out when the operator leaves the truck. Operators must be trained not to use trucks unless authorized. Training should follow the operating instructions included in the truck manufacturer's manual. General instructions are:

1. Do not operate the truck with wet or greasy hands.
2. Lead the truck from right or left of the handle. Face the direction of travel and keep one hand on the handle.
3. When entering an elevator, back the truck in to keep it from getting caught between the handles and the elevator walls. Operate the truck in reverse whenever it must be run close to a wall or other obstruction.
4. Always give pedestrians the right of way.
5. Stop at blind corners, doorways, and aisle intersections to prevent collisions.
6. Never operate the truck faster than normal walking pace.
7. Handle flammable or corrosive liquids only when they are in approved containers.
8. Never ride on the truck, unless it is specifically designed for the driver to ride (Figure 13-4).
9. Never permit others to ride on the truck.
10. Do not indulge in horseplay.

These general rules for operators of powered hand trucks should also be followed by operators of power trucks. All such training must be documented.

Powered Industrial Trucks

To comply with OSHA regulations, only qualified and trained operators should be allowed to operate powered industrial trucks. Training programs must include safe driving practices and supervised experience driving on a training course. As supervisor, be sure you emphasize workers' safety awareness. Trained and authorized drivers should have badges or other clear authorization to drive, which they should display at all times. These steps will help the company to meet OSHA requirements.

Power trucks have either a battery-powered motor or an internal combustion engine. Trucks should be maintained according to the manufacturers' recommendations. All trucks acquired on or after February 15, 1972, must meet the design and construction requirements established in ANSI/ASME B56.1–1988, *Low Lift and High Lift Trucks*. Modifications and additions that affect capacity and safe operation should not be made by the customer or user without the manufacturer's prior written approval. All nameplates and markings must be accurate, in place, and legible. Eleven different designations of trucks or tractors are authorized for use in various locations (*Powered Industrial Trucks, Including Type Designations, Areas of Use, Maintenance, and Operation*, ANSI/NFPA 505–1987).

Battery-charging installations. This equipment must be located in areas designated for that specific purpose. The company must provide facilities for flushing and neutralizing spilled electrolyte, and fire protection to prevent trucks from damaging the charging apparatus. There must also be adequate ventilation for dispersal of gases or vapors from gassing batteries. Racks used to support batteries must be made of spark-resistant materials or be coated or covered to prevent spark generation.

A conveyor, overhead hoist, or equivalent equipment must be used for handling batteries. Reinstalled batteries must be properly positioned and secured in the truck. Use a carboy tilter or siphon

Figure 13-4. Riders should be allowed on powered materials handling trucks only when they are specifically designed to be ridden.

for handling electrolyte. Teach workers that acid must always be poured into water; water must *never* be poured into acid—it overheats and splatters. During charging operations, vent caps must be kept in place to avoid electrolyte spray. Make sure that vent caps are functioning. Battery or compartment cover or covers must be open to dissipate heat.

Take all precautions to prevent open flames, sparks, or electric arcs in battery-charging areas. Tools and other metallic objects must be kept away from the tops of uncovered batteries. Employees charging or changing batteries must be authorized to do this work, trained in proper handling techniques, and required to wear protective clothing, including face shields, long sleeves, rubber boots, aprons, and gloves. Smoking is prohibited in the charging area.

Refueling. All internal combustion engines must be turned off before refueling. Refueling should be in the open or in specifically designated areas where ventilation is adequate to carry away fuel vapors. Liquid fuels, such as gasoline and diesel fuel, must be handled and stored in accordance with the National Fire Protection Association's *Flammable and Combustible Liquids Code*, NFPA 30. Liquefied petroleum gas fuel (or LP-gas) must be stored in accordance with *Storage and Handling of Liquefied Petroleum Gases*, NFPA 58. LP-gas tanks must be secured with both straps while on the truck so as not to shake loose and possibly catch fire. Smoking must be strictly prohibited in the service areas; make sure signs to that effect are prominently displayed.

Train employees to follow general rules for driving powered industrial trucks in your operations. For each job, you should draw up a specific set of rules (referenced from OSHA 29 *CFR* 1910.178).

1. All traffic regulations must be observed, especially plant speed limits. (N1—Traveling)
2. Safe distances must be maintained approximately three truck lengths from the truck in front. Trucks must be kept under control at all times so that an emergency stop, if necessary, can be made in the clear distance ahead. (N1)
3. The right of way must be yielded to ambulances, fire trucks, and other vehicles in emergency situations. (N2)
4. Other trucks traveling in the same direction must not be passed at intersections, blind spots, or other dangerous locations. (N3)
5. Drivers are required to slow down and sound horns at cross aisles and other locations where vision is obstructed. If the forward view is obstructed because of the load being carried, drivers are required to travel with the load trailing behind them. (N4)
6. Railroad tracks must be crossed diagonally whenever possible, and drivers must never park within 8 ft (2.5 m) of the center of the nearest railroad track bed. (N5)
7. Drivers are required to keep their eyes on the direction of travel and to have a clear view of the path ahead at all times. They should never back up without looking. Be especially careful on loading docks (Figure 13-5). (N6)

Figure 13-5. Forklift truck operators should be well trained in safety procedures, including keeping eyes in the direction of travel, having a clear view of the path ahead, and checking that the dock plate is secured. (Courtesy Bader Rutter.)

8. Grades are to be ascended or descended slowly and loaded trucks must be driven with the load upgrade on grades in excess of ten percent. Unloaded trucks must be operated on all grades with their load-engaging means downgrade. On all grades, load and load-engaging means are to be tilted back if applicable and raised only as far as necessary to clear the road surface. (N7)

9. Trucks must always be operated at a speed that will permit them to stop safely. Drivers are required to slow down on wet and slippery floors. Stunt driving and horseplay are not tolerated. (N8, N9, N10)

10. Dockboards or bridge plates are to be driven over carefully and slowly and only after they have been properly secured. Their rated weight capacity must never be exceeded. (N11)

11. Elevators shall be approached slowly and then entered squarely after the elevator car is properly leveled. Once on the elevator, the controls should be neutralized, power shut off, and the brakes set. (N12)

12. Never run over loose objects on the roadway surface. (N14)

13. While negotiating turns, speed must be reduced to a safe level by means of turning the hand steering wheel in a smooth, sweeping motion. When maneuvering at a low speed, the hand steering wheel must be turned at a moderate and even rate. (N15)

14. Only stable or safely balanced loads shall be handled, and extra caution must be exercised when handling off-center loads. (O1—Loading)

15. Only loads within the rated load capacity of the truck shall be handled, and long or high (including multiple-tiered) loads that may affect capacity must be adjusted. (O2, O3)

16. Load-engaging means must be placed under the load as far as possible; the mast is then

carefully tilted backward to stabilize the load. (O5)

17. Extreme care must be used when tilting the load forward or backward, particularly when placing items on high tiers. Tilting forward with the load-engaging means elevated is not permitted except to pick up a load. Elevated loads must not be tilted forward, except when the load is in a deposit position over a rack or stack. When stacking or tiering, use only enough backward tilt to stabilize the load (Figure 13-6). (O6)

18. When operating in close quarters, keep hands out of the way so they cannot be pinched between steering controls and projecting stationary objects. Keep legs and feet inside the guard or the operating stations of the truck. (M4—Truck Operations)

19. Do not use the reverse control on electric trucks for braking.

20. Park trucks only in designated areas—never in an aisle or doorway or where they will obstruct equipment or material. Fully lower the load-engaging means, neutralize the

Figure 13-6. When using a forklift truck to place loads on a stack or tier, use only enough backward tilt to stabilize the load. Do not travel with loads elevated.

controls, shut off the power, and set the brakes. Remove the key (or connector plug) when leaving a truck unattended. If the truck is parked on an incline, block the wheels. (M5)

Research on safety belts in power trucks conducted during 1982 and 1983 convinced some manufacturers that safety belts should be installed for the protection of the drivers. In tests, it was found that if a truck tipped over, the driver was safer sitting on the seat rather than attempting to jump clear. A driver jumping clear could be struck by some part of the truck body or overhead guard and be seriously injured or killed.

Depending on its use, a power truck should have a dead-man control and be equipped with guards, such as a hand-enclosure guard or a canopy guard. A lift truck should also have upper and lower limit switches to prevent overtravel. No power truck should be used for any purpose other than the one for which it was designed.

A forklift should not be used to elevate employees (for example, in servicing light fixtures or when stacking materials) unless a safety platform meeting OSHA requirements, with standard railing and toeboards, is fastened securely to the forks (Figure 13-7). Failure to use safety platforms has resulted in many serious accidents.

Combustion by-products. When internal-combustion engine trucks are operated in enclosures, the concentration of carbon monoxide should not exceed the limits specified by local or state laws. In no case should the time-weighted average concentration exceed 35 ppm (parts per million) for an 8-hour exposure. The atmosphere must contain a minimum of 19 percent oxygen, by volume—air usually contains 20.8 percent oxygen, by volume.

Many companies equip their fuel-powered trucks with catalytic exhaust purifiers to burn the carbon monoxide before it reaches the atmosphere. Some power trucks are approved and marked only for use in specific locations. These include areas where flammable gases or vapors are present in quantities sufficient to produce explosive or ignitable mixtures, or areas where combustible dust or easily ignitable fibers or flyings are present. (See ANSI/NFPA 505–1987, *Powered Industrial Trucks*, for classification of types.)

Loads. Operators should refuse to carry unstable loads. If material of irregular shape must be transported, make sure it is securely placed and blocked, tied, or otherwise strapped down. If possible, crosstie the load after neatly stacking it. Never place an extra counterbalance on a forklift to help handle overloads because it puts added strain on the

Figure 13-7. When a forklift is used to elevate employees, a safety platform with standard railing and toeboards must be securely fastened to the forklift.

truck and endangers the operator and nearby employees. It is difficult to operate an overloaded truck safely.

Operators of powered industrial trucks should not drive up to anyone standing in front of a bench or other fixed object, or allow anyone to stand or pass under the elevated portion of any truck, whether loaded or empty. Unauthorized personnel are not permitted to ride on trucks; even if authorized, they are permitted only when they are in the space provided. No one should be allowed to place arms or legs between uprights of the mast or outside the running lanes of the truck.

Raise or lower loads at the point of loading or unloading, not during travel. Operators should be trained to look before raising a load, so that they

will not strike structural members of the building, electric wiring or cables, or piping (especially ones carrying gases or flammable fluids).

Operators must make sure that there is sufficient headroom under overhead installations, such as lights, pipes, and sprinkler systems, before driving underneath them. Overhead guards must be used as protection against falling objects. However, keep in mind that an overhead guard is intended to offer protection from the impact of small packages, boxes, bagged materials, and so forth, but not to withstand the impact of a falling capacity load. Load backrest extensions must be used whenever necessary to minimize the possibility of the load or part of the load falling toward the rear.

When using a scoop or shovel attachment on a

front-end loader, operators should be told to work the front of the pile and not to undercut. Overhangs are dangerous.

Floors on which power trucks travel should be kept in peak condition. They should have a load capacity great enough to support both truck and load. The combined weight of the heaviest truck and load should not exceed one-fifth of the design strength of the floor. For example, a floor originally designed to bear 1,000 pounds per square foot (psf)—or 5,000 kg/m²—should not be allowed to bear more than 200 pounds per square foot. Maximum floor loadings should be prominently displayed (ANSI standard A58.1-1982, *Minimum Design Loads for Buildings and Other Structures*).

The 5 to 1 ratio, although not excessively conservative, is important in establishing a low risk. A conservative safety factor provides an adequate margin of protection to offset possible miscalculations due to inadequate structural data, and misinformation about age, condition of floor members, type of floor, and other details.

Dock Plates

Dock plates (bridge plates) that are not built in should be secured to prevent them from "walking" or sliding when used. Secure by drilling holes and dropping pins or bolts through them.

Plates should be large and strong enough to hold the heaviest load, including the equipment. A safety factor of five or six should be used. Plates can have the edges turned up, or have angle-iron welded to the edges, to keep trucks from running over their edges. Keep plates in good condition, and make sure they do not have curled corners or bends. They should be kept clean and dry—clear of snow and ice, and free of oil and grease. Canopies can be used to protect them from the weather.

Dock plates should be positioned with a forklift or other mechanical equipment to avoid back injuries and pinched fingers. The dock plate must be kept under control and not allowed to bounce. Handholds or other effective means must be provided on portable dock plates to permit safe handling.

Permanent, adjustable dock levelers are the safest and most efficient method of traversing between the dock and a vehicle. They should be of sufficient capacity to handle present and future loads, and be long enough to reduce operating grades. Dock levelers should be inspected and maintained on a regular basis (Figure 13-8.)

The wheels of trucks or trailers backed up to the dock should be blocked or chocked to prevent movement when power trucks are loading or unloading. When not in use, dock plates should be stored safely to avoid acting as tripping hazards. They should not be placed on edge so they can fall over. For details of loading highway trucks and cars, see later sections in this chapter.

Conveyors

The primary rule regarding conveyors is simple: no one is allowed to ride them. Post prominent warning signs making this rule clear. If people must cross over conveyors, the company should provide bridges so that they do not have to crawl across or under the conveyor.

In the case of certain roller conveyors, it is possible to provide a hinged section that opens up and permits people to walk through without having to climb over. The hinged section should be returned to its original position as soon as the individual has reached the other side. Tell workers to make sure that no merchandise is on the roller conveyor when they use the hinged section.

Gravity conveyors, usually the chute or roller type, should be equipped with warning or protective devices to prevent workers' hands from being caught in descending material or from being jammed between material and the receiving table. If the conveyor jams up, first try to free it from the top side. If someone must enter the chute to free a jam, the person should wear a lifeline and have another worker stationed at the top of the chute to assist in case of emergency. Inform your employees of these hazards and train them to handle such hazards safely.

Pinch points on conveyors—gears, chain drives, and revolving shafts—should be guarded. Tell workers not to remove these guards. Should the conveyor need servicing, the power should be shut off and the switch locked out.

Pneumatic or blower conveyors must be shut off and locked before inspections or repairs are made to keep material from being blown into the workers' faces or into the working area. If an inspection or service door must be opened while a system is operating, the inspector should wear goggles and, if necessary, respiratory equipment. Inspection ports, equipped with transparent coverings made of durable, nonflammable, shatter-resistant material, permit inspections without stopping the operation.

Screw conveyors should be completely enclosed whenever possible. Their principal danger is that workers may try to dislodge material or free a jam with their hands or feet, and be caught in the conveyor. Screw conveyors should be repaired only when the power has been shut off and the switch locked out. Any exposed sections should be covered with a metal grate with openings no larger than ½ in.

Figure 13-8. Mechanical dock levelers, which work on a spring and counterbalance system, are first raised and then "walked down" to the proper truck bed height.

(13 mm), or with a solid plate—ANSI/ASME standard B20.1-1987, *Safety Standards for Conveyors and Related Equipment.* The cover should be strong enough to withstand abuse. Before removing any cover, grating, or guard, padlock the main disconnect in the OFF position. Do not rely on interlocks. Never step or walk on covers, gratings, or guards.

Cranes

All overhead and gantry cranes must meet the design specifications of the ANSI/ASME B30.17-1985, *Overhead and Gantry Cranes*; all mobile, locomotive, and truck cranes must meet ANSI/ASME B30.5—1989, *Mobile and Locomotive Truck Cranes.*

Although many types of cranes are used in industry, safe operating procedures are much the same. Standard signals (Figures 13-9 and 13-10) should be thoroughly understood by both operator and signalman. You should assign only one signalman for each crane, and tell the crane operator to obey only this person's instructions. People working with or near a crane should keep out from under the load, be alert at all times, and watch warning signals closely. When a warning signal sounds, they should move to a safe place immediately.

No crane should be loaded beyond its rated load capacity. The weight of all auxiliary handling devices, such as hoist blocks, hooks, and slings, must be considered as part of the load rating. Substantial and durable rating charts with clearly legible letters and figures are to be fixed to the crane cab in a location clearly visible to the operator while seated at the control station.

Hoist chain or hoist rope must be free of kinks and twists and must not be wrapped around the load. Loads must be attached to the load block hook by means of a sling or other approved device. Care must be taken to make certain that the sling clears all obstacles. The load must be well secured and properly balanced in the sling or lifting device before it is lifted more than a few inches. Before starting to hoist, make sure that multiple-part lines are not twisted around each other.

Hooks should be brought over the load slowly to prevent swinging. During hoisting operations, see to it that there is no sudden acceleration or deceleration of the moving load and that the load does not come in contact with any obstruction.

Cranes must not be used for side pulls, unless specifically authorized by a responsible person who

HOIST. With forearm vertical, forefinger pointing up, move hand in small horizontal circle.

LOWER. With arm extended downward, forefinger pointing down, move hand in small horizontal circles.

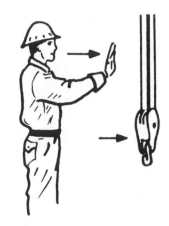

BRIDGE TRAVEL. Arm extended forward, hand open and slightly raised, make pushing motion in direction of travel.

TROLLEY TRAVEL. Palm up, fingers closed, thumb pointing in direction of motion, jerk hand horizontally.

STOP. Arm extended, palm down, move arm back and forth.

EMERGENCY STOP. Both arms extended, palms down, move arms back and forth.

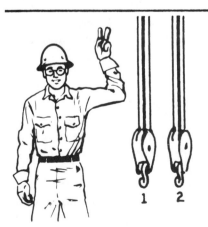

MULTIPLE TROLLEYS. Hold up one finger for block marked "1" and two fingers for block marked "2". Regular signals follow.

MOVE SLOWLY. Use one hand to give any motion signal and place other hand motionless in front of hand giving the motion signal. (Hoist slowly shown as example.)

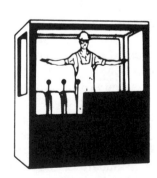

MAGNET IS DISCONNECTED. Crane operator spreads both hands apart palms up.

Figure 13-9. Standard hand signals for controlling operation of overhead and gantry cranes. (Reprinted with permission from ANSI/ASME B30.5–1989.)

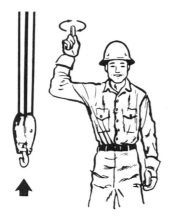

HOIST. With forearm vertical, forefinger pointing up, move hand in small horizontal circle.

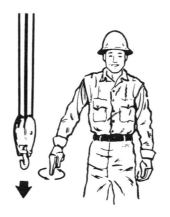

LOWER. With arm extended downward, forefinger pointing down, move hand in small horizontal circles.

USE MAIN HOIST. Tap fist on head; then use regular signals.

USE WHIP LINE. (Auxiliary Hoist) Tap elbow with one hand; then use regular signals.

RAISE BOOM. Arm extended, fingers closed, thumb pointing upward.

LOWER BOOM. Arm extended fingers closed, thumb pointing downward.

MOVE SLOWLY. Use one hand to give any motion signal and place other hand motionless in front of hand giving the motion signal. (Hoist Slowly shown as example)

RAISE THE BOOM AND LOWER THE LOAD. With arm extended thumb pointing up, flex fingers in and out as long as load movement is desired.

LOWER THE BOOM AND RAISE THE LOAD. With arm extended, thumb pointing down, flex fingers in and out as long as load movement is desired.

Figure 13-10. Standard hand signals suitable for crawler, locomotive, and truck boom cranes. (Reprinted with permission from ANSI/ASME B30 series.)

with bolts, and gangplanks and skids securely placed before using. Operators of mechanized equipment should avoid bumping doors or door posts. Metal bands should be removed or cut to avoid dangling ends. Stack special equipment, such as DF bars, at the ends of a car when empty, and close and latch the doors before moving the car. Specially equipped cars with movable bulkheads have instructions in the cars that must be followed. When bulkhead partitions are difficult to move, examine the rollers to determine the reason for binding to avoid having the bulkhead fall on those working in the car.

Highway Trucks

When loading or unloading highway trucks, set the brakes and place wheel chocks under the rear wheels or use vehicle restraints to prevent the trucks from rolling (Figure 13-11). With vehicle restraints, trailers are held in place by their underride protection bar, or ICC bar. Unlike wheel chocks, vehicle restraints hold the truck safely to the dock if the landing gear or jack collapses, or if the driver tries to pull away from the dock before the loading or unloading is completed. They are also effective in icy and wet environments (Figure 13-12).

If a powered industrial truck is used inside a semitrailer, place a fixed jack to support the trailer if it is not coupled to a tractor, to prevent upending the semitrailer. Make sure jacks are snug against the trailer bottom. For more details of truck and truck terminal safety, see the National Safety Council's *Motor Fleet Safety Manual.*

Motorized Equipment

Heavy-duty trucks, mobile cranes, tractors, bulldozers, and other motorized equipment used in the production of stone, ore, and similar materials and also in construction work have been involved in frequent, and often serious, accidents. In general, prevention of these accidents requires:

- Safe equipment
- Systematic maintenance and repairs
- Safety training for operators
- Safety training for repair personnel

Operation. Manufacturers' manuals contain detailed information on the operation of equipment. Recommended driving practices are similar to those necessary for the safe operation of highway vehicles. However, off-the-road driving involves certain hazards that require special training and safety measures. The modern heavy-duty truck or off-the-road vehicle is a carefully engineered and expensive piece of equipment and warrants operation only by drivers who are qualified physically and mentally. For safety, driving standards should be especially high.

The time required for prospective drivers or operators to become thoroughly familiar with the mechanical features of the equipment, safety rules, driver reports, and emergency conditions varies. In no case should even an experienced driver be permitted to operate equipment until the instructor or supervisor is satisfied with his or her abilities. After workers have been trained, supervisors have the im-

Figure 13-11. Place wheel chocks in front of the rear wheels of trucks at a loading dock.

Figure 13-12. Vehicle restraints can prevent accidents and injuries at loading docks by holding trucks in place. They are often considered safer than wheel chocks, especially in wet, slippery environments. (Courtesy Bader Rutter.)

portant responsibility of seeing that drivers continue to operate vehicles as they were instructed.

Equipment. Safe operation of heavy equipment begins with the purchase specifications. A good policy is to specify safeguards over exposed gears, safe oiling devices, handholds, and other devices when the order is placed with the manufacturer. In any case, before equipment is put into operation, it should be thoroughly inspected and necessary safety devices installed and checked.

Both operators and repair personnel, whether experienced or not, should know and follow the recommendations of the manufacturer pertaining to lubrication, adjustments, repairs, and operating practices for equipment. A preventive maintenance program is essential to ensure worker safety and efficiency. Frequent and regular inspections and prompt repairs provide effective preventive maintenance.

In general, the operator is responsible for inspecting such mechanical conditions as hold-down bolts, brakes, clutches, clamps, hooks, and similar vital parts. Wire ropes should be lubricated (see manufacturer's instructions); for inspection guidelines, see the section that follows.

Many of the basic safety measures recommended for trucks also apply to motor graders and other earth-moving equipment. All machines should be inspected regularly by the operator, who should report any defects promptly. Equipment safety and efficiency of equipment are increased by adherence to periodic maintenance.

ROPES, CHAINS, AND SLINGS

Special safety precautions apply to using ropes, rope slings, wire rope, chains and chain slings, and storing chains. You should know both the properties of these various types and the precautions pertaining to their use and maintenance. For additional help regarding ropes, chains, or slings, consult the supplier.

Fiber Ropes

Fiber rope is used extensively in handling and moving materials. The rope is generally made from manila (abaca), sisal, or nylon. Manila or nylon ropes give the best uniform strength and service. Other types of rope include those made from polyester or polypropylene or a composite of both that is adaptable to special uses.

Sisal is not as satisfactory as manila because the strength varies in different grades. Sisal rope is about 80 percent as strong as manila. Manila rope is yellowish with a somewhat silvery or pearly luster. Sisal rope is also yellowish, but often has a green tinge and lacks the luster of manila. Its fibers tend to splinter, and it is rather stiff. The safe working loads and breaking strengths of the rope that you are using should be known and the limits must be observed.

Maintenance tips. Precautions should be taken to keep ropes in good condition. Kinking, for example, strains the rope and may overstress the fibers. It may be difficult to detect a weak spot made by a kink. To prevent a new rope from kinking while it is being uncoiled, first lay the rope coil on the floor with the bottom end down. Then pull the bottom end up through the coil and unwind the rope counterclockwise. If it uncoils in the other direction, turn the coil of rope over and pull the end out on the other side.

Rope should not be stored unless it has first been cleaned. Dirty rope can be hung in loops over a bar or beam and then sprayed with water to remove the dirt. The spray should not be so powerful that it forces the dirt into the fibers. After washing, the ropes should be allowed to dry and then shaken to remove any remaining dirt. Rope must be thoroughly dried out after it becomes wet, otherwise it will deteriorate quickly. A wet rope should be hung up or laid in a loose coil until thoroughly dried. Rope will deteriorate more rapidly if it is alternately wet and dry than if it remains wet. Furthermore, rope should not be allowed to freeze.

Store rope in a dry place where air circulates freely. Small ropes can be hung up, and larger ropes can be laid on gratings so air can get underneath and around them. Rope should not be stored or used in an area where the atmosphere contains acid or acid fumes, because it will quickly deteriorate. Signs of deterioration are dark brown or black spots.

Sharp bends should be avoided whenever possible because they create extreme tension in the fibers. When cinching or tying a rope, be sure the object has a large enough diameter to prevent the rope from bending sharply. A pad can be placed around sharp corners.

If at all possible, ropes should not be dragged because dragging abrades the outer fibers. If ropes pick up dirt and sand, abrasion within the lay of the rope will rapidly wear them out. Lengths of rope should be joined by splicing. A properly made short splice will retain approximately 80 percent of the rope's strength. A knot will retain only about 50 percent. (See details in Chapter 9, under Lifelines.)

Ropes should be inspected at least every 30 days, more often if they are used to support scaffolding on which people work. The following procedure will assist in the inspection of natural fiber rope:

1. Check for broken fibers and abrasions on the outside.
2. Inspect fibers by untwisting the rope in several places. If the inner yarns are bright, clear, and unspotted, the strength of the rope has been preserved to a large degree.
3. Unwind a piece of yarn 8 in. (20 cm) long and ¼ in. (0.6 cm) diameter from the rope. Try to break this with your hands. If the yarn breaks with little effort, the rope is not safe.
4. Inspect rope used around acid or caustic materials daily. If black or rusty brown spots are noted, test the fibers as described in steps 1, 2, and 3, and discard all rope that fails the tests.
5. As a general rule, rope that has lost its pliability or stretch, or in which the fibers have lost their luster and appear dry and brittle, should be viewed with suspicion and replaced. This is particularly important if the rope is used on scaffolding or for hoisting where, if the rope breaks, workers may be seriously injured or property may be damaged.

Rope Slings

Because of the high tensile strength needed for a sling, fiber rope slings should be made of manila only. The following are some precautions to be observed in using manila-rope slings:

1. Make sure that the sling is in good condition and of sufficient strength. Take into account the factors of the splice and the leg angles of the sling. Hooks, rings, and other fittings must be properly spliced.
2. Reduce the load by one-half after the sling has been in use for six months, even though the sling shows no sign of wear.
3. If the sling shows evidence of cuts, excessive wear, or other damage, destroy it.

Whether fiber or wire rope slings are used, certain general precautions should be taken. (These were given earlier under Fiber Ropes—Maintenance tips.) In addition, it is best to keep the legs of slings as nearly vertical as possible (Figure 13-13).

Wire Ropes

Wire rope is used widely instead of fiber because it has greater strength and durability for the same diameter and weight. Its strength is constant either wet or dry and it maintains a constant length under varying weather conditions. Wire rope is made according to its intended use. A large number of wires give flexibility, while fewer strands with fewer wires are less flexible.

A wire rope used for general hoisting should not be subjected to a working load greater than one-eighth its breaking strength—a safety factor of five. Greater factors of safety (six or seven, for example) are always desirable for extra security. Ropes should be thoroughly inspected at least once a month; and written, dated, and signed reports kept on file. OSHA regulations spell out the details of this inspection. A tag attached to a sling can identify it and also indicate load capacity.

Wire rope should be lubricated at regular intervals to prevent rust and excessive wear. Sheaves for rope should be as large as possible. The less flexible the rope, the larger the sheave or drum diameter must be, otherwise the rope will be bent too sharply (Table 13-1). Rope should be wound in one layer only. Using several layers will mash or jam the rope and shorten its life.

Sheaves and drums should be aligned as much as possible to prevent excessive wear of rope. Reverse bending of rope, in which rope is bent first in one direction and then in the other, should be avoided, as this wears out wire rope faster than any other abuse. Fittings should be properly selected and attached if they are to be safe. Some of the principal wire rope attachments are illustrated in Figure 13-14, and include:

- Babbitt- or zinc-coated connections

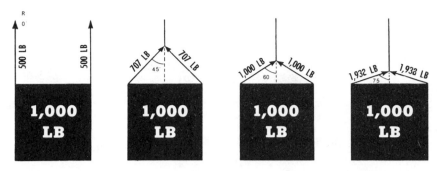

Figure 13-13. Increasing the sling angle (the angle of the sling leg with the vertical) increases the stress on each leg of the sling, even though the load remains constant. The stress on each leg can be calculated by dividing the load weight by the cosine of the angle.

TABLE 13-1. Tread Diameters for Sheaves and Drums

Rope Classification	Average Recommended	Minimum
6 × 7	72 times rope diameter	42 times rope diameter
6 × 19	45 times rope diameter	30 times rope diameter
6 × 37	27 times rope diameter	18 times rope diameter
8 × 19	31 times rope diameter	21 times rope diameter

Source: ANSI/ASME A17.1–1987, *Elevators and Escalators.*

- Wedge sockets
- Swagged attachments
- Thimble-with-clip connections
- Three-bolt clamps
- Spliced eye-and-thimble connections
- Crosby clips

The crosby clip is probably the most commonly used fitting. It is important that these clips be installed properly. Wedge sockets are dependable and will develop approximately 75 to 90 percent of the strength of the rope. Swagged attachments are satisfactory for small-diameter (¼ to 1 in. or 0.6 to 2.5 cm) ropes.

Wire rope should be spliced only by workers who have been trained to do it properly. Splices should be tested to twice the load they are expected to carry. Again, any questions should be referred to the supplier or manufacturer.

Chains and Chain Slings

Alloy steel chain, approximately twice as strong—size for size—as wrought-iron chain, has become standard material for chain slings. One advantage is that it is suitable for high-temperature operations. Continuous operation at temperatures of 800 F (425 C) (the highest temperature for which continuous operation is recommended) requires a reduction of 30 percent in the regular working load limit. For intermittent service, these chains can be used at temperatures as high as 1,000 F (540 C), but at only 50 percent of the regular working load limit (Table 13-2).

An alloy steel chain should never be annealed when the chain is being serviced or repaired. The process reduces the hardness of the steel; and this, in turn, reduces the strength of the chain. Wrought-iron chain, on the other hand, should be annealed periodically by the manufacturer (or persons specially trained in the operation) if it is used where failure might endanger human life or property.

Impact loads, caused by faulty hitches, bumpy crane tracks, and slipping hookups, can add significant stress in the chain. The impact resistance of heat-treated alloy steel chain does not increase in proportion to the strength of the chain. Under a full working load, it will fail before a fully loaded wrought-iron or heat-treated carbon steel chain will.

Chain slings preferably should be purchased complete from the manufacturer. Whenever they need repair, they should be returned to the manufacturer. Query suppliers and manufacturers about details of strength and specifications of the slings.

Load and wear. Chains should not be overloaded. The working load limit and break test limit for alloy steel chains are given in Table 13-2. Some of the principal causes of chain failure are:

1. Brittleness caused by cold working of the metal surface
2. Failure of the weld
3. Repeated and severe bending or deformation of the links
4. Metal fatigue

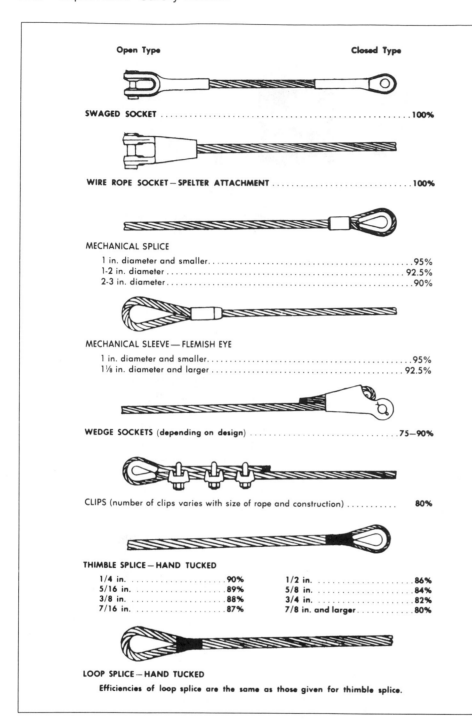

Open Type   Closed Type

SWAGED SOCKET 100%

WIRE ROPE SOCKET — SPELTER ATTACHMENT 100%

MECHANICAL SPLICE
     1 in. diameter and smaller......................................95%
     1-2 in. diameter 92.5%
     2-3 in. diameter 90%

MECHANICAL SLEEVE — FLEMISH EYE
     1 in. diameter and smaller......................................95%
     1⅛ in. diameter and larger 92.5%

WEDGE SOCKETS (depending on design) 75–90%

CLIPS (number of clips varies with size of rope and construction) 80%

THIMBLE SPLICE — HAND TUCKED

1/4 in.	90%	1/2 in.	86%
5/16 in.	89%	5/8 in.	84%
3/8 in.	88%	3/4 in.	82%
7/16 in.	87%	7/8 in. and larger	80%

LOOP SPLICE — HAND TUCKED
Efficiencies of loop splice are the same as those given for thimble splice.

Figure 13-14. Typical efficiencies of attaching wire rope to fittings in percentages of strength of rope. (Adapted from Wire Rope Users' Manual.)

5. Brittleness caused by defects in the metal
6. Tensile failure—full elongation of the links

How much wear a chain can stand is determined by its use, the original factor of safety, and the loads the chain is expected to carry. A regular schedule of checking and calipering the links should be set up for each chain, depending on usage (Table 13-3).

To check elongation, chain should be calipered in gage lengths of 1 to 3 (preferably 5) ft when new,

and a record of that test should be kept. The chain (or sling) should be discarded if it shows a stretch (includes link wear) of more than 5 percent. Some states specify that a chain must be destroyed or discarded when any 3-ft (0.9-m) length is found to have been stretched one-third of a linklength.

A broken chain should never be spliced by inserting a bolt between two links. Nor should one link be passed through another and a nail or bolt inserted to hold them.

Use. Before making a lift, workers should check

TABLE 13-2. Working Load Limits and Break Test Limits for Alloy Steel Chain

Nominal Size of Chain Bar in Inches	Working Load Limits In Pounds	Minimum Break Test in Pounds
¼	3,250	10,000
⅜	6,600	19,000
½	11,250	32,500
⅝	16,500	50,000
¾	23,000	69,500
⅞	28,750	93,500
1	38,750	122,000
1⅛	44,500	143,000
1¼	57,500	180,000
1⅜	67,000	207,000
1½	80,000	244,000
1¾	100,000	325,000

1 in. = 2.54 cm; 1 kg = 2.2 lb.

*Source: *Specification for Alloy Steel Chain*, American Society for Testing and Materials, A391–1975. The strengths for wrought-iron chain are slightly less than half these figures.

the chain to be sure it has no kinks, knots, or twists. Workers should lift the load gradually and uniformly and make certain that it is not merely attached to the top of the hook. They should never hammer a link over the hook as this will stretch the hook and the links of the chain. When lowering a crane load, brakes should be applied gradually to bring the load to a smooth stop.

Each chain should be tagged to indicate its load capacity, date of last inspection, date of purchase, and type of material. This information must not be stamped on the lifting links, however, as stamping can cause stress points that weaken the chain (Figure 13-15).

Chain attachments (rings, shackles, couplings, and end links) should be made of the same material to which they are fastened. Hooks should be made of forged or laminated steel, and should be equipped with safety latches. Replace hooks that have been overloaded, loaded on the tips, or have a permanent set greater than 15 percent of the normal throat opening.

Storing chains. Proper chain storage serves two purposes. It preserves the chain and promotes good housekeeping. Chains are best stored on racks inside dry buildings that have a fairly constant temperature. Chains should be stored where they will not be run over by trucks or other mobile equipment, nor exposed to corrosive chemicals.

Storage racks and bins should be constructed in such a way that air circulates freely around the chain and keeps it dry. Use power equipment to lift large and heavy chains. Attempting to lift a chain manually can cause an injury.

TABLE 13-3. Correction Table for the Reduction of Working Load Limits of Chain Due to Wear

Reduce working load limits of chain by the following percent when diameter of stock at worn section is as follows:

Nominal Original Chain Size, Inches		5 Percent	10 Percent	Remove from Service
¼	0.250	0.244	0.237	0.233
⅜	0.375	0.366	0.356	0.335
½	0.500	0.487	0.474	0.448
⅝	0.625	0.609	0.593	0.559
¾	0.750	0.731	0.711	0.671
⅞	0.875	0.853	0.830	0.783
1	1.000	0.975	0.949	0.895
1⅛	1.125	1.100	1.070	1.010
1¼	1.250	1.220	1.190	1.120
1⅜	1.375	1.340	1.310	1.230
1½	1.500	1.460	1.430	1.340
1⅝	1.625	1.590	1.540	1.450
1¾	1.750	1.710	1.660	1.570
1⅞	1.875	1.830	1.780	1.680
2	2.000	1.950	1.900	1.790

1 in. = 2.54 cm

Source: ANSI/ASME B30.17–1985, *Overhead and Gantry Cranes* (Top Running Bridge, Single Girder, Underhung Hoist).

The supplier or manufacturer can arrange a time to give you on-location instructions on the care, use, inspection, and storage of chains. By instructing some of your maintenance workers on chains, you will be in a better position to evaluate your own needs.

MATERIALS STORAGE

Both temporary and permanent storage of all materials must be neat and orderly. Materials piled haphazardly or strewn about increase the possibility of accidents. The warehouse supervisor must direct the storage of raw materials (and sometimes processed stock) kept in quantity lots for a length of time. The production supervisor is usually responsible for storing of limited amounts of materials and stock, for short periods, close to the processing operations. Planned materials storage minimizes (a) handling required to bring materials into production, and (b) removing finished products from production to shipping.

When supervisors plan materials storage, they should make sure that materials do not obstruct fire-alarm boxes, sprinkler system controls, fire extinguishers, first-aid equipment, lights, electric

Figure 13-15. In a typical double-chain sling, all components, such as the oblong master link, the body chain, and the hook, are carefully matched for compatibility. Note the permanent identification tag. (Courtesy Columbus McKinnon Corporation.)

switches, and fuse boxes. Exits and aisles must be kept clear at all times.

There should be at least 18 in. (0.5 m) clearance below sprinkler heads to reduce interference with water distribution. This clearance should be increased to 36 in. (0.9 m), if the material being stored is flammable.

Aisles designed for one-way traffic should be at least 3 ft wider than the widest vehicle, when loaded. If materials are to be handled from the aisles, the turning radius of the power truck also needs to be considered. Employees should be told to keep materials out of the aisles and out of loading and unloading areas, which should be marked with white lines.

Storage is facilitated and hazards are reduced by using bins and racks. Material stored on racks, pallets, or skids can be moved easily and quickly

from one workstation to another. When possible, crosstie material piled on skids or pallets.

In an area where the same type of material is stored continuously, it is a good idea to paint a horizontal line on the wall to indicate the maximum height to which material may be piled. This will help to keep the floor load within the proper limits and the sprinkler heads in the clear.

Containers and Other Packing Materials

Because packing containers vary considerably in size and shape, height limitations for stacking these materials vary. The weight of the container itself must also be figured when keeping within the proper floor-loading limit.

Sheets of heavy wrapping paper placed between layers of cartons will help to prevent the pile from shifting. Better still, crosstying will prevent shifting and sagging and permit higher stacking. Wirebound containers should be placed so that sharp ends do not stick into the passageway. When stacking corrugated paper cartons on top of each other, remember that increased humidity may cause them to slump.

Bagged materials can be crosstied with the mouths of the bags toward the inside of the pile, so contents will not spill if the closure breaks. Piles over 5 ft (1.5 m) high should be stepped back one row. Step back an additional row for each additional 3 ft (0.9 m) of height. Sacks should be removed from the top of the pile, not halfway up or from a corner. Remove sacks from an entire layer before the next layer is started. This keeps the pile from collapsing and possibly injuring people and damaging material.

Small-diameter bar stock and pipe are usually stored in special racks, located so that when stock is removed, passersby are not endangered. The front of the racks should not face the main aisle, and material should not protrude into the aisles. When the storage area is set up, the proper floor load should be considered.

Large-diameter pipe and bar stock should be piled in layers separated by strips of wood or by iron bars. If wood is used, it should have blocks at both ends; if iron bars are used, the ends should be turned up. Again, the floor load must be considered.

Lumber and pipe have a tendency to roll or slide. When making a pyramid pile, operators should (a) set the pieces down rather than dropping them, and (b) not use either hands or feet to stop rolling or sliding material; this can result in serious injury.

Sheet metal usually has sharp edges and should be handled with care, using leathers, leather gloves, or gloves with metal inserts. Large quantities of sheet metal should be handled in bundles by power equipment. If sheet-metal piles are separated by strips of wood, they will be easier to handle when the material is needed for production and will be less likely to shift or slide.

Tinplate strip stock is heavy and razor sharp. Should a load or partial load fall, it could badly injure anyone in its way. Two measures can be taken to prevent spillage and injuries: (a) band the stock after shearing and (b) use wooden or metal stakes around the stock tables and pallets that hold the loads. It is the responsibility of the supervisor and all who handle the bundles to band loads properly and make sure the stakes are in place when the load is on the table.

Packing materials, such as straw, excelsior, or shredded paper, should be kept in a fire-resistant room equipped with sprinklers and dustproof electric equipment. Materials received in bales should be kept baled until used. Remove only enough material for immediate use or for one day's supply. Packing materials should be taken to the packing room and placed in metal (or metal-lined wood) bins. Bin covers should have fusible links that close automatically in case of fire. To prevent injury in case a counterweight rope should break, the weight should be boxed in.

Steel and plastic strapping, which may be flat or round, require that the worker be trained in both application and removal (Figure 13-16). In all cases, the worker should wear safety goggles or glasses with

Figure 13-16. The worker should always stand to one side when tensioning strapping to be out of the direct line of the strapping if it slips or breaks and recoils. (Courtesy Signode.)

side shields, and leather-palm gloves. Heavy strap may require the wearing of steel-studded gloves.

Barrels or kegs stored on their sides should be placed in pyramids, with the bottom rows securely blocked to prevent rolling. If piled on end, the layers should be separated by planks. If pallets are used, they should be large enough to prevent overhang.

Hazardous Liquids

Acid carboys. These are best handled with special equipment, for example, carboy trucks. Boxed carboys should generally be stacked no higher than two tiers and never higher than three. Not more than two tiers should be used for carboys of strong oxidizing agents, such as concentrated nitric acid or concentrated hydrogen peroxide. The best method of storage is in specially designed racks.

Before handling carboy boxes, inspect them thoroughly to make sure that nails have not rusted or that the wood has not been weakened by acid. Empty carboys should be completely drained and their caps replaced.

The safest way to draw off liquid from a carboy is to use suction from a vacuum pump or a siphon started by either a rubber bulb or an ejector. Another method is to use a carboy inclinator that holds the carboy at the top, sides, and bottom, and returns the carboy to an upright position automatically when released. Pouring by hand or starting pipettes or siphons by mouth suction to draw off the contents of a carboy should never be permitted.

Portable containers. When liquids are handled in portable containers, the contents should be clearly labeled. Substances should be stored separately in safety storage cabinets that meet pertinent OSHA and NFPA 30 requirements (Figure 13-17). Purchasing agents should specify to suppliers that all chemicals must be properly labeled showing contents, hazards, and precautions. Labels are discussed in Chapter 8, Industrial Hygiene.

Drums. Storage drums containing hazardous liquids should be placed on a rack, not stacked. It is best to have a separate rack for each type of material. Racks permit good housekeeping, easy access to the drums for inspection, and easy handling. Racks allow drums or barrels to be emptied through the bung. Self-closing spigots should be used, particularly where individuals are allowed to draw off their own supplies. If there is a chance that spigots might be hit by material or equipment, drums should be stored on end and a pump used to withdraw the contents. Drums containing flammable liquid, and the racks that hold them, should be

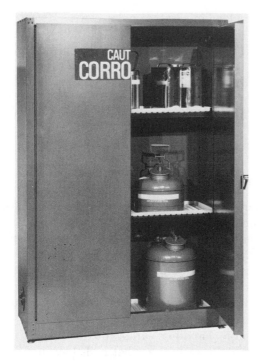

Figure 13-17. If the safety storage cabinets meet the requirements of NFPA 30, *Flammable and Combustible Liquids Code*, they are suitable for storing containers of up to five-gallon capacity of flammable liquids. This cabinet is also safe for acids and corrosives. (Courtesy Eagle Manufacturing Company.)

grounded (Figure 13-18). In addition, the smaller container into which the drum is being emptied should be bonded to the chime of the drum by a flexible wire with a C-clamp at each end.

Storage areas for liquid chemicals should be naturally well ventilated. Natural ventilation is better than mechanical because the mechanical system

may fail. The floor should be made either of concrete or of some other material treated to reduce absorption of the liquids. The floor should be pitched toward one or more corrosion-resistant drains that can be cleaned easily.

Where hazardous chemicals are stored, handled, or used, emergency flood showers or eyewash fountains should be available. Operators should wear chemical goggles, rubber aprons, boots, gloves, and other protective equipment necessary to handle the particular liquid (see Chapter 9, Personal Protective Equipment).

Each type of hazardous material requires either special handling techniques or special protective clothing. In all cases, the supervisor or the purchasing agent should consult the safety professional or the safety department for handling precautions.

Additional measures. . When handling hazardous liquids, take these extra precautions:

1. Wear proper protective equipment.
2. Keep floors clean; do not allow them to become slippery.
3. Drain all siphons, ejectors, and other emptying devices completely before removing them from carboys.
4. To dilute acid, always add it to water, never add water to acid. Add the acid slowly and stir constantly with a glass implement.
5. In case of an accident, give first aid: flush burned areas with lots of water. Call a doctor immediately.
6. Do not force hazardous chemicals from containers by injecting compressed air. The container may burst or the contents may ignite.

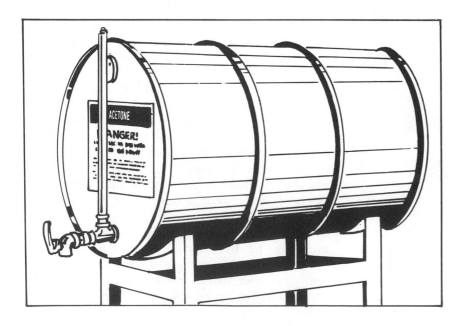

Figure 13-18. A storage rack reduces harzards, aids identification of the drum contents, and permits easy grounding.

7. Do not try to wash or clean a container unless its label specifically requires that it be cleaned before it is returned.

8. Do not store hazardous chemicals in glass or other containers near heat or steam pipes, or where strong sunlight will strike them. The contents may expand and may cause a fire or explosion. Bottles may focus sunlight to ignite a fire in nearby combustibles.

9. Do not stir acids with metal implements.

10. Do not store chemicals or solvents in dark or poorly lit areas because the wrong chemical may be selected.

11. Do not pour corrosives from a carboy by hand.

12. Do not move a carboy unless it is securely capped.

Pipelines to outside storage tanks are safer if large quantities of liquid chemicals or solvents are used, because they reduce and localize any spillage. When hazardous liquids are used in small quantities, only enough for one shift should be kept on the job. In many cases, local ordinances limit the amount of hazardous materials that can be stored or processed inside a building. The main supply should be stored in tanks located in an isolated place.

Pipelines. These should be color-coded and labeled to identify their contents. Make sure workers understand the company's use of identifying colors and labels. Should outside contractors work in the area, they should be adequately informed of piping system contents and any hazards.

When a valve is to be worked on, it should be closed. The section of pipe in which it is located should be thoroughly drained before the flanges are loosened. Appropriate personal protective equipment must be worn by workers.

As a temporary measure, the opening between faces of a flange may be covered with a piece of sheet lead, and this wrapped around to make a sleeve while the flange bolts are being loosened and the faces separated. If the flanges do not separate readily, a strong nail can be driven through the lead shield into the flange joint to part it.

Workers should loosen the flange bolts farthest away first, so that the pipe will drain away from them. A blind (or blank) should be inserted between the flanges as soon as they are parted. If a line is opened often, a two-ended metal blank can be permanently installed on a flange bolt so that the blank can be pivoted on the bolt when the flanges are parted. One end of the blank is shaped in the form of a gasket, to permit flow of liquid. The other end is blind and is swung to cover the pipe opening when the flanges are unbolted and parted.

Tank cars. These storage cars should be protected on sidings by derails and by blue stop flags or blue lights before they are loaded or unloaded. Hand brakes should be set and wheels chocked. Before the car is opened, it should be bonded to the loading line. The track and the loading or unloading rack should be grounded, and all connections checked regularly. Chemical tank cars should be unloaded through the dome rather than through the bottom connection.

Gas Cylinders

There are many regulations regarding the storage and handling of gas cylinders. Several government agencies and private organizations and their standards should be consulted—OSHA, National Fire Protection Association, Compressed Gas Association, and others. Care should be taken to comply with all local government codes as well.

Compressed gas cylinders should be stored on end on a smooth floor. All cylinders should be chained or otherwise fastened firmly against a wall, post, or other solid object. Different kinds of gases should either be separated by aisles or stored in separate sections of the building or storage yard. Empty cylinders must be stored apart from full cylinders.

Set up storage areas away from heavy traffic. Containers should be stored in a place where there is minimal exposure to excessive temperature, physical damage, or tampering. Never store cylinders of flammable gases near flammable or combustible substances. Before cylinders are moved, check to make certain all valves are closed. Always close the valves on empty cylinders. Never let anyone use a hammer or wrench to open valves. As stated before, cylinders should never be used as rollers to move heavy equipment. Handle all cylinders with care—a cylinder marked "empty" may not be.

To transport cylinders, use a carrier that does not allow excessive movement, sudden or violent contacts, and upsets (Figure 13-19). When a two-wheeled truck with rounded back is used, chain the cylinder upright. Never use a magnet to lift a cylinder. For short-distance moving, a cylinder may be rolled on its bottom edge, but never dragged. Cylinders should never be dropped or permitted to strike one another. Protective caps must be kept on cylinder valves when cylinders are not being used. When in doubt about how to handle a compressed gas cylinder, or how to control a particular type of gas once it is released, ask the safety professional or the safety department for advice.

Figure 13-19. Using the proper carrier, one person can handle this liquefied gas container.

Combustible Solids

Bulk storage of grains and granular or powdered chemicals or other materials presents fire and explosion hazards. Many materials not considered hazardous in solid form often become combustible when finely divided. Some of these are carbon, fertilizers, food products and by-products, metal powders, resins, waxes and soaps, spices, drugs and insecticides, wood, paper, tanning materials, chemical products, hard rubber, sulfur, starch, and tobacco. This list is by no means complete, mentioning only the general categories of materials that may generate explosive dusts.

Dust. To avoid dust explosions, prevent formation of an explosive mixture or eliminate all sources of ignition. Either the dust must be kept down, or enough air supplied to keep the mixture below the combustible limit. Good housekeeping and dust-collection equipment will go far toward preventing disaster. Make sure dust does not accumulate on floors and other structural parts of a building. A dust explosion is a series or progression of explosions. The initial one is small, but it shakes up dust that has collected on rafters and equipment so that a second explosion is created. This chain reaction usually continues until the result is the same as if a single large explosion had taken place.

Wearing protective equipment is important when dusts are toxic. Such equipment may range from a respirator to a complete set of protective clothing. The Council, or a state or local safety agency, can recommend the proper type of protective clothing for specific jobs. See the recommendations in Chapter 9, Personal Protective Equipment.

Bins. Bins in which solids are stored should have a sloping bottom to allow the material to run out freely and to prevent arching. For some materials, especially if they are in process, the bin should have a vibrator or agitator in the bottom to keep the material flowing.

If large rectangular bins, open at one side, are entered with power shovels or other power equipment, workers should take care not to undercut the material and thereby endanger themselves if arched material should suddenly give way. Where practical, tops of bins should be covered with a 2-in. (5-cm) mesh screen, or at least by a 6-in. (15-cm) grating (or parallel bars on 6-in. centers) to keep people from falling into the bins.

Tests for oxygen content and for presence of toxic materials should be made before anyone enters a bin or tank for any purpose. If such tests indicate the need for protection, people should use air-supplied masks, in addition to a safety belt and lifeline, before entering. Filling equipment should be made inoperative so that it cannot be started again, except by the worker after leaving the bin or by the immediate supervisor after checking the worker out of the bin. When a person is working in a bin, he or she should have a companion employee, equipped to act in case of emergency, stationed on top of the bin. Bins should be entered from the top only.

If flammable vapors, dusts, or other flammable air contaminants are found, workers should use only spark-resistant equipment and electrical equipment approved by a recognized testing laboratory.

SUMMARY OF KEY POINTS

This chapter covered the following key points:

- Manual handling of materials accounts for an estimated 25 percent of all occupational injuries. These injuries are caused by improper lifting, carrying too heavy a load, incorrect gripping, failing to observe proper foot or hand clearances, failing to use or wear proper safety equipment, and poor job design. Supervisors should look at their operations to see if manual handling can be eliminated, how materials injure workers, whether handling aids can be used, and whether protective clothing is needed.
- Because the largest number of injuries occur to fingers and hands, workers must be taught how to pick up and put down heavy, bulky, or long objects. Protective gloves, eyewear, and safety shoes should be worn to prevent injuries. If manual lifting cannot be eliminated entirely, then workers must be taught correct lifting practices. Researchers have found that no single lifting method is best for everyone; instead, workers should choose the method that is most comfortable for them. Workers must be carefully taught how to handle specific shapes—box, carton, crate, drum, or barrel—to minimize the risk of injury to themselves and damage to the container. Special equipment or tools can make the job easier and safer.
- Materials handling equipment include hand tools, nonpowered and powered hand trucks, powered industrial trucks, dock plates, conveyors, cranes, railroad cars, highway trucks, and motorized equipment. Hand tools, such as chisels, crowbars, rollers, hooks, and shovels, should be in good condition and used safely to prevent injuries.
- Nonpowered hand trucks include two-wheel and four-wheel trucks. Four major hazards associated with these trucks are (a) running wheels off bridge plates or platforms, (b) colliding with other trucks or obstructions, (c) jamming hands between the truck and other objects, and (d) running wheels over feet. Only qualified personnel should be allowed to handle these trucks, and they should be used only for the specific purpose they were designed for. Loads should be carefully placed so they do not shift, slide or obstruct the handler's view of the path ahead.
- Major hazards associated with powered hand trucks are (a) catching the operator between the truck and another object, and (b) collisions with objects or people. The truck should be equipped with a dead-man control, wheel guards, and an ignition key that can be removed when the operator leaves the truck. Only qualified workers should be allowed to use these vehicles.
- Powered industrial trucks are either battery operated or run by internal combustion engines. These trucks must meet stringent standards and be operated only by qualified drivers. Battery-charging installations must be in special areas, with adequate precautions for handling spilled electrolyte and dealing with fire hazards. Refueling areas must be adequately ventilated and secured against fire. Loads must be carefully stacked and secured on these trucks to avoid spilling the load or overturning the truck. All employees must be carefully trained in the use and maintenance of this equipment.
- Dock plates not built in must be fastened down to prevent them from moving or sliding when used. Plates should be large and strong enough to hold the heaviest load—a safety factor of five or six should be adequate. They should be inspected frequently and kept in good condition.
- The primary rule in using conveyors is simple: never allow anyone to ride on them. Workers should know the specific safety rules for working on or around gravity, pneumatic or blower, or screw conveyors. Supervisors should train employees in how to maintain guards and other safety equipment on conveyors, free conveyor jams, inspect conveyors for hazards, and cross safely from one side of a conveyor to the other.
- Although many types of cranes are used in industry, safe operating procedures are much the same. Supervisors should assign only one signalman for each crane and tell crane operators to obey only this person's instruc-

tions. Loads must be properly balanced and carefully lifted before being moved from one location to another. Cranes and their chains and ropes must be inspected regularly for signs of wear, corrosion, or deterioration. Supervisors should pay particular attention to all control mechanisms, all safety devices, crane hooks, ropes, and electrical apparatus.

- Moving any railroad car should be done after considering the facilities or equipment available to move the car, number of cars and distance they are to be moved, and whether the track is level or on a grade. The supervisor may need power equipment or may have enough workers to move the car by hand. Workers should be alert to the dangers involved in opening and closing doors, transferring material in and out of the cars, and securing car doors.

- When loading or unloading highway trucks, workers should set the brakes and place wheel chocks under the rear wheels or use a vehicle restraint. They should use a fixed jack to support a semitrailer not attached to a tractor.

- All motorized equipment requires special handling to prevent accidents. Workers must be taught to operate this equipment according to detailed instructions in the manufacturers' manuals. Equipment must be thoroughly inspected and given safety clearance before it is put into service. A preventive maintenance program is necessary to ensure worker safety and efficiency.

- Ropes, chains, and slings are common materials handling equipment that require careful inspection and maintenance to remain in good working order. Supervisors should become familiar with the properties of various types of ropes, chains, and slings and the precautions for their use and maintenance. This equipment should never be used to lift a load if it has a kink, knot, or twist in the rope or chain. Fiber rope, generally made from manila, sisal, or nylon, must be kept clean and dry and stored in a dry, warm place. If the rope becomes frayed or worn, it should be replaced immediately. Wire ropes have greater durability and strength, and are used to handle heavier loads than are fiber ropes. They must be carefully lubricated to prevent rust and excessive wear.

- Chains and chain slings must be inspected frequently for wear, failure of a weld, metal fatigue, deformation, or stress in the links. They should not be overloaded or forced onto a hook as this weakens the links. Each chain should be tagged to indicate its load capacity, and chains should be stored in dry buildings with fairly constant temperatures.

- Both temporary and permanent storage of all materials must be neat and orderly. Supervisors should make sure that materials do not obstruct fire-alarm boxes, sprinkler system controls, fire extinguishers, first-aid equipment, lights, electric switches, fuse boxes, and exits and aisles. The use of bins and racks, pallets, or skids can reduce hazards associated with materials storage.

- Materials storage includes the careful stacking and placement of containers and other packing materials, hazardous liquids, gas cylinders, and combustible solids. Supervisors must be familiar with the hazards and precautions associated with each of these materials. Containers and other packing materials can be stacked safely by taking into consideration height limitations and weight to keep within safe floor-loading limits. Materials should be stacked to prevent loads from sliding or shifting and kept in fire-resistant rooms.

- Hazardous liquids are best handled with special equipment. Acid carboys, portable containers, and drums must be moved only by trained personnel who are equipped with the proper tools and protective gear. Each type of hazardous material requires special clothing or handling techniques.

- Pipelines, tank cars, and gas cylinders also require special handling when used for storage. Pipelines should be color-coded and labeled to identify their contents. Tank cars should be protected on sidings by derails and by blue stop flags or blue lights before they are loaded or unloaded. Gas cylinders must be carefully chained or fixed to a stationary object. Empty cylinders should be stored apart from full cylinders and kept away from flammable or spark-generating materials. Workers must be trained thoroughly in safe work practices when dealing with these materials.

- Bulk storage of grains and granular or powdered chemicals or materials presents fire and explosion hazards. To avoid dust explosions, dust must be kept down or enough air supplied to keep the mixture below the combustible limit. Good housekeeping and dust-collection equipment will go far toward preventing disaster. Workers should wear protective equipment when working near toxic dust.

- If bins are used for storage, they should have a sloping bottom and, in some cases, agitators or vibrators to keep material moving. Supervisors should always check for oxygen content and for presence of toxic materials before allowing workers to enter a bin or tank for any reason. Workers should wear protective clothing and have a companion worker stationed outside in case of emergency.

14

Electrical Safety

After reading this chapter, you will be able to:

- Explain the action of electricity on the human body and how to use grounding to prevent shocks and shock hazards
- Describe how to use plug- and cord-connected equipment and extension cords safely
- Understand how to test branch circuits and electrical equipment
- Explain the uses and limitations of ground-fault circuit interrupters
- Understand the classes and divisions of hazardous locations and the proper equipment to use for each classified location
- Instruct workers in the proper safety standards and procedures to handle electrical equipment safely on or off the job

Electrical safety is one of a supervisor's most important responsibilities in ensuring safe production. Although everyone knows high voltage is dangerous, low alternating current levels can also cause fatalities by affecting the heart. An estimated 31 percent of the known electrocution fatalities occur in the home, 24 percent in general industry, and the remaining in the generation and distribution of electrical power. This means that approximately 55 percent of electrocutions are due to contact with so-called low voltage circuits, that is, 600 volts and under.

The misuse of electricity can also claim lives from fire-related accidents. The U.S. Product Safety Commission estimates that there are 169,000 house fires of electrical origin each year, claiming 1,100 lives and injuring 5,600. Property losses are estimated at $1.1 billion a year.

MYTHS AND MISCONCEPTIONS ABOUT ELECTRICITY

Misunderstandings workers have about electricity can lead to serious accidents and property damage. Make sure your workers know the facts. Some of the more common myths and misconceptions are as follows:

1. Electricity takes the path of least resistance. This myth implies that current only takes low-resistance paths. Actually, current will take any path, high or low resistance, in order to return to the source that provides it power.
2. Electricity wants to go to ground. A person is led to believe that electricity wants to go to ground and simply disappear. In reality, ground serves as just one of the electrical loops that misdirected current can use to get back to the grounded power source.
3. If an electrical appliance or tool falls into a sink or tub of water, the item will short out and trip the circuit breaker. When an electrical tool or appliance falls into water, it does not short out. If the switch is ON, the item will continue to operate. If the switch is OFF, it will simply get wet.

   However, in both cases all persons should be warned not to reach into the water to retrieve the appliance or tool. The water serves as an electrical conducting path for the electricity to flow out of the item. Any person reaching into the water with one hand when some other part of the body is touching a grounded object could receive a severe or even fatal shock.

4. AC reverse polarity is not hazardous. The *National Electrical Code* (*NEC*) requires that power tools, attachment plugs, and receptacles be properly wired so as not to reverse the designated polarity. Many tools have switches in only one of the two conductors supplying power to the item. The switch is supposed to be on the "hot" conductor supplying the power. If the tool (e.g., a drill) is plugged into a receptacle with reverse polarity and accidentally dropped into water, all of the internal wiring will be energized.

The increased energized surface area in the water will allow more fault current to flow and increase the hazard of serious or fatal shock. If the tool and the receptacle are properly polarized, this would minimize the shock hazard. Only the terminal on the switch would be energized, not the entire tool, as in the case of reverse polarity.

5. It takes high voltage to kill; 120 volts AC is not dangerous. Current is the culprit that kills. Voltage just determines by what means. Under the right conditions, AC voltage as low as 60 volts can kill. At higher voltages, the body can be severely burned and yet the victim could live. Respect all AC voltages as having the potential to kill.

6. Double-insulated power tools are doubly safe and can be used in wet and damp locations. Read the manufacturer's operating instructions carefully. Double-insulated power tools can be hazardous if dropped into water. Electrical current can flow out of the power tool into the water.

Every piece of equipment is a potential source of electrical shock. Even an electrical shock too small to cause injury can trigger an involuntary reaction that results in physical injury; for example, after touching a live wire, a person may lose his or her balance and fall off a ladder. Obviously, design, layout, installation, and a preventive maintenance program can minimize live or "hot" electrical equipment. An effective electrical safety policy coupled with an employee training and hazard awareness program can further prevent electrical shock. It is also recommended that an effective electrical inspection program be implemented and conducted periodically as conditions warrant.

This chapter provides the basic information that will enhance your ability to recognize, understand, and control electrical hazards. This will help you to eliminate and/or prevent electrical shock injuries. Make sure that your employees do not believe in, or cling to, any of the myths and misconceptions about electricity. This chapter will further clarify these issues so that you may train others in the correct truth about electrical action. The major topics covered in this chapter are as follows: electrical fundamentals review, branch circuit and grounding concepts, plug- and cord-connected equipment and extension cords, branch circuit and equipment testing methods, ground-fault circuit interrupters, hazardous locations, common electrical deficiencies, and electrical inspection guidelines.

ELECTRICAL FUNDAMENTALS REVIEW

Ohm's Law

Before discussing electrical hazards, a brief discussion of Ohm's Law is necessary. Important terms are voltage (v), current, amperes (amp), milliamperes (m), watts (w), ohms, resistance (R), impedance, and power.

Ohm's Law simply states that one volt will cause a current of one ampere to flow through a resistance of one ohm. As a formula, the relationship is represented by E (volts) = I (amps) times R (resistance). An easy way to remember this formula, and the different ways it can be expressed, is to put the symbols in a circle (Figure 14-1).

1. Put your finger on I-I = E/R.
2. Put your finger on R-R = E/I.
3. Put your finger on E-E = I × R.

With this basic formula, you can better understand and explain the effects that electrical current has on the human body.

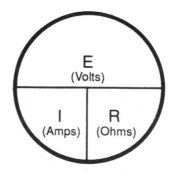

Figure 14-1. Ohm's Law.

The above applies to direct current (DC); however, in alternating current (AC) usage, Ohm's Law is E = I × Z, where Z is called the impedance. Impedance is the sum of the resistance and the capacitive and inductive reactance, expressed this way:

$$Z = \sqrt{R^2 + (X_L - X_C)^2}$$

Capacitance (C) is the ability to hold an electrical charge. Inductance is the coefficient of self-introduction of a circuit. Since the body is essentially resistive, the explanations and examples that follow will treat the body as a resistive factor only.

To demonstrate an understanding of Ohm's Law, try to solve this problem. Assume a person is working and perspiring and has a hand-to-hand resistance of 1,000 ohms. This person contacts 100 volts with one hand and touches a ground surface with the other, completing a loop to the voltage source. What would be the current going through the body?

$$I = E/R = 100/1{,}000 = 0.1 \text{ amp (100 ma)}$$

Another important formula is the relationship of voltage, current, and power utilization (watts). Power is measured in watts and is equal to E (volts) x I (current). In other words, one watt would be equal to one ampere of current flowing through a resistor with one volt of potential difference. Another way to put this relationship to practical use would be to consider a 6-watt electric bulb (about the size of a night light or panel-board light). Determine the current flowing through the filament of a 6-watt bulb that is being used in a 120-volt light socket.

$$I = P/E = 6/120 = 0.05 \text{ amp} = 50 \text{ ma}$$

This current is enough to affect the human heart and induce a fatal irregular heart beat, which can result in a heart attack. These examples of current levels from 50 ma to 20,000 ma are used to illustrate and clarify misconceptions about electricity and its effect on the human body.

Body Resistance Model

The human body impedance, or resistance to electric current, has three distinct parts: internal body resistance and the two skin impedances associated with contact with surfaces of different voltage potential. The internal body impedance is reported to be essentially resistive with no reactive components. If the voltage applied is high enough, the skin offers virtually no impedance to the electric current.

A body resistance model (Figure 14-2) would indicate approximately 1,000 ohms from hand to hand, or about 1,100 ohms from hand to foot. It is

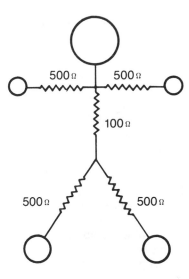

Figure 14-2. Human body resistance model.

believed that the greatest number of injurious shocks involve a current pathway that is either hand to hand or hand to feet. Figure 14-2 illustrates that wet contact with an energy source (such as 120 volts AC) in one hand and a ground loop in the other produces current through the chest area equal to: I = E/R = 120/1,000 = 0.12 amp or 120 ma. This current level is definitely hazardous.

It is easy to analyze electrical shock hazards today. A small amount of current (50 ma) can cause the heart to start beating rapidly and irregularly—a potentially fatal condition known as ventricular fibrillation. The example given previously, that a 6-watt bulb used 50 ma of current when energized by 120 volts, simply means that merely the current required to light a 6-watt bulb is capable of causing death.

For instance, once ventricular fibrillation occurs, death will ensue in a few minutes. Resuscitation techniques, if applied immediately, can save the victim. Thus, contrary to many other beliefs, low AC current is dangerous and it doesn't take high voltage to kill, as shown in Table 14-1.

Electrical Current Danger Levels

Since experiments on ventricular fibrillation could obviously not be performed on humans, animals have been substituted. The first work of this type was that of Ferris, Spence, King, and Williams of the Bell Telephone Laboratories in 1936. Since that time, researchers at John Hopkins University and the USSR Academy of Sciences have expanded the knowledge of the effects of electrical currents on humans. As a result, the danger threshold for ventricular fibrillation was established for typical adult workers.

TABLE 14-1. Current and Its Effect on the Human Body

	Current in Milliamperes					
	Direct		Alternating			
			60 Hz		10,000 Hz	
Effect	Men	Women	Men	Women	Men	Women
Slight sensation on hand	1	0.6	0.4	0.3	7	5
Perception threshold	5.2	3.5	1.1	0.7	12	8
Shock—not painful, muscular control not lost	9	6	1.8	1.2	17	11
Shock—painful, muscular control not lost	69	41	9	6	55	37
Shock—painful, let-go threshold	76	51	16	10.5	75	50
Shock—painful and severe, muscular contractions, breathing difficult	90	60	23	15	94	63
Shock—possible ventricular fibrillation effect from 3-second shocks	500	500	100	100		
Short shocks lasting t seconds			$165/\sqrt{t}$	$165/\sqrt{t}$		
High voltage surges	50*	50*	13.6*	13.6*		

*Energy in watt-seconds or joules

Note: Data are based on limited experimental tests, and are not intended to indicate absolute values.

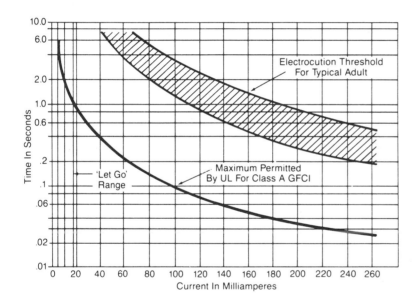

Figure 14-3. This chart shows the effects of electric current intensity versus the time it flows through the human body.

Figure 14-3 plots the range of electrical values that affect humans. Also shown is the requirement of Underwriters Laboratories for a Class A ground-fault circuit interrupter (GFCI) intended for protection. Note the time-current relationship at much lower currents and shorter time periods. The electrical current range that causes muscles to freeze (called the "let go" range) is also illustrated.

From Figure 14-3, one can determine that a 100 ma current flowing for 2 seconds through a human adult body will cause death by electrocution. This doesn't seem like much current when you consider that a small, light-duty portable electric drill draws 30 times that much. But, because a current as small as 3 ma (3/1,000 amp) can cause painful shock, it is imperative that all electrical plugs and cords and extension cords be kept in good condition and connected to a properly wired grounded circuit.

Factors Enhancing Electrical Shock

The route that electrical currents take through the human body determines the degree of injury, and voltage determines how much current flows. In cases where individuals come in contact with distribution lines, high voltage and high current can cause the moisture in the body to heat so rapidly that body parts can literally explode. This extreme expansion is the result of the body fluids changing to steam. This could result in a person's being injured severely, but not electrocuted.

When investigating shock incidents—whether near misses or those that have caused injury—the factors of voltage, current, and body resistance can be used as an investigative tool. Any report of a shock hazard should be checked. The combination of one or more of these factors can be analyzed and action taken to eliminate or to control the hazard potential. Other factors that enhance electrical shock potential are as follows:

- Wet and/or damp locations
- Ground/grounded objects
- Current loop from power source back to power source
- Path of current through body and duration of contact
- Area of body contact and pressure of contact
- Physical size, condition, age of person
- Type and/or amount of voltage
- Personal protective equipment, gloves, shoes
- Metal object such as watches, necklaces, rings
- Miscellaneous
  - Poor workplace illumination
  - Color blindness
  - Lack of safety training and knowledge of electricity
  - No safe work procedures

Basic Rules of Electrical Action

There are four basic rules of electrical action that everyone should know. They are as follows:

1. Electricity isn't "live" until current flows.
2. Electrical current won't flow until there is a complete loop, out from and back to the power source.
3. Electrical current always returns to its source, that is, the transformer that created it.
4. When current flows, work (measured in watts) is accomplished.

Rule 1. This rule explains why faulty power tools with live metal cases (due to a defective equipment-grounding conductor) can be carried around and used without causing a shock. Only when a person comes in contact with a ground loop with low enough resistance will current flow through the body and produce a shock.

Rule 2. When a current loop is formed, the electric current now has a path to return to its source. If the loop is broken, the current will stop flowing. To minimize electrical shock hazards, look for ways to prevent ground loops. Insulated mats, electrical gloves, and insulated tools are some methods that can be used.

Rule 3. Current will find any available loop or path to return to its source. In a grounded system the *NEC* requires that the power source be electrically attached to an earth-driven ground rod as well as to the water piping. All other noncurrent-carrying metal parts must also be grounded. This creates many potential ground loops for current to get back to its source. Water pipes, metal ventilation ducts, metal door frames, T-bars holding suspended ceiling panels, and metal ridge rolls on a roof with a grounded lightning system are but a few of the many ground loops available.

Rule 4. If current flows through the body, it can do harm depending on the amount of current available. As noted in Table 14-1, the effect on the human body can be anything from a mild shock to irreversible damage.

These rules should help you analyze how and why someone received an electrical shock. Use the loop concept and Ohm's Law to investigate actual or potential electrical shock hazards in the workplace and to design proper safeguards.

BRANCH CIRCUITS AND GROUNDING CONCEPTS

Single-Phase Service

A typical pole-mounted transformer (single phase) is shown in Figure 14-4. Industrial plants may have their transformer vaults and the feeder lines and branch circuits distributed throughout the building in conduit systems. This system is used in residential, commercial, and industrial operations to provide 120 volts for lighting and small appliances. The 240-volt, single-phase power is used for heavy appliances and small motor loads. It is important that an understanding of the simple, single-phase, three-element, 120-volt electrical system be understood, and that some misconceptions about grounding, bonding, and reverse polarity be clarified.

The *National Electrical Code (NEC)* requires that electrical wiring design ensure continuity of the system. In Figure 14-4, the ungrounded conductor from the service entrance panel (SEP) is wired to the small slot of the receptacle. The terminal screw and/or the metal connecting to the small slot is required to be a brass color. The insulation on the ungrounded conductor is generally black or red. Remember the phrase "black to brass" or the initials "B & B." The insulation on the grounded conductor is either white or gray. This conductor fastens to the large slot side of the receptacle and is defined by the code as the "grounded conductor." An easy

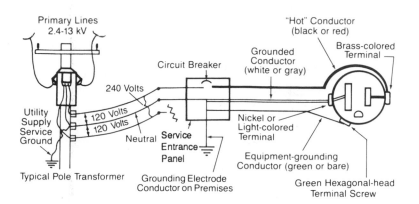

Figure 14-4. Note how the elements are identified in this single phase service-branch circuit wiring.

way to remember this wiring method is to think "white to light."

It is important that the receptacle be wired in this manner. If it is wired in reverse, that is, the white insulated conductor to the brass screw and the black insulated wire to the silver screw, it produces a condition known as reverse polarity, mentioned previously in this chapter.

Reverse Polarity (AC)

Many individuals experienced with electrical wiring and appliances think that reverse polarity is not hazardous. An example of one hazardous situation would be an electric hand lamp. Figure 14-5 illustrates a hand lamp improperly wired and powered.

When the switch is turned off, the shell of the lamp socket is still energized. If a person accidentally touched the shell with one hand and a ground loop back to the transformer, a shock could result. If the lamp had no switch and was plugged in as shown, the lamp shell would be energized when the plug was inserted into the receptacle. Most 2-prong plugs have blades that are the same size and the situation described is just a matter of chance. If the plug is reversed, Figure 14-6, the voltage is applied to the bulb center terminal and the shell is at ground potential. Contact with the shell and ground would not create a shock hazard in this situation.

Another example is the case of electric hair

dryers or other plastic-covered electrical appliances. Figure 14-7 illustrates a hair dryer properly plugged into a receptacle with the correct polarity. You will notice that the switch is a single pole throw (SPT). When the appliance is plugged in with the switch in the hot or 120-volt leg, the voltage stops at the switch when it is in the OFF position. If the hair dryer was accidentally dropped into water, current could flow out of the plastic housing using the water as the conducting medium.

The water does not short the appliance since it has resistance that creates current flow, but in the 100-ma range. This fault current is not sufficient to trip a 20-amp circuit breaker. In order to trip the circuit breaker, there would have to be a line-to-line short that would cause an excess of 20 amps (20,000 ma). When the appliance is polarized or plugged in, as shown in Figure 14-7, the switch connection is the only surface area where the hot conductor is fastened and is so small that high resistance allows only a small current to be available ($I = E/R$). A person could retrieve the appliance from the water while it is still plugged into the outlet. In this configuration, the fault current would be extremely low (unless the switch is in the ON position).

If the appliance is plugged in so that its polarity is reversed, as shown in Figure 14-8, when the switch is off, voltage will be present throughout all the internal wiring of the appliance. Now if it is dropped into water, the $I = E/R$ ratio will have increased

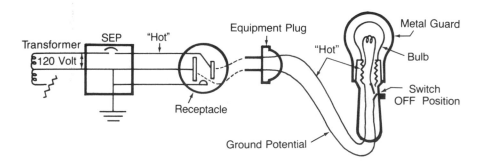

Figure 14-5. When this hand lamp is plugged in so that the polarity is reversed, the shell is still energized even when the switch is turned off. (SEP is the service entrance panel.)

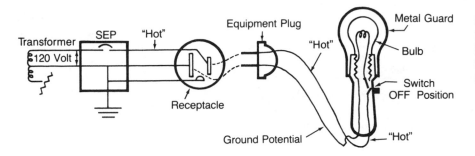

Figure 14-6. When this hand lamp is plugged in with the plug in correct polarity, current flows to the bulb center terminal and the shell is at ground potential.

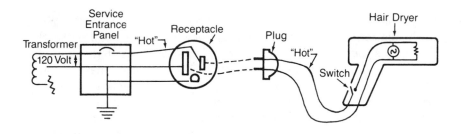

Figure 14-7. This hair dryer is plugged in correctly so that the switch is on the hot conductor.

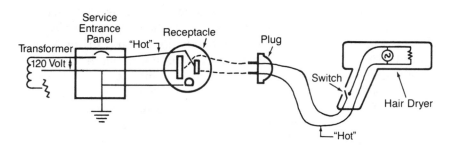

Figure 14-8. If the hair dryer is plugged in so that the polarity is reversed, voltage will be present throughout the internal wiring of the appliance even when the switch is turned off.

current available in the water, since the heating element has a large surface area, decreasing the resistance and increasing the available fault current. A person who accidentally tries to retrieve the dryer is now in a hazardous position because the voltage in the water could cause current to flow through the body if another part of the body contacts a ground loop. This illustrates the concept that reverse polarity is a problem whenever appliances are used with plastic housings in areas near sinks, or where rain or water can wet down the appliance. Remember, most motorized appliances have air passages for cooling. Wherever air can go, so can moisture and water.

If the appliance had a double pole-double throw switch (DPDT), then it would make no difference how the plug was positioned in the outlet. The hazard would be minimized since the energized contact surface would be extremely small, resulting in a high resistance contact in the water and a resulting low available fault current. Later, we will see how ground-fault circuit interrupters (GFCIs) can be used to protect against shock hazards when using appliances with nonconductive housing around water.

Grounding Concepts

The terms "grounded conductor," "equipment-grounding conductor," "neutral," and "ground" are probably the cause of many difficulties in understanding basic 120-volt AC system-design concepts. Grounding falls into two safety categories. One of these is systems safety grounding, and the other is equipment safety grounding.

System grounding provides protection of the power company equipment and the consumer. The system safety grounding loop activates protective devices during lightning strikes and when higher voltage lines fall on secondary lines during storms. *Equipment safety grounding* provides a system where all noncurrent-carrying metal parts (such as the metal frame of a drill press or electric refrigerator) will be bonded together and kept at the same potential. Grounding concept definitions are as follows:

Grounded conductor. The *NEC* defines grounded conductor as "a system or circuit that is intentionally grounded." Figure 14-9 shows the conductor from the service entrance panel (SEP) to the outlet as the grounded conductor. As shown in the

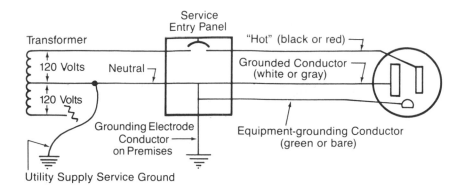

Figure 14-9. Note the wiring terminology, particularly the relationship of the neutral to the grounded conductor and the equipment-grounding conductor.

illustration, for 120-volt AC, 3-conductor outlets, the grounded conductor (white or gray insulation) is fastened to the silver screw, which, in turn, is the large parallel slot.

The grounded conductor is a current-carrying conductor, but is also a connected ground at the SEP. From the SEP back to the power source, the grounded conductor is called neutral. On 120-volt circuits, refer to the grounded conductor from the SEP to the receptacle as just that, the *grounded conductor*, not the neutral.

Neutral. The term "neutral" connotes a neutral conductor that carries only unbalanced current from the other conductors in the case of normally balanced circuits. From the SEP to the power source, the "neutral" term is used. In circuits from the SEP to large 240-volt AC appliance receptacles, the grounded conductor would also be referred to as the neutral.

Equipment-grounding conductor. Figure 14-9 shows the equipment-grounding conductor going from the SEP to the appliance outlet. This path is sometimes also called the grounding wire. This grounding conductor, if insulated, is colored green or green with a yellow stripe. This conductor may also be a bare wire like those used in plastic sheathed cable. The *NEC* allows the conduit system to be the grounding conductor path also.

The grounding conductor normally does not carry a current. Only when an equipment malfunction occurs—allowing current to flow from the motor or electrical device to the equipment case—does the grounding conductor carry the fault current back to the power source and keeps the equipment case at ground potential. (More on this in the discussion on grounded power tools and equipment.)

Portable equipment and portable power tools can be grounded (equipment grounding) by the use of a separate grounding wire from the equipment housing to a *known* ground. This method is satisfactory for stationary equipment as long as conditions do not change, but is less than satisfactory for portable equipment where the integrity of the

grounding wire may be questionable and the quality of the ground is unknown. However, it is better than no grounding provision at all. If used, the grounding wire should be at least 1 in. longer than the power cord, so that the plug can be removed from the receptacle before the ground is disconnected.

Ground. The *NEC* defines this term as a "conducting connection, whether intentional or accidental, between an electrical circuit or equipment and earth, or to some conducting body that serves in place of earth." Ground, then, means an electrical connection to a water pipe or to a driven ground rod, or, where the pipe is not available, to the rod alone. To "ground" something means to connect it electrically to the ground. Since the power source and the SEP are both connected to ground, the earth or ground loop becomes an electrical loop hazard.

Figure 14-10 illustrates that accidental contact with the hot conductor, while a person is standing on the ground, could allow a current to return to the power source through the ground loop formed by the person and the earth. This can cause a shock.

Remember that grounded objects are many. The metal studs used in new construction are usually grounded from the electrical conduit system. The T-bar hangers for suspended ceilings can be a ground loop since the electric lighting fixtures are grounded and electrically may cause the T-bars to be ground potential. Ventilation duct work may be ground potential since electric switches, dampers, and other devices may be mounted on the duct work.

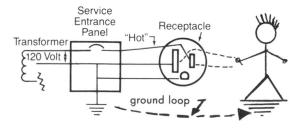

Figure 14-10. A shock is caused when a person contacts the hot wire while standing on the ground.

PLUG- AND CORD-CONNECTED EQUIPMENT AND EXTENSION CORDS

Grounded Power Tools and Equipment

The earlier discussion of two-conductor appliances with plastic housings revealed the hazard of dropping ungrounded electrical devices in water. Now it is important to explain how appliances with metal cases properly grounded provide safety if accidentally immersed in water.

Virtually all portable power tools, except double-insulated tools, used in industry and on construction sites are equipped with a 3-prong plug. Unfortunately, not all receptacles are equipped with provisions for a 3-prong plug.

This fact can lead a worker, who is unwilling to take the time or effort to use an adapter, to break off the U-blade, or ground, and defeat the protection of an equipment-grounding circuit. This is a highly dangerous practice on construction sites and in other areas that are often damp and wet. Cutting off the equipment-grounding conductor U-blade not only destroys the grounding path, but also may allow the plug to be inserted in the receptacle in a reverse polarity mode (Figure 14-11).

As shown in this figure, the grounding pin is cut off. If, in this situation, a fault developed when the power tool was plugged into an outlet, the case would have leakage voltage. The power tool would operate and the person using the power tool would be safe—but only until a ground loop was encountered. Then, current could flow through the person, through the ground loop, and back to the power source. This would cause an electrical shock to the person; the loop resistance would determine the seriousness of the shock.

Double-Insulated Power Tools

Double-insulated tools and equipment generally do not have an equipment-grounding conductor. Protection from shock depends upon the dielectric properties of the internal protective insulation and the external housing to insulate the user from the electrical parts. External metal parts, such as chucks and saw blades, are internally insulated from the electrical system.

Today, many electrical power tool manufacturers produce a line of double-insulated tools. As a result, many of their advertisements lead us to believe that double-insulated power tools are "doubly safe." However, you should take the following precautions when using double-insulated power tools:

- Double-insulated tools are designed so that the inner electrical parts are isolated physically and electrically from the outer housing. The housing is nonconductive. Particles of dirt and other foreign matter from the drilling and grinding operations may enter the housing through the cooling vents and become lodged between the two shells, thereby voiding the required insulation properties.
- Double insulation does not protect against defects in the cord, plug, and receptacle. Continuous inspection and maintenance are required.
- A product with a dielectric housing—for example, plastic—protects users from shock should the interior wiring contact the housing. Immersion in water, however, can allow a leakage path that may be either high or low resistance. Handling the product with wet hands, in high humidity, or outdoors after a rain storm can be hazardous.

The best indication of the safety of a dou-

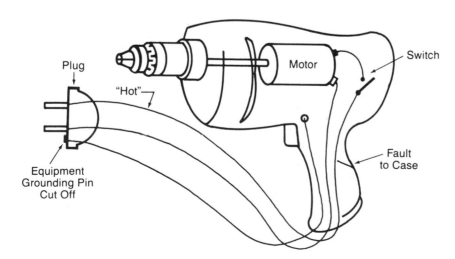

Plug

"Hot"

Motor

Switch

Equipment Grounding Pin Cut Off

Fault to Case

Figure 14-11. Removing the grounding pin destroys the grounding path and can allow the plug to be inserted in the receptacle in reverse polarity.

ble-insulated tool is the Underwriters Laboratories (or other recognized testing laboratory's) label attached to the housing. The UL listing is evidence that the tools meet minimum standards. A tool marked "double insulated," but without a UL label, may not be safe.

Double-insulated tools and equipment should be inspected and tested, like all other electrical equipment. Remember, do not use double-insulated power tools or equipment where water and a ground loop are present. They should not be used in any situation where dampness, steam, or potential wetness can occur, unless protected by a GFCI.

Attachment Plugs and Receptacles

Attachment plugs are devices fastened to the end of a cord so that electrical contact can be made between the conductor in the equipment cord and the conductors in the receptacle. The plugs and receptacles are uniquely designed for different voltages and currents, so that only matching plugs will fit into the correct receptacle. In this way, a piece of equipment rated for one voltage/current combination cannot be plugged into a power system that is of a different voltage or current capacity (Figure 14-12).

The polarized 3-prong plug is also designed with the equipment-grounding blade slightly longer than the two parallel blades. This provides equipment grounding before the equipment is energized. Conversely, when the plug is removed from the receptacle, the equipment-grounding pin is the last to leave, assuring a grounded case until power is removed. The parallel line blades may be the same width on some appliances since the 3-prong plug can only be inserted in one way.

Figure 14-12 illustrates the National Electrical Manufacturers Association (NEMA) standard plug and receptacle connector blade configurations. Each has been developed to standardize the use of plugs and receptacles for different voltages, amperes, and phases from 115 through 600 volts; from 10 through 60 amps; and for single- and three-phase systems.

You should watch for workers who may jury-rig adapters, for example, trying to match a 50-amp attachment plug to a 20-amp receptacle configuration. The danger is getting the voltage and current ratings mixed up and causing fire and/or shock hazards to personnel using the equipment. Equipment attachment plugs and receptacles should match in voltage and current ratings to provide safe power to meet the equipment ratings.

Electrical Extension Cord Deficiencies

Electrical extension cords are used in manufacturing, maintenance, construction, and even in office operations. Defective, worn, or damaged extension

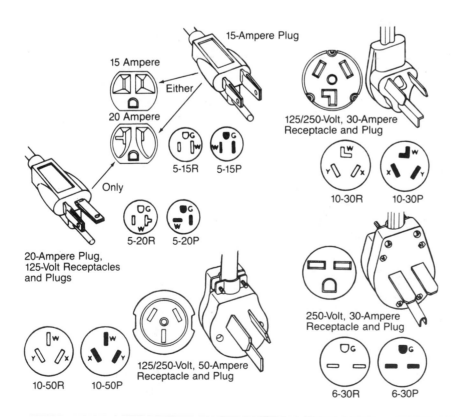

Figure 14-12. Some NEMA plug and receptacle blade configurations. (From OSHA Publication 3073, *An Illustrated Guide to Electrical Safety.*)

cords have been the cause of electrical equipment damage, injury, fire, and even death. Often the rush to meet schedules on time or the lack of funds result in jury-rigged cords being fabricated and used. Purchasing staff may buy the cheapest units available, and workers end up with inferior extension cord sets that fail the contact tension test after a few cycles of use. Many times neither the receiving department nor the workers test extension cords prior to using. This results in the possibility of brand-new defective units being put into service.

Other problems resulting from extension cord use include overloading, cutting off the grounding blade, and incorrect maintenance and repair. The following safety criteria should be strictly observed and used for training and inspection purposes:

- Before new or repaired extension cords are put into use, they must be tested. In addition, always purchase UL-listed extension cords. One supplier purchased and distributed several thousand units that were incorrectly wired (reverse polarity). These defective units were not UL listed.

    If extension cords are repaired by your company, electrical shop personnel, be sure they test these units prior to returning them to service. Follow the Testing Extension Cords procedures described later in this chapter.

- A visual and electrical inspection should be made of the extension cord each time it is used. Cords with cracked or worn insulation or damaged ends should be removed from service immediately. Cords with the grounding prong missing or cut off must also be removed from service.

- If extension cords are fabricated, only qualified electricians should do the work. The use of jury-rigged receptacles fabricated from junction boxes should be prohibited. Only UL-listed attachment plugs, cords, and receptacle ends should be used when fabricating extension cords.

- Use of 2-conductor extension cords should be prohibited. Only grounding-type extension cord sets should be used.

- When cords must cross passageways, whether vehicular and/or personnel, the cords should be protected and identified with appropriate warnings. Cords should be used in this manner only for temporary or emergency use. They must never be used where they pass through exit doorways, hazardous storage areas, smoke barriers, or fire barriers.

- Extension cords must be of continuous length without splices or taps. Personnel should be instructed to remove from service any cord sets that have exposed wires, cut insulation, loose connections at the plug or receptacle ends, and jury-rigged splices.

- Extension cords must not be connected or disconnected until all electrical load has been removed from the extension cord receptacle end.

- Extension cords used in hazardous locations must be approved for the applicable hazards (see the Hazardous Location section in this chapter).

- When the extension cord is not in use, it should be disconnected and neatly stored. Cords should be stored in a dry, room-temperature area and not exposed to excessively cold or hot temperatures.

- Never overload an extension cord electrically. If it is warm or hot to the touch, have it and/or the appliance being used checked by a qualified person to determine the problem. The wire size of the extension cord may be too small for the amperage of the appliance being used.

- Extension cords must not be draped over hot surfaces such as steam lines, space heaters, radiators, or other heat sources.

- Extension cords must not be run through standing water, wet floors, or other wet areas. If the extension cord is to be used to provide electrical power to equipment used in damp or wet locations, use GFCI protection.

- Extension cords used outdoors should be listed and marked for outdoor use. Do not allow damaged cord sets to be used outside.

BRANCH CIRCUIT AND EQUIPMENT TESTING METHODS

Testing Branch Circuit Wiring

Branch circuit receptacles should be tested periodically. The frequency of testing should be established on outlet usage. In shop areas, quarterly testing may be necessary, while office areas may need only annual testing. A preventive maintenance program should be established, particularly because it is not unusual to find outlets as old as the facility. For some reason, a popular belief exists that outlets never wear out. This is obviously false. For example, outlets take severe abuse from employees disconnecting the plug from the outlet by yanking on the cord. This can put severe strain on the contacts inside the outlet, as well as on the plastic face.

The electrical outlet is a critical electrical system component. It must act as a strong mechanical connection for the appliance plug and provide a continuous electrical circuit for each of the prongs. The receptacle must be electrically correct or serious accidents can result. For this reason, a three-step testing procedure is recommended. The first two steps can easily be done by most collateral duty safety inspectors. The third step is one that should be performed by electrical maintenance personnel prior to accepting contractor wiring jobs. Also, the third step should be done on any circuit where electrical maintenance has been performed. Some inspectors use the third step as a spot check of ground-loop quality.

Step 1. Plug in a 3-prong receptacle circuit tester and note the combination of indicator lights (Figure 14-13). (The circuit tester has three lights; the flashing light indicates which of three problems is present in the wires.) The tester checks the receptacle for proper connection of ground wire, correct polarity, or fault in any of the three wires. If these are normal, proceed to the next test. If the tester indicates a wiring problem, contact electrical maintenance and have it corrected as soon as possible. Retest after the problem is corrected.

Step 2. Once you are sure the outlet is wired correctly, test the receptacle contact tension. A typical tension tester is shown in Figure 14-14. The tension should be 8 oz or more. If it is less than 8 oz, have it replaced. The first receptacle function that loses its contact tension is usually the ground-

Figure 14-14. Use a receptacle tension tester to be sure receptacle contact tension is 8 oz or more.

ing contact circuit. Because this is a critical circuit, it illustrates the importance of checking for receptacle contact tension.

In determining the frequency of testing, the interval must be based on receptacle usage. All electrical maintenance personnel should be equipped with receptacle testers and tension testers. Other maintenance employees could also be equipped with testers and taught how to use them. In this manner, the receptacles can be tested before maintenance workers use them.

Step 3. The ground loop impedance tester (GLIT) should be used next. This step may be conducted by the electrical maintenance person whenever any electrical circuit is reworked or repaired. An electrical preventive maintenance program could be established to perform annual ground loop testing in high usage areas. The safety inspector can use the GLIT for spot-checking (Figure 14-15).

To perform this test, simply insert the 3-prong plug of the GLIT into the receptacle. Press the grooved portion of the red bar and observe the meter reading. As you have learned in previous sections, a reading of 1 ohm or less is required. Any reading of more than 1 ohm should be reported to electrical maintenance for analysis and corrective action.

WARNING: The GLIT must never be used to test a branch circuit when patient monitoring devices or computer equipment of any kind is in use.

The GLIT should not be used until steps 1 and 2 have been completed successfully. The tester will

Figure 14-13. A receptacle circuit tester checks the receptacle for proper connection of ground wire, correct polarity, and faults in any of the three wires.

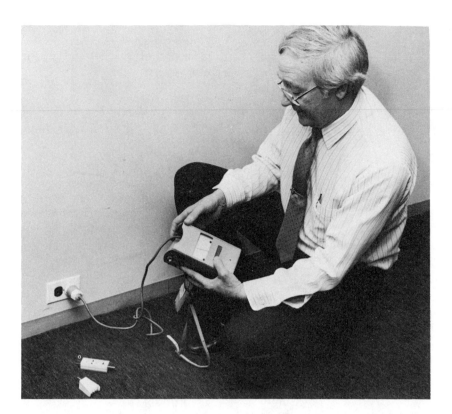

indicate the wiring fault. Prior to making the circuit operational, all wiring faults must be corrected and the circuit retested.

Testing Extension Cords

Extension cord testing and maintenance is critical to worker safety. The extension cord conducts electrical energy from a fixed outlet or source to a remote location. It must be wired correctly, or the cord can become the critical fault path.

Many electrical accidents have been caused by faulty or incorrectly repaired extension cords. Many times the male plug on a 3-conductor extension cord becomes damaged and the repair person installs a 2-prong plug (Figure 14-16). This obviously means the receptacle end has no grounding path. It also means the plug can be inserted with correct polarity or reverse polarity.

Sometimes the repair may be made in such a way that the hot and ground are interchanged, as shown in Figure 14-17. This could happen on repair of the plug or the receptacle. In this configuration, when a grounded power tool is plugged into the extension cord, voltage will be applied to the tool casing. This could cause a dangerous or fatal electrical shock.

The same dangerous condition is illustrated in Figure 14-18. A repair was made replacing the broken 3-prong plug with a 2-prong plug. The repair person connected the white- and green-insulated conductors on one prong, then connected the black-insulated conductor to the other prong. Depending on how the plug was inserted into the outlet, in one mode, the case would be grounded; and in the other mode, the case would be energized.

With a "hot" case, the user would have to find

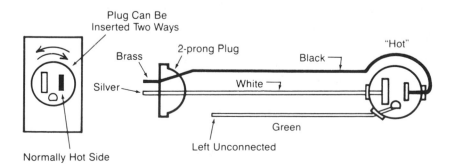

Figure 14-16. A 3-conductor extension cord repaired with a 2-prong plug.

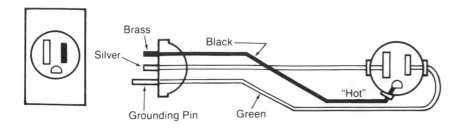

Figure 14-17. A 3-conductor extension cord with hot and ground wires interchanged.

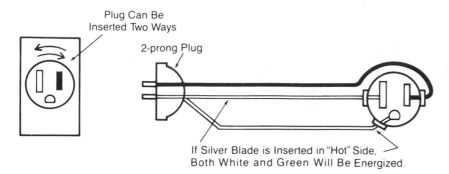

Figure 14-18. This 3-conductor extension cord is repaired with a 2-prong plug that has the white and green wires tied together.

only a ground loop to the source and a fatal shock could occur. Interestingly enough, the power tool would operate in either mode.

Figure 14-19 illustrates a repaired or a jury-rigged extension cord fabricated with a 3-prong plug and receptacle, but the connecting cord is only 2-conductor. This configuration would obviously show an open ground when tested. Because at times the cord may look like a 3-conductor cord, the tester should be used to make sure the extension cord is safe.

New extension cords must be tested before being put into service. Many inspectors have found that new extension cords have open ground or reverse polarity. Don't assume that a new extension cord is good—test it!

Testing extension cords must follow the same three steps used for electrical outlet testing. These three steps are:

Step 1. Plug the extension cord into an electrical outlet that has successfully passed the three-step outlet testing procedure. Plug any 3-prong receptacle circuit tester into the extension receptacle and note the combination of indicator lights. If the tester shows the extension cord is safe, proceed to the next step. If the tester indicates a faulty con-

dition, return the extension cord to electrical maintenance for repair. Once the cord is correctly repaired and passes the 3-prong circuit tester test, it must be tested according to step 2.

Step 2. Plug a reliable tension tester into the receptacle end. The parallel receptacle contact tension and the grounding contact tension should check out at 8 oz or more. If the tension is less than 8 oz, the receptacle end of the extension cord must be repaired. As with fixed electrical outlets, the receptacle end of an extension cord loses its grounding contact tension first. This path is the critical human protection path and must be both electrically and mechanically in good condition.

Step 3. Once an extension cord has passed steps 1 and 2, the ground-loop impedance tester may be used. The extension cord must be plugged into a properly connected outlet (as noted in step 1). Plug the GLIT into the receptacle end of the extension cord. A reading of 1 ohm or less is required. If you obtain a reading of more than 1 ohm, return the cord to maintenance for analysis and repair or replacement.

If a GLIT is not available, use an ohmmeter to test the continuity from the U-shaped grounding pin to the grounding outlet hole in the receptacle end

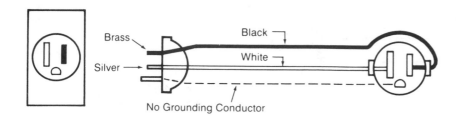

Figure 14-19. This extension cord with a 3-prong plug and receptacle is connected to a 2-conductor cord.

of the extension cord. This test is done with the extension cord de-energized. **The extension cord should not be plugged into an electrical outlet when using an ohmmeter. Damage to the ohmmeter could result.**

Testing Plug- and Cord-Connected Equipment

The last element of the electrical systems test is the cord- and plug-connected equipment. The electrical inspector should be alert for jury-rigged repairs made on electric power tools and appliances. Many times a visual inspection will disclose 3-prong plugs with the grounding prong broken or cut off. Obviously, the grounding path to the equipment case has been destroyed. (Also, the plug can now be plugged into the outlet in the reverse polarity configuration.)

Double-insulated equipment generally has a nonconductive case and will not be tested using the procedures in this section. Some manufacturers who have UL listed double-insulation ratings may also provide a 3-prong plug to ground any exposed non-current-carrying metal using the grounding path conductor. In these cases, the grounding path continuity can be tested.

A common error from a maintenance standpoint is the installation of a 3-prong plug on a 2-conductor cord to the appliance. Obviously, there will be no grounding path if there are only 2 conductors in the cord. Anything can be jury-rigged, but normally the maintenance error is simply to hook up the black- and white-insulated conductors to the brass and silver screws, respectively, and leave the third screw unconnected. Some hospital-grade plugs have transparent bases that allow visual inspection of the electrical connection to each prong. Even in this situation, you should still perform an electrical continuity test.

Field testing using an ohmmeter only. Many electrical power tool testers require the availability of electric power. The ohmmeter can be used in the field and in locations where electric power is not available or is not obtained easily. An additional safety feature is that the plug- and cord-connected equipment tests are made on de-energized equipment. Testing of de-energized equipment in wet and damp locations also can be done safely.

The plug- and cord-connected equipment test using a self-contained battery-powered ohmmeter is simple and straightforward. The two-step testing sequence that can be performed on 3-prong plug grounded equipment is as follows:

Step 1. *Ground pin-to-case test.* Put the ohmmeter selector switch to the lowest scale (such

as R × 1). Zero the meter. Place one test lead (tester probe) on the grounding pin of the de-energized equipment (Figure 14-20). While holding that test lead steady, take the other test lead and make contact with an unpainted surface on the metal case of the appliance or power tool. You should get a reading of less than 1 ohm. If the grounding path is open, the meter will indicate infinity. If no continuity, then the appliance must be tagged out and removed from service. If the appliance grounding path is safe, proceed to step 2.

Step 2. *Appliance leakage test.* Place the ohmmeter selector switch on the highest ohm test position (such as R × 1,000). Zero the meter. Place one test lead on an unpainted surface of the appliance case, then place the other test lead on one of the plug's parallel blades. Observe the reading. The ideal is close to infinity (Figure 14-21). (If a reading of less than 1 mega ohm is noted, return the appliance to maintenance for further testing.) With one test lead still on the case, place the other test lead on the remaining parallel blade and note the ohmmeter reading. A reading approaching infinity is required. (Again, anything less than 1 mega ohm should be checked by a maintenance shop.)

Voltage Detector Testing

When conducting an electrical inspection, a voltage detector (Figure 14-22) should be used in conjunction with the circuit tester, ohmmeter, and tension tester. There are several types of these devices on the market, varying in price from $10 to $40. The unit shown in Figure 14-22 costs approximately $10. These testers are battery-powered and are constructed of nonconducting plastic. Being lightweight and self-contained, these testers make an ideal inspection tool.

The voltage detector works like a radio receiver in that it can receive or detect the 60-hertz electromagnetic signal from the voltage waveform surrounding an ungrounded ("hot") conductor. Figure 14-23 illustrates the detector being used to detect the hot conductor in a cord connected to a portable hand lamp. Whenever the front of the detector is placed near an energized (hot or ungrounded) conductor, the tester will provide an audible as well as visual warning. In Figure 14-23, if the plug was disconnected from the receptacle, the detector would not give an alarm since there would not be any voltage waveform present.

The detector can also be used to test for properly grounded equipment. Whenever the tester is positioned on a properly grounded power tool (e.g., the electric drill shown in Figure 14-24), the tester will not give a warning indication. The reason is

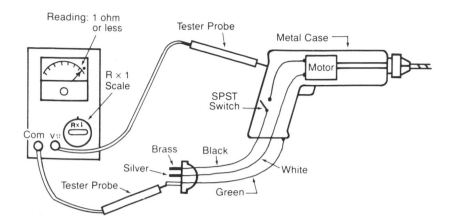

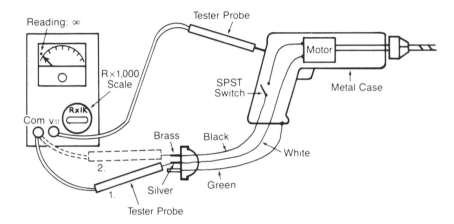

Figure 14-20. In a continuity test for checking the grounding path of an appliance, the reading on the ohmmeter should be less than 1 ohm.

Figure 14-21. The leakage test should read near infinity on the ohmmeter when checking an appliance for current-to-case leakage.

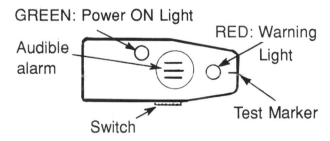

Figure 14-22. Use a voltage detector during an electrical inspection to find the "hot" conductor in a cord.

that the electromagnetic field is shielded from the detector so that no signal is picked up. If the grounding prong had been removed and the drill was not grounded, the electromagnetic waveform would radiate from the drill and the detector would receive the signal and give an alarm.

As illustrated in Figure 14-24, the detector can be used to test for many things during an inspection. Receptacles can be checked for proper AC polarity. Circuit breakers can be checked to determine if they are ON or OFF. All fixed equipment can be checked for proper grounding. If the detector gives a warning indication on any metal-enclosed equipment, have it tagged out and repaired quickly. Ungrounded

equipment can be just that or, in addition, there may be current leakage to the enclosure, making it hot with respect to ground. An experienced inspector could use the voltmeter to determine if there is voltage on the enclosure. The voltage detector should be used as an indicating tester, while qualitative testing should be done using other testing devices such as a multimeter. The use of the detector can speed up the inspection process by allowing you to check equipment grounding quickly and safely.

Another unique feature of the voltage detector shown in Figure 14-22 is that it is voltage sensitive. Table 14-2 lists distances from a hot conductor, at which the detector red light turns on, versus the strength of the voltage waveform. As an example, conductors energized with 120 volts AC can be detected from 0 to 1 in. from the conductor. On the other hand, power source wiring on a furnace automatic ignition system rated at 10,000 volts can be detected from 6 to 7 ft away. You can use this feature to assist in making judgments regarding the degree of hazard and the urgency in obtaining corrective action on any appliance or power tool.

When using the voltage detector, you must understand how it operates. If you put the tester on an energized, double-insulated power tool, you will

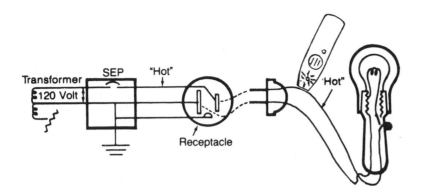

Figure 14-23. To test for a "hot" conductor, place the front of the voltage detector near an energized conductor; the tester gives a visible and audible warning.

see the red warning light turn on. In this situation, the voltage waveform is detected because there is no metal enclosure to shield the waveform. The same would occur if you were to test many of the typewriters used in offices. This does not mean that the plastic- or nonmetallic-enclosed equipment is unsafe, only that it is energized and not a grounding-type piece of equipment. By understanding the operational features of a voltage detector, you can see the benefits it offers and the many electrical hazards it can detect that other equipment may overlook.

GROUND-FAULT CIRCUIT INTERRUPTERS

Operational Theory

The ground-fault circuit interrupter (GFCI) is a fast-acting device that monitors the current flow to a protected load. It can sense any leakage current returning to the power supply by any electrical loop other than through the white (grounded) and the black (hot) conductors.

When any leakage current 5 ma or over is sensed, the GFCI, in a fraction of a second, shuts off the current on both the hot and grounded conductors, thereby interrupting the fault current to the appliance and the fault loop. This action is illustrated in Figure 14-25. As long as I_1 is equal to I_2 (normal appliance operation with no ground fault leakage), the GFCI switching system remains closed (N.C.). If a fault occurs between the metal case of an appliance and the hot conductor, fault current I_3 will cause an imbalance (5 ma or greater for personnel protection), allowing the GFCI switching system to open (as illustrated) and remove power from both the white and hot conductors.

Another type of ground fault can occur when a person contacts a hot conductor directly or touches an appliance with no (or faulty) equipment-grounding conductor. In this case, I_4 represents the fault current loop back to the transformer. This ground fault is generally the type that personnel are exposed to.

The GFCI is intended to provide worker protection by de-energizing a circuit, or portion of a circuit, in approximately $1/40$th of a second when the ground fault current exceeds 5 ma. The GFCI should not be confused with ground fault protection (GFP) devices that protect equipment from damaging line-to-line fault currents. Protection provided by GFCIs is independent of the condition of the equipment-grounding conductor.

The GFCI can provide protection to personnel even when the equipment-grounding conductor is accidentally damaged and made inoperative. When replacing receptacles in older homes or business buildings (equipped with nongrounding-type receptacles), an exception in the *National Electrical Code* allows GFCIs to be used when a grounding means does not exist for installation of a grounding-type receptacle. The existing nongrounding receptacles are permitted to be replaced with grounding-type receptacles when power is provided to them through a GFCI-style receptacle or circuit breaker.

It is important to remember that a fuse or circuit breaker cannot provide hot-to-ground loop protection at the 5-ma level. The fuse or circuit breaker is designed to trip or open the circuit if a line-to-line or line-to-ground fault occurs that exceeds the circuit protection device rating. For a 15-amp circuit breaker, a short in excess of 15 amps or 15,000 ma would be required. The GFCI will trip if 0.005 amp or 5 ma starts to flow through a ground fault in a circuit it is protecting. This small amount of current (5 ma) flowing for the extremely short time required to trip the GFCI will not electrocute a person but will shock the person in the magnitude noted in Table 14-1.

GFCIs are available in several different types. Figure 14-26 illustrates three of the types available. Circuit breakers can be purchased with built-in GFCI protection. This combination unit can be secured in the circuit-breaker panel to prevent unauthorized persons from having access to the GFCI.

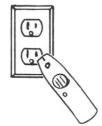

Safely and quickly check for voltage at AC outlets

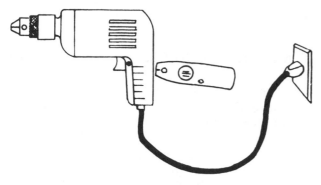

Easily check power tools for proper grounding

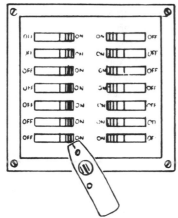

Quickly check circuit breakers

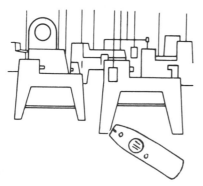

Check fixed equipment grounding

Figure 14-24. A voltage detector can be used to check outlet voltage, equipment grounding, and circuit breaker condition.

TABLE 14-2. Voltage Versus Detection Distance

To Determine Approximate Voltage: Slowly approach the circuitry being tested with the front sensor end of the unit and watch at what distance the red LED light glows along with the "beep" sound. Use the chart below to determine the approximate voltage in the circuitry.

Voltage (V)	100	200	600	1K	5K	9K
Distance (inch)	0–1	1–2	3–5	15	5 ft	6–7 ft up

These figures may vary due to conditions governing the testing, i.e., static created by your standing on grounded material, carpets, etc.

The receptacle GFCI is the most convenient type in that it can be tested and/or reset at the location where it is being used. It can also be installed so that it is closest to the circuit-breaker panel and will provide GFCI protection to all receptacles on the load side of the GFCI. The portable type can be carried in the toolbox of a maintenance person for instant use at a work location where the available AC power is not GFCI-protected.

Other versions of GFCIs are available including multiple outlet boxes for use by several power tools and for outdoor applications. The obvious difference between a regular electrical receptacle and a GFCI receptacle is the presence of the TEST and RESET buttons. The user should follow the manufacturer's instructions regarding the testing of the GFCI. Some recommend pushing the TEST button and resetting monthly. Defective units should be replaced immediately.

Be aware that a GFCI will not protect the user from line-to-line electrical contact. For example, suppose a person is standing on a surface insulated from ground (e.g., a dry, insulated floor mat) while holding a faulty appliance with a hot case in one hand. The worker then reaches with the other hand to unplug the appliance and contacts an exposed grounded (white) conductor. The worker has made a line-to-line contact. However, the GFCI will not protect him or her because there is no ground loop. To prevent this type of accident from occurring, it is essential to have an equipment-grounding conductor inspection program in addition to a GFCI program. Supervisory personnel should train and encourage employees to follow the electrical safety policy near the end of this chapter.

GFCIs do not replace an ongoing electrical equipment inspection program. Rather, GFCIs should be considered an additional personnel safety factor for protection against the most common form of electrical shock and electrocution, the line (hot)-to-ground fault. It is recommended that GFCI pro-

GFCI Protection Device

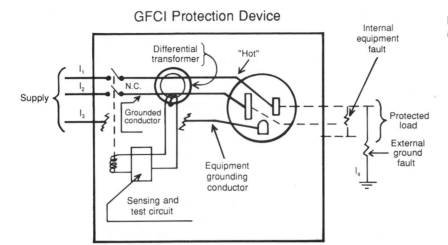

Figure 14-25. A circuit diagram for a GFCI.

Figure 14-26. Three types of GFCI: (top) a circuit breaker with a GFCI, (middle) a receptacle GFCI, (bottom) a portable GFCI.

tection be provided in any working area where damp or wet conditions can or do exist.

GFCI Uses

GFCIs should be used in dairies, breweries, canneries, steam plants, construction sites, and inside metal tanks and boilers, or when workers are exposed to humid or wet conditions and may come in contact with ground or grounded equipment. It is recommended that they be used in any work environment which is or can become wet and in other areas that are highly grounded.

You can also stress off-the-job electrical safety. The *National Electrical Code* requires GFCIs on 15-amp, 20-amp, 120-volt outside receptacles, garage circuits, and bathrooms. The *NEC* also requires that GFCIs be installed in crawl spaces at or below grade level and in unfinished basements. Receptacles installed within 6 ft of a kitchen sink to serve countertop surfaces must also be GFCI protected. Other locations where the *NEC* requires GFCI-protection include the following: pools, fountains, commercial garages, pipeline heating, marinas and boat yards, mobile homes, and recreational vehicles. It is recommended that you consult the *NEC* for the specific requirements applicable to any of these occupancies.

Nuisance GFCI Tripping

When GFCIs are used in construction activities, the GFCI should be located as close as possible to the electrical equipment it protects. Excessive lengths of temporary wiring or long extension cords can cause ground fault leakage current to flow by capacitive and inductive coupling. The combined leakage current can exceed 5 ma, causing the GFCI to trip.

Other nuisance tripping may be caused by one or several or the following items:

- Wet electrical extension cord to tool connections
- Wet power tools
- Outdoor GFCIs not protected from rain or water sprays
- Bad electrical equipment with case-to-hot-conductor fault
- Too many power tools on one GFCI branch
- Resistive heaters
- Coiled extension cords (long lengths)
- Poorly installed GFCI
- Defective or damaged GFCI
- Electromagnetic-induced current near high-voltage lines
- Portable GFCI plugged into a GFCI-protected branch circuit

Remember that a GFCI does not prevent shock. It simply limits the duration so the heart is not affected. The shock lasts about 1/40 second (.025 second) and can be intense enough to knock a person off a ladder or otherwise cause an accidental injury. Emphasize to your workers that they should not consider a GFCI adequate protection against electrical hazards and that they must follow all safety steps rigorously.

HAZARDOUS LOCATIONS

Overview of Classes and Divisions

This section provides only a summary of electrical equipment requirements as a guide for inspection or system analysis of various hazardous locations. For more in-depth design and engineering requirements, refer to subpart S, Electrical, in the OHSA Standards (29 *CFR* 1910) and Chapter 5 of the *NEC*. Hazardous locations are areas where flammable liquids, gases, or vapors, combustible dusts, or other easily ignitable materials exist, or can exist accidentally, in sufficient quantities to produce an explosion or fire. In hazardous locations, specially designed equipment and special installation techniques must be used to protect against the explosive and flammable potential of these substances.

Hazardous locations are classified as Class I, Class II, or Class III, depending on what type of hazardous substance is or may be present. In general, Class I locations are those in which flammable vapors and gases may be present. Class II locations are those in which combustible dusts may be found. Class III locations are those in which there are ignitable fibers and flyings.

Each of these classes is divided into two hazard categories, Division 1 and Division 2, depending on the likelihood of a flammable or ignitable concentration of a substance. Division 1 locations are designated as such because a flammable gas, vapor, dust, or easily ignitable material is normally present in hazardous quantities. In Division 2 locations, the existence of hazardous quantities of these materials is not normal, but they occasionally exist either accidentally or when material in storage is handled. In general, the installation requirements for Division 1 locations are more stringent than for Division 2 locations.

Additionally, Class I and Class II locations are also subdivided into groups of gases, vapors, and dusts having similar properties. Table 14-3 summarizes the various hazardous (classified) locations.

Equipment Requirements

General-purpose electrical equipment can cause explosions and fires in areas where flammable vapors, liquids, and gases, and combustible dusts or fibers are present. Hazardous areas require special electrical equipment designed for the specific hazard involved. This includes explosion-proof equipment for flammable vapor, liquid, and gas hazards and dust-ignition-proof equipment for combustible dust. Other kinds of equipment are nonsparking equipment, intrinsically safe equipment, and purged and pressurized equipment.

Many pieces of electrical equipment have parts, such as circuit controls, switches, or motors, that arc, spark, or produce heat under normal operating conditions. These energy sources can produce temperatures high enough to cause ignition. Thus, when electrical equipment must be installed in hazardous areas, the sparking, arcing, and heating nature of the equipment must be controlled.

Installations in hazardous locations must be (a) intrinsically safe; (b) approved for the hazardous location; or (c) of a type and design that provides protection from the hazards arising from the combustibility and flammability of the vapors, liquids, gases, dusts, or fibers present. Installations can have one or more of these options. Each option is described in the following paragraphs.

Intrinsically safe. Equipment and wiring approved as intrinsically safe are acceptable in any hazardous (classified) location for which they are designed. Intrinsically safe equipment cannot release sufficient electrical or thermal energy under normal or abnormal conditions to ignite specific flammable or combustible materials present in the location.

Make sure that flammable gases and vapors do not come into contact with or pass through the equipment. Additionally, evaluate all interconnec-

TABLE 14-3. Summary of Class I, II, and III Hazardous Locations

Classes	Groups	Divisions	
		1	2
I Gases, Vapors, and Liquids (Art. 501)	A: Acetylene B: Hydrogen, etc. C: Ether, etc. D: Hydrocarbons, Fuels, Solvents, etc.	Normally explosive and hazardous	Not normally present in an explosive concentration (but may accidentally exist)
II Dusts (Art. 502)	E: Metal dusts (conductive* and explosive) F: Carbon dusts (some are conductive,* and all are explosive) G: Flour, Starch, Grain, Combustible Plastic or Chemical Dust (explosive)	Ignitable quantities of dust normally is or may be in suspension, or conductive dust may be present	Dust not normally suspended in an ignitable concentration (but may accidentally exist). Dust layers are present.
III Fibers and Flyings (Art. 503)	H: Textiles, woodworkings, etc. (easily ignitable, but not likely to be explosive)	Handled or used in manufacturing	Stored or handled in storage (exclusive of manufacturing)

*Note: Electrically conductive dusts are dusts with a resistivity less than 10^6 ohm-centimeters.
Source: *An Illustrated Guide to Electrical Safety*, OSHA Pub. No. 3073.

tions between circuits to be sure that an unexpected source of ignition is not introduced through other nonintrinsically safe equipment. Separation of intrinsically safe and nonintrinsically safe wiring may be necessary to make sure that the circuits in hazardous (classified) locations remain safe.

Approved for the hazardous (classified) location. Under this option, equipment must be approved for the class, division, and group of locations. Two types of equipment are designed specifically for hazardous (classified) locations: explosion proof and dust-ignition proof. Explosion-proof apparatus is intended for Class I locations, while dust-ignition-proof equipment is primarily intended for Class II and III locations. Equipment listed specifically for hazardous locations should have a recognized testing laboratory's label or mark indicating the class, division, and group of locations where it may be installed. Equipment approved for use in a Division 1 location may be installed in a Division 2 location of the same class and group.

Explosion proof. Generally, equipment installed in Class I locations must be approved as explosion proof. Since it is impractical to keep flammable gases outside of enclosures, arcing equipment must be installed in enclosures that are designed to withstand an explosion. This minimizes the risk of having an external explosion that occurs when a flammable gas enters the enclosure and is ignited by the arcs.

Not only must the equipment be strong enough to withstand an internal explosion, but the enclosures must be designed to vent the resulting explosive gases. This venting must ensure that the gases are cooled to a temperature below that of ignition temperature of the hazardous substance involved before being released into the hazardous atmosphere.

There are two common enclosure designs: threaded-joint enclosures (see Figure 14-27) and ground-joint enclosures (see Figure 14-28). When hot gases travel through the small openings in either of these joints, they are cooled before reaching the surrounding hazardous atmosphere.

Other design requirements, for example, sealing, prevent the gases, vapors, or fumes from passing into one portion of an electrical system from another. Motors, which typically contain sparking brushes or commutators and tend to heat up, must also be designed to prevent internal explosions.

Dust-ignition proof. In Class II locations,

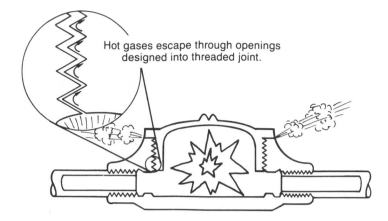

Hot gases escape through openings designed into threaded joint.

Figure 14-27. This threaded-joint explosion-proof exclosure is made of cast metal strong enough to withstand the maximum explosion pressure of a specific group of hazardous gases or vapors. Small openings designed into the threaded joint cool the hot gases as they escape. (From OSHA Publication 3073, *An Illustrated Guide to Electrical Safety.*)

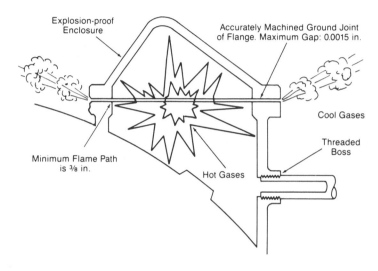

Explosion-proof Enclosure

Accurately Machined Ground Joint of Flange. Maximum Gap: 0.0015 in.

Cool Gases

Threaded Boss

Minimum Flame Path is ⅜ in.

Hot Gases

Figure 14-28. This ground-joint enclosure cools gases as they escape through the ground joint of the flanges. (From OSHA Publication 3073, *An Illustrated Guide to Electrical Safety.*)

equipment must generally be dust-ignition proof. Section 502-1 of the *NEC* defines dust-ignition proof as equipment,

> enclosed in a manner that will exclude ignitable amounts of dust or amounts that might affect performance or rating and that, where installed and protected in accordance with the Code, will not permit arcs, sparks, or heat otherwise generated or liberated inside the enclosure to cause ignition of exterior accumulations or atmospheric suspensions of a specified dust on or in the vicinity of the enclosure.

Dust-ignition-proof equipment is designed to keep ignitable amounts of dust from entering the enclosure. In addition, dust can accumulate on electrical equipment, causing overheating as well as dehydration or gradual carbonization of organic dust deposits. Overheated equipment may malfunction and cause a fire. Dust that has carbonized is susceptible to spontaneous ignition or smoldering. Therefore, equipment must also be designed to operate below the ignition temperature of the specific

dust involved even when blanketed. The shape of the enclosure must be designed to minimize dust accumulation when fixtures are out of reach of normal housekeeping activities, e.g., lighting fixture canopies.

Class II hazardous locations are comprised of three groups—E, F, and G. (See NFPA 497M for an explanation of hazardous atmosphere groups A–G.) Special equipment designs are required for these locations to prevent dust from entering the electrical equipment enclosure. For example, assembly joints and motor shaft openings must be tight enough to prevent dust from entering the enclosure. The design also must take into account the insulating effects of dust layers on equipment and must make sure that the equipment will operate below the ignition temperature of the dust involved. If conductive combustible dusts are present, the design of equipment must take the special nature of these dusts into account.

In general, explosion-proof equipment is not designed for, and is not acceptable for use in Class II locations, unless specifically approved. For example, since grain dust has a lower ignition temperature than that of many flammable vapors, equipment approved for Class I locations may operate at a temperature that is too high for Class II locations. In contrast, equipment that is dust-ignition proof is generally acceptable for use in Class III locations, since the same design considerations are involved.

Safe design for hazardous locations. Under this option, equipment installed in hazardous locations must be of a type and design that provide protection from the hazards arising from the combustibility and flammability of vapors, liquids, gases, dusts, or fibers. The employer has the responsibility of demonstrating that the installation meets this requirement. Guidelines for installing equipment under this option are contained in the *NEC* in effect at the time of installation of that equipment. Compliance with these guidelines is not the only means of complying with this option; however, the employer must demonstrate that installation is safe for the hazardous (classified) location.

Equipment Marking

Approved equipment must be marked to indicate the class, group, and operating temperature range (based on 40 C ambient temperature) for which it is designed to be used. Furthermore, the temperature marked on the equipment must not be greater than the ignition temperature of the specific gases or vapors in the area. There are, however, four exceptions to this marking requirement:

- First, equipment that produces heat but that has a maximum surface temperature of less than 100 C (or 212 F) is not required to be marked with operating temperature range. The heat normally released from this equipment cannot ignite gases, liquids, vapors or dust.
- Second, any permanent lighting fixtures that are approved and marked for use in Class I, Division 2 locations do not need to be marked to show a specific group. This is because these fixtures are acceptable for use with all of the chemical groups for Class I, that is, for Groups A, B, C, and D.
- Third, general-purpose equipment, other than lighting fixtures, considered acceptable for use in Division 2 locations, does not have to be labeled according to class, group, division, or operating temperature. This type of equipment does not contain any devices that might produce arcs or sparks and, therefore, is not a potential ignition source. For example, squirrel-cage induction motors with-

out brushes, switching mechanisms, or similar arc-producing devices are permitted in Class I, Division 2 locations (see *NEC*, Section 501-8(b)); therefore, they need no marking.
- Fourth, for Class II, Division 2, and Class III locations, dust-tight equipment (other than lighting fixtures) is not required to be marked. In these locations, dust-tight equipment does not present a hazard so it need not be identified.

ELECTRICAL STANDARDS MOST OFTEN VIOLATED

To aid the new or even experienced inspector to do a more thorough job of electrical inspecting, a listing of the most often noted electrical standard violations is provided (Table 14-4). This table is a compilation of federal and private sector inspection experience and primarily concerns electrical personnel safety. The average inspector should become adept at recognizing the items in this listing. These items should be included in any electrical inspection checklist you may be using.

Table 14-4 also provides a cross reference to either the *NEC*, 1990 edition, and/or the OSHA standard paragraph. Each electrical standard noted in Table 14-4 is discussed below. Consult the *NEC* or the OSHA standard for additional details and interpretations.

- **110-3.** *Examination, Identification, Installation, and Use of Equipment.* This standard provides operational characteristics for practical employee and facility safeguarding. You should evaluate suitability, mechanical strength of enclosures, insulation, heating and wiring effects, proper use of listed or labeled equipment, and any other factor that would provide personnel safeguarding. Fabricating and using extension cords with junction box receptacle ends would be a violation of this standard.
- **110-12(a).** *Unused Openings.* This reference requires that electrical equipment be installed in a neat and professional manner. All openings in junction boxes and electrical equipment must be effectively closed to prevent metal objects from entering the enclosure and causing arcing or shorting of the supply conductors. Personnel protection is also provided by preventing contact with electrically live parts.
- **110-13(a).** *Secure Mounting of Equipment.*

TABLE 14-4. Most Often Violated Electrical Standards

NEC-NFPA 70–1990 Reference	Subject	OSHA Standard 29 CFR 1910
110-3	Suitability for safe installation and use in accordance with listing or labeling	.303(b)(1)
110-12(a)	Unused openings in cabinets, boxes, and fittings	.305(b)(1)
110-13(a)	Secure mounting of equipment	—
110-14(a)	Electrical terminal connections	—
110-14(b)	Electrical splices	.303(c)
110-16	Working space about electric equipment	.303(g)(1)
110-17	Guarding of live parts	.303(g)(2)
110-22	Disconnect and circuit identification	.303(f)
200-11	Reverse polarity	.304(a)(2)
210-7 & 410-58	Grounding-type receptacles, cord connectors, and attachment plugs	.305(j)(2)(i)
210-8	Ground-fault circuit interrupters	—
210-63	Heating, air-conditioning, and refrigeration equipment outlet	—
250-42	Grounding fixed equipment—general	.304(f)(5)(iv)
250-45	Founding of cord- and plug-connected equipment	.304(f)(5)(v)
250-51	Effective grounding path	.304(f)(4)
250-59	Methods of grounding cord-connected equipment	—
400-7	Flexible cord & cable uses permitted	.305(g)(1)(i)
400-8	Flexible cord & cable uses not permitted	.305(g)(1)(iii)
400-9	Flexible cord & cable splices	.305(g)(2)(ii)
400-10	Pull at joints and terminals	.305(g)(2)(iii)

All fixed electrical equipment must be firmly secured to the surface on which it is installed. Many times you may discover a conduit hanging loose or equipment boxes not secured to the wall. These are examples of violations of this standard.

- **110-14(a).** *Electrical Terminal Connections.* Loose or improperly tightened terminal connections have been determined as the cause of many electrical fires and equipment burnouts. It is important to be alert to this problem when intermittent equipment operation or flickering lights are observed.

- **110-14(b).** *Splices.* Splices are required to be joined by suitable splicing devices or methods. The three elements of a proper splice are (a) mechanical strength, (b) electrical conductivity, and (c) insulation quality. These factors must at least be equivalent to the conductors being spliced.

- **110-16.** *Working Space About Electrical Equipment* (600 volts, nominal, or less). This paragraph requires that sufficient access and working space be provided and maintained around all electrical equipment. It is important that the electrical equipment clearance space does not become storage space. It may be necessary to make the clearance space obvious by using floor stripes or other methods.

- **110-17.** *Guarding of Live Parts* (600 volts, nominal, or less). This reference requires that the live parts of electrical equipment operating at 50 volts or more must be guarded against accidental contact. Approved enclosures are recommended. When the alternatives of location (e.g., 8-ft elevation above the floor) are used, employee safeguarding should be carefully evaluated.

- **110-22.** *Disconnect and Circuit Identification.* This reference requires that the disconnecting means for motors and appliances, and each service, feeder, or branch circuit at the point where it originates be legibly marked to indicate its purpose unless located and arranged so the purpose is evident. Many times a contractor will install new wiring and leave circuit-breaker panels with blank circuit identification cards. Whenever electrical modifications are completed, the circuit identifications cards should be updated.

- **200-11.** *Polarity of Connections.* No grounded conductor should be attached to any terminal or lead so as to reverse designated polarity. This usually can be determined by using the 3-prong circuit tester. Be sure to test not only wall receptacles but also extension cord receptacles.

- **210-7 & 410-58.** *Grounding-Type Receptacles, Cord Connectors, and Attachment Plugs.* These references require that receptacles installed on 15- and 20-amp branch circuits be of the grounding type. Grounding-type attachment plugs and mating cord connectors must also be used on electrical equipment where grounding-type receptacles are provided. An exception would be the use of UL-listed double-insulated tools or appliances.

- **210-8.** *Ground-Fault Circuit-Interrupter Protection (GFCI) for Personnel.* Whereas this section applies to dwelling units, hotels, and motels, be reminded that GFCI protection may be provided for other circuits, locations, and occupancies where personnel can be protected against line-to-ground shock hazards. GFCI-protection requirements for other specific applications are as follows: 305-6 Construction Sites; 426-6 Fixed Heating for Pipelines; 426-31 Fixed Outdoor De-Icing Equipment; 511-10 Commercial Garages; 551-41 (c) Recreational Vehicles; and 680 Swimming Pools, Tubs, and Fountains. The various codes and regulations are minimal requirements and in many cases not retroactive. From an accident-prevention standpoint, you should look for areas where potential line-to-ground shock hazards could occur and recommend GFCI protection regardless of the lack of retroactivity.

- **210-63.** *Heating, Air-Conditioning, and Refrigeration (HVAC) Equipment Servicing Outlet.* It is now a requirement that a 125-volt, 15- or 20-amp receptacle be installed on rooftops and in attic spaces for use by maintenance personnel for servicing HVAC equipment. These receptacles must be on the same level and within 75 ft of the HVAC equipment. Although the Code does not require it, you should consider providing GFCI-protected receptacles if maintenance work must be done during wet weather or in roof areas where water may be present.

- **250-42.** *Grounding Fixed Equipment—General.* This paragraph requires exposed noncurrent-carrying metal parts of fixed equipment to be grounded. By using the noncontact voltage detector, you can quickly determine whether the equipment metal housing is properly grounded.

- **250-45.** *Grounding of Cord- and Plug-Connected Equipment.* This paragraph requires that exposed noncurrent-carrying metal parts of cord- and plug-connected equipment (listed) shall be grounded. Exceptions to this requirement are tools and appliances listed as double-insulated or that are supplied through an isolating transformer with an ungrounded secondary of not more than 50 volts.

- **250-51.** *Effective Grounding Path.* The path to ground from circuits, equipment, and conductor enclosures must (a) be permanent and continuous, (b) have capacity to conduct safely any fault current likely to be imposed on it, and (c) have sufficiently low impedance to limit the voltage to ground and to facilitate the operation of the circuit-protection devices. If a portable power tool has the equipment-grounding prong broken off, the 250-51 (1) requirement would not be met since the grounding path is not continuous. Extension cords being used with defects in the equipment-grounding conductor path would also not meet this requirement. Finally, if the grounding path tested out at more than 10 ohms, obviously 250-51 (3) would not be met. It is recommended that the equipment-grounding conductor loop impedance be less than 1 ohm. This can be determined only by using a ground-loop impedance tester.

- **250-59.** *Cord- and Plug-Connected Equipment.* This section describes three methods that can be used to ground equipment effectively. One method is to use the electrical conduit in conjuction with approved connectors and receptacles. The second method is to run the equipment-grounding conductor along with the power-supply conductors in a cable assembly or flexible cord. The third method is to use a separate flexible wire or cable. The main purpose of alternative methods is to ensure that all metal enclosures and frames are at the same ground potential.

- **400-7.** *Flexible Cord and Cable Uses Permitted.* Specific applications where flexible cords and cables are permitted include pendants, wiring of fixtures, portable lamps and appliances, and stationary equipment that may have to be moved frequently for maintenance or other purposes. Another important use is to prevent the transmission of noise and vibration. Special attention should be given to the proper installation and protection of flexible cords and cables.

- **400-8.** *Flexible Cord and Cable Uses Not Permitted.* Flexible cords must not be used as a substitute for fixed wiring. Additionally, they must not be run through holes in walls, ceilings, or floors nor through doorways or windows.

- **400-9.** *Splices.* This section prohibits the use of flexible cords that have been spliced or tapped. Using splices to repair hard-service cord No. 14 or larger is permitted if done in accordance with the previously covered Section 110-14(b) Splices.

• **400-10.** *Pull at Joints and Terminals.* This reference requires that flexible cords be connected to devices and to fittings so that tension will not be transmitted to joints or terminal screws. This requirement applies to the points where the cord connects to the appliance and to the attachment plug.

The above list is not intended to be comprehensive, but rather identifies some of the most common electrical safety problem areas. To determine the exact requirements, consult the *NEC* or the OSHA standard noted. Using the testing methods noted in the Branch Circuit and Equipment Testing Methods section of this chapter will help you to find and document many of these hazards.

INSPECTION GUIDELINES AND CHECKLIST

Before you conduct an electrical inspection, check your test equipment to be sure it works properly. As a minimum, you should have a circuit tester, a GFCI tester (combination circuit/GFCI testers are available), a contact tension tester, and a volt-ohmmeter. Avoid wearing metal jewelry or loose hanging necklaces. Your footwear should have synthetic soles (do not wear leather soles). A clipboard, writing material, and the Table 14-5 checklist should also be included. A camera is optional; if used, remember that before and after photos or slides make valuable visual aids for training use.

Table 14-5 provides general guidelines that will

TABLE 14-5. Inspection Guidelines

1. **Service Entrance Panel**—Circuit I.D., Secure Mounting, Knockouts, Connectors, Clearances, Live Parts, Ratings.
2. **System Grounding**—Secure Connections, Corrosion, Access, Protection, Proper Size.
3. **Wiring (General)**—Temporary, Splices, Protected, J Box Covers, Insulation, Knockouts, Fittings, Workmanship.
4. **Electrical Equipment/Machinery**—Grounding, Wiring Size, Overcurrent/Disconnect Devices, Installation, Protection.
5. **Small Power Tools**—Attachment Plugs, Cords, Cord Clamps, Leakage, Grounding.
6. **Receptacles**—Proper Polarity, Adequate Number, Mounting, Covers, Connections, Protection, Adapters.
7. **Lighting**—Grounding, Connections, Attachment Plugs and Cords, Cord Clamps, Live Parts.
8. **GFCI Protection**—Bathrooms, Wet Locations, Fountains, Outdoor Circuits, Testing.

assist you in checking for electrical hazards from the service entrance panel(s) to the equipment using the power. Further explanation of these guidelines is as follows:

1. Service Entrance Panel—Check the branch circuit identification. It should be up to date and posted on the panel door. Be sure the panel and cable or conduit connectors are secure. There should be no storage within 3 ft of the panel. No flammable materials of any kind should be stored in the same area or room. Look for corrosion and water in or around the area. Missing knockouts, covers, or openings must be covered properly to eliminate workers' exposure to live parts.

2. System Grounding—Check connection of the grounding electrode conductor to the metal cold-water pipe and to any driven ground rod. Also check any bonding jumper connections and any supplemental grounding electrode fittings. These items should not be exposed to corrosion but should be accessible for maintenance and visual inspection.

3. Wiring (General)—Temporary wiring used for an extended time should be replaced with fixed wiring. Conduit and/or cable systems must be protected from damage by vehicles or other mobile equipment. All fittings and connections to junction boxes and other equipment must be secure. No exposed wiring can be allowed. Check for missing knockouts and cover plates. Jury-rigged splices on flexible cords and cables should be repaired correctly. Electrical equipment should be installed in a neat and professional manner. Check for damaged insulation on flexible cords and pendant drops.

4. Electrical Equipment/Machinery—The most important item is to test for proper grounding. All electrical equipment and machinery must be grounded effectively so that no potential difference exists between the metal enclosures. Use the voltage detector to find discrepancies and use other test equipment to determine the corrective action required.

   Disconnects should be identified clearly regarding the specific machinery they shut off and should be placed near the machinery for emergency use. The disconnects should be exercised periodically to be sure they are operable. All electrical connections to the equipment must be secure so that no wiring tension is transmitted to the electrical terminals within the equipment. The wiring installation should be protected from damage at all times while the machinery is being used.

5. Small Power Tools—Attachment plugs should be checked for defective cord clamps and broken or missing blades. Connection of the cord to the power tool should be secure. Use your ohmmeter to check for leakage and an effective equipment-grounding conductor.

6. Receptacles—The receptacles should be tested for proper wiring configuration. Be sure to install enough receptacles to eliminate the use of extension cords as much as possible. Covers should be in place and not broken. Multiple outlet adapters on a single outlet should be discouraged to prevent overloading. Surface-mounted receptacle boxes should be protected from damage by housekeeping or other mobile equipment.

7. Lighting—Cord- and plug-connected metal lamps and fixtures should be tested for grounding. Use the ohmmeter to do this. Check all cord clamps for secure connections. Frayed or old cords should be replaced.

8. GFCI Protection—Generally, GFCI protection is not required by the *NEC* on a retroactive basis. However, where employees are exposed to potential line-to-ground shock hazards, GFCI protection is recommended. This is especially important in workplaces where portable electrical equipment is being used in wet or damp areas in contact with earth or grounded conductive surfaces.

Typically, areas where GFCIs should be installed include bathrooms, outdoor fountains, tubs, countertop sinks, crawl spaces under buildings, outdoor circuits, swimming pools, and other wet locations. Use your with GFCI tester to be sure that the GFCI is operable. After years of use, they can become defective and will need to be replaced. Check to make sure that receptacles being provided with GFCI protection are labeled as such.

SAFEGUARDS FOR PORTABLE HOME ELECTRICAL APPLIANCES

This section is primarily provided for use in off-the-job safety training. There are instances where home appliances are used on the job, and the following general precautions would be applicable to those situations. It is recommended that any electrical train-

ing include discussions of safeguards for small, portable, electrical home appliances. You should recommend installation of GFCIs in bathrooms and other locations required by the *NEC* for new dwelling construction.

When using electrical appliances, especially when children are present, basic safety precautions should always be followed. Appliance manufacturers provide written procedures for safe use of their appliances. Their instructions should be read by adults and children alike. Safety precautions should be understood. Strong emphasis should be given on the danger of using small electrical appliances in or near water. The following list established a commonsense baseline for the safe handling, use, and storage of electrical appliances in the home. To reduce risk of electrocution:

- Install GFCI-protected receptacles in bathrooms, near kitchen sinks, outdoors, and in other areas near water or where wet or damp conditions may exist.
- Always unplug and store appliances after using.
- Do not use any electrical appliance while bathing. For examples of warning labels, see Figure 14-29.
- Do not place or store appliances where they can fall or be pulled into a sink or tub (such as on towel bar on tub door).
- Do not place or drop any appliance into water or other liquid.
- Do not reach for an appliance that has fallen

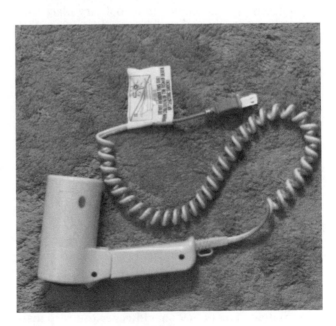

Figure 14-29. Do not remove hair dryer labels warning against using them in bathtubs.

into water. Unplug the electrical cord immediately.

To reduce the risk of burns, electrocution, fire, or injury:

- Never leave an appliance unattended when it is plugged in.
- Closely supervise an appliance used by, on, or near children or invalids.
- Use the appliance only for its intended use as described in the manufacturer's manual. Do not use attachments not recommended by manufacturer.
- Never operate an appliance if it has a damaged cord or plug, is not working properly, has been dropped or damaged, or has been immersed in water. Return the appliance for examination and electrical or mechanical adjustment and repair.
- Keep the cord away from heated surfaces.
- Never block the air openings of an appliance. Keep the air openings free of lint, hair, paper, or other materials.
- Never use appliances when you are drowsy.
- Never drop or insert any object into any appliance opening.
- Do not use an appliance where flammable liquid vapors or aerosol (spray) products are being used or where oxygen is being administered.
- Do not attempt to repair appliances yourself. Most small electrical appliances do not lend themselves to normal repair procedures.
- Keep your appliance in a cool, dry place and out of the reach of children. When handling or storing your unit, do not put any stress on the line cord and the point where it enters the unit. Excessive stress could cause the cord to fray and to break.
- Never plug an appliance into a damaged receptacle or one that has a missing face plate.
- Keep all electrical appliances away from hot tubs, bath tubs, circulating water tubs, sinks, and swimming pools.

SAFETY PROGRAM POLICY AND PROCEDURES

Safety Policy for Electrical Equipment

Each facility should have an electrical safety program policy. The policy should cover the responsibilities of all employees, including supervisors, workers, and the specialists who inspect, install, and

maintain the electrical systems and equipment. The policy should stress management's concern and support of safety issues. Individuals responsible for applying and enforcing the electrical policy should have standards of performance which include periodic assessment of their electrical safety performance.

In addition to policy and implementation procedures, an electrical safety program should also include four basic areas of concern: training and education, hazardous condition reporting, work practices, and housekeeping. The National Safety Council Data Sheets (Table 14-6) provide an excellent source of technical and training information to support any facility electrical safety program. In addition, the professional electrician responsible for installing, repairing, and maintaining electrical

TABLE 14-6. National Safety Council Data Sheets that Pertain to Electrical Safety Programs

Data Sheet No.	Title
I-240	Cathode-Ray Tubes
I-248	Auxiliary Electrical Systems and Emergency Lighting
I-316	Low-Voltage Extension Light Cords and Systems
I-385	Electric Cords and Fittings
I-498	Live Line Tools
I-515	Temporary Electric Wiring for Construction Sites
I-544	Electrical Switching Practices
I-546	Maintenance of Electric Motors for Hazardous Locations
I-547	Static Electricity
I-579	Applications of Electric Plug and Receptacle Configurations
I-581	Silicon Diode Grounding Devices
I-598	Flexible Insulating Protective Equipment for Electrical Workers
I-607	Direct Buried Utility Cables
I-620	Care, Fitting, and Maintenance of Lineman Climbers
I-624	Electrical Controls for Mechanical Power Presses
I-635	Lead-Acid Storage Batteries
I-636	GFCIs for Personnel Protection
I-641	Electrical Testing Installations
I-644	Treatment of Extraneous Electricity in Electric Blasting
I-657	Undergound Residential Distribution of Electricity
I-660	Electrical Safety in Health Care Facilities
I-675	Electric Hand Saws, Circular Blade Type
I-684	Equipment Grounding

equipment and systems should be familiar with NFPA 70E, *Electrical Safety Requirements of Employee Workplaces.*

Management should support an effective, preventive maintenance program. Use NFPA 70B, *Electrical Equipment Maintenance,* as a guide to implement or fine tune this type of program. All employees are responsible for detecting and reporting unsafe electrical equipment. The following discussion provides suggestions for ensuring that these responsibilities are carried out effectively.

Supervisory Responsibilities

Training and education. Supervisors should be given training courses to assist them in discharging their electrical safety program responsibilities for their specific areas. If any employees under your supervision use, install, repair, or modify electrical equipment and/or appliances, you must ensure that they have received the proper training. You should also monitor employees and assess their performance against the established facility safety program policy.

Hazardous condition reporting. A written procedure promoting the observation and reporting of electrical hazards should be implemented. You might also institute an employee recognition program in conjunction with hazard reporting. This will reward employees who help to locate electrical hazards so they can be controlled in a timely manner.

Work practices. You must make sure that employees follow safe work practices. A sample of suggested work practices is included in this section under Employee Responsibilities. Employees should be rated on their performance in following safe work practices. You should also be familiar with OSHA standards as they apply to the workplace under your responsibility.

Housekeeping. It is important that housekeeping be monitored closely by the supervisor. Floor-area problems always present challenges to the supervisor. Areas around electrical equipment such as circuit-breaker panels, disconnects, and fixed power tools should be kept free from stored items, debris, and any liquids or material that would create slippery floors or obstruct access to the equipment for maintenance or emergencies. When hazards of this nature are reported to the supervisor, they should be logged in and necessary work orders issued for corrective action.

Employee Responsibilities

Training and education. Employees must receive training in electrical safety work practices and equipment operation. Any changes in job duties will require additional safety training. Many accidents are caused by employees' lack of knowledge about the equipment or its operation. Sometimes blame for accidents is placed on employees when, in reality, specific training was not provided.

Hazardous condition reporting. The employee should always report unsafe equipment, conditions, or procedures. A team effort is required by both the employee and the supervisor. Getting equipment repaired should receive priority, even if it requires rescheduling a process or project. Under no conditions should defective electrical equipment causing electrical shocks to employees be used to get a job done. The Electrical Safety Policy should be followed and deviations reported immediately.

Work practices. Employees are responsible for following safe work practices, procedures, and safety policy as established by the employer. Workers should also be familiar with OSHA regulations as they apply to workplace safety.

Housekeeping. In the process of routine housekeeping, employees should be observant and report conditions that could cause virtually any type of accident or electrical shock hazards. For example, the use of improperly grounded electrical equipment in areas that have water on the floor can expose workers to the risk of electrical shock. Storing tools or other materials around electrical panels or equipment disconnects can create hazards for others as well as prevent immediate disconnection in an emergency. Cleaning tools and electrical equipment with solvents can create health and physical safety problems. Also, discarding these rags into trash could create fire hazards.

Electrical Safety Policy

Supervisors must know all facets of their employers' electrical safety policy and make sure workers understand what is expected of them. You should stress that, as a minimum, the following items must be observed:

- Plug power equipment into wall receptacles with power switches in the OFF position.
- Unplug electrical equipment by grasping the plug and pulling. Do not pull or jerk the cord to unplug the equipment.
- Do not drape power cords over hot pipes, radiators, or sharp objects.
- Check the receptacle for missing or damaged parts. Do not plug equipment into defective receptacles.
- Check for frayed, cracked, or exposed wiring on equipment cords. Also check for defective cord clamps at locations where the power cord enters the equipment or the attachment plug.

- Extension cords should not be used in office areas. Generally, extension cords should be limited to use by maintenance personnel.
- "Cheater plugs," extension cords with junction box receptacle ends, or other jury-rigged equipment should not be used.
- Consumer electrical equipment or appliances should not be used if not properly grounded.
- Personnel should know the location of electrical circuit-breaker panels that control equipment and lighting in their respective areas. Circuits and equipment disconnects must be identified.
- Temporary or permanent storage of any materials must not be allowed within 3 ft of any electrical panel or electrical equipment.
- When defective electrical equipment is identified by personnel, it should be tagged immediately and removed from service for repair or replacement.
- Any electrical equipment causing shocks or with high leakage potential must be tagged with a DANGER—DO NOT USE label or equivalent.

ELECTRICAL DISTRIBUTION SYSTEM REVIEW

The electrical distribution system should be designed, installed, operated, and maintained to provide electrical power for all required operations in a safe and reliable manner. The facility should be analyzed to determine that the following requirements are documented and observed regarding the overall electrical system operation and maintenance. This questionnaire should be used to determine compliance to OSHA and *NEC* requirements. If there are any No answers, obtain professional electrical services to correct the situation and to eliminate potential hazards.

- There is a program of preventive maintenance and periodic inspection to ensure that the electrical distribution system operates safely and reliably. Inspections and corrective actions are documented.
  Yes   No
- The facility identifies components of the electrical distribution system to be included in the program. Consideration is given to reliability of receptacles, wiring, fuse and/or breaker panels, transformers, switch gear, and service equipment.
  Yes   No
- There is a current set of documents which

indicate the distribution of and controls for partial or complete shutdown of each electrical system.
  Yes   No
- Electrical maintenance and operating personnel are provided with appropriate job training, and written records of this training are maintained.
  Yes   No
- The capacity of electrical feeds and transformers is adequate for the electrical demands of the facility.
  Yes   No
- Any special-purpose electrical subsystems or devices installed are maintained as required.
  Yes   No
- The electrical distribution system is designed with enough receptacles and circuits to power devices used in each area of the facility without the use of extension cords.
  Yes   No

SUMMARY OF KEY POINTS

This chapter covered the following key points:

- Electrical safety is one of a supervisor's most important responsibilites to ensure safe production. Supervisors must know the facts about electricity and be able to counter the myths and misconceptions that workers may hold.
- The four basic rules of electrical action are (1) electricity isn't "live" until current flows, (2) current won't flow until there is a complete loop out from and back to the power source, (3) current always returns to its source, and (4) when current flows, work measured in watts is performed. These four rules can help supervisors to analyze electrical shock hazards or investigate how and why electrical injuries occurred in their department.
- According to Ohm's Law, it is current that can injure or kill a worker, while voltage simply determines how—either by burns or electrocution. The human body represents resistance to current. Should a worker establish an electrical loop between himself or herself and a ground, the current will pass through the body on the way to the ground. Even a small amount of current can cause the heart to go into ventricular fibrillation, which can produce sudden death if the heart is not resuscitated within a few minutes.

  Factors that can enhance shock potential

include wet or damp locations, nearby grounded objects that the worker or electrical tool may touch, lack of worker training, and type and amount of voltage and current.

- Thus, the keys to preventing electrical shock hazards are (1) to train workers thoroughly in safe work practices; (2) to wire, insulate, and ground all electrical equipment properly so that there are no ground loops between an electrical tool or machine and the worker; and (3) routinely inspect, maintain, and replace such equipment, along with all cords, outlets, plugs, attachments, and power sources.

- Supervisors should understand the single-phase, three-element, 120-volt electrical system that powers all industrial and commercial operations in the United States. The 240-volt, single-phase power is used for heavy-duty appliances and small motor loads. If the system is not wired properly, or if a tool or appliance is not plugged in properly, it can create a condition known as reverse polarity. In reverse polarity the entire casing of a tool or machine is energized and creates a shock hazard, particularly if accidentally immersed in water.

- A ground is a conducting connection, whether intentional or accidental, between an electrical circuit or equipment and earth, or to some conducting body that serves in the place of earth. A ground loop can become an electrical hazard if workers accidentally form part of the loop. Grounding falls into two safety categories—system grounding and equipment safety grounding.

  System grounding provides protection of the power company equipment and the consumer. Equipment safety grounding provides a system where all noncurrent-carrying metal parts are bonded together and kept at the same potential. Grounding concepts include the grounded conductor, neutral conductor, and equipment-grounding conductor.

- Power tools and equipment can be designed to protect workers against shock hazards by being double insulated or by being equipped with 3-pronged plugs in which one prong acts as a ground. However, these safeguards are not foolproof, and workers must remember to inspect and test their equipment regularly. Double-insulated equipment should not be used in damp or wet conditions nor immersed in water.

- Attachment plugs and receptacles are designed for specific voltage/current combinations and should not be used for any other combination. To do so can cause fires or shock

hazards to personnel using the equipment. Likewise, all electrical extension cords should be the proper gage to carry the current load required. These cords should be UL listed, tested before being put into use, and inspected regularly.

- Supervisors can use several testing procedures for branch circuits and electrical equipment to guard against shock hazards and ensure safe production. Branch circuit receptacles should be tested periodically to ensure that electrical outlets are safe and in good working order. Inspectors often use a three-step process using a 3-prong receptable circuit tester, a receptacle contact tension tester, and a ground-loop impedance tester. Extension cords should be tested using the same three-step procedure to make sure the cords are in good repair and can carry the required current safely.

  Finally, cord- and plug-connected equipment should be tested using a two-step procedure of ground-to-pin testing and an appliance leakage test. Inspectors should be alert for jury-rigged repairs made on electric power tools and appliances and for signs of wear or deterioration. An ohmmeter can be used where electric power is not available for testing equipment. A voltage detector can be used in conjunction with the circuit tester, ohmmeter, and tension tester to identify electrical hazards. It can detect voltage waveforms surrounding an ungrounded, or hot, conductor. It is also voltage sensitive and can detect voltage from a distance.

- The ground-fault circuit interrupter (GFCI) is a fast-acting device that monitors the current flow to a protected load. If it senses a leakage current, it will shut off the current on grounded and hot conductors. The GFCI protects workers against electrical hazards caused by ground faults, but it will not protect workers from line-to-line electrical contact. GFCIs should be used in areas where workers are exposed to wet or humid conditions and where they may come in contact with ground or grounded equipment. GFCIs do not prevent shock but only limit the duration so the heart is not affected.

- Hazardous locations are classified as Class I (flammable vapors and gases present), Class II (combustible dusts present), and Class III (ignitable fibers and flyings present). Each class is divided into two hazard categories: Division 1—flammable gas, vapor, dust, or easily ignitable material is present in hazardous quantities, and Division 2—hazardous quantities of these materials is not normal but may exist occasionally. Installations in hazardous locations must consist of equipment and wiring that is (1) intrinsically safe, (2) approved for hazardous locations, or (3) of a type or design that provides protection from the hazards present. Generally, equipment installed in Class I locations must be approved as explosion proof, while Class II equipment must be dust-ignition proof. With some exceptions, equipment must be clearly marked to indicate the class, group, and operating temperature range for which it is designed.

- Supervisors and inspectors must be familiar with the electrical standards most often violated on the job. This knowledge will help them identify and correct electrical hazards frequently occurring in the workplace. Guidelines established for inspecting electrical equipment and wiring include the following items: service entrance panel, system grounding, electrical equipment machinery, small power tools, receptacles, lighting, and GFCI protection.

- Supervisors should also instruct workers in off-the-job safety training to guard against electrical hazards in the home. These precautions can prevent risk of burns, electrocution, fire, or injury to the workers or their family members.

- Finally, supervisors must establish ongoing safety program policies and procedures for their departments and enforce the program consistently. An electrical safety program should include training and education, hazardous condition reporting, safe work practices, and housekeeping. Management personnel and employees should know clearly what their responsibilities are and what is expected of them. Part of the safety program involves setting up an electrical distribution system review to provide safe, reliable electric power for all required operations.

15

Fire Safety

After reading this chapter, you will be able to:

- Understand the basic principles of fire safety, including the chemistry of fires
- Identify fire hazards, causes of fires, and safeguards required to prevent fires
- Conduct regular, periodic inspections to ensure that work areas remain in a fire-safe condition
- Understand the use and operation of fire protection equipment and systems
- Instruct employees in the procedures for reporting fires, fighting fires, and evacuating work areas
- Develop a sound fire protection education program for on and off the job

An effective fire protection program must, in the long run, depend on you, the supervisor. Although the overall program may be under the authority of the director of safety, fire protection, security, engineering, or maintenance, each supervisor also has a direct interest in, and responsibility for, the program's success.

Fire protection is a science in itself. This manual obviously cannot cover all aspects of fire prevention and extinguishment, but this chapter does present condensed, basic information to help supervisors conduct fire-safe operations. More specialized information is available, and sources are cited throughout this chapter.

BASIC PRINCIPLES

Fire is an oxidation process that emits light and heat. In order to explain or understand fire development, experts have devised various fire models. One of the earliest is shown as a tetrahedron in Figure 15-1. To sustain most fires, four elements must be available at the same time: elevated temperature, oxygen, fuel, and an uninhibited chain reaction. Fire extinguishing agents act by removing one or more of the tetrahedron's sides. However, as Figure 15-1 shows, an uninhibited chain reaction is not necessary for a deep-seated, surface-glowing type of fire to continue. As the supervisor, you need to understand that under the proper conditions, many materials will burn. The best example is steel, which in the form of steel wool burns quite readily.

Fire spreads from an ignition source to a fuel source to other fuel sources by conduction, convection, and radiation. Conduction transfers heat through contact with solid material. Convection transfers heat through heated air. Radiation heat transfers through electromagnetic waves given off by flames.

Supervisors, because of their knowledge of day-to-day operations, are in an excellent position to determine necessary fire prevention measures. You should be able to recognize the need for specific fire protection equipment and see to it that such equipment is provided. Take the time to become thoroughly familiar with the fire equipment best suited to your particular operations.

Departmental housekeeping is also under the

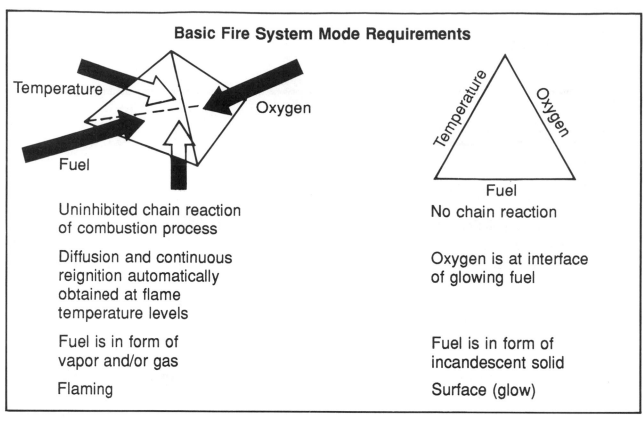

Basic Fire System Mode Requirements

Uninhibited chain reaction of combustion process

Diffusion and continuous reignition automatically obtained at flame temperature levels

Fuel is in form of vapor and/or gas

Flaming

No chain reaction

Oxygen is at interface of glowing fuel

Fuel is in form of incandescent solid

Surface (glow)

Figure 15-1. In the presence of oxygen, fuel, and elevated temperature, an un-inhibited chain reaction can produce a flaming fire, but without a chain reaction you can still produce a deep-seated, surface-glowing fire.

supervisor's control. Make sure your employees follow safe housekeeping practices to prevent fires. Continuous training in fire-safe work procedures, regular inspections of work areas, and close supervision of employee job performance are the primary requirements of a successful fire prevention program. Although fire protection equipment may be maintained by others, you have ultimate responsibility for the safety of employees, materials used in processes, and production equipment.

Understanding Fire Chemistry

Supervisory expertise, in regard to fire prevention and control problems, begins with an understanding of basic fire chemistry. Every ordinary fire (one that does not produce its own oxygen supply) results when a substance (fuel) in the presence of air (oxygen) is heated to a critical temperature, called its "ignition temperature."

This important concept is best illustrated by the fire tetrahedron. By looking at this model, you can understand how most fires are extinguished. The methods used include:

• **Oxygen removal.** removing or lowering ox-

ygen levels is difficult because a fire needs about the same amount for burning (percentage of oxygen in the air) that humans need for breathing. In some cases, oxygen levels can be reduced below the minimum percentage needed for combustion by purging and rendering the atmosphere inert in closed containers or processing systems. Firefighting foam extinguishes fires by smothering (and cooling) action.

• **Fuel removal.** In many cases, it is neither possible nor practical to remove all fuels (solids and liquids). However, try to keep the quantity of stored combustible materials at a minimum. Storage containers must be placed in orderly piles with adequate aisle space. Good housekeeping, with the frequent removal of waste materials, is also a crucial factor in keeping a small, accidental fire from rapidly spreading.

• **Heat source control.** Eliminating and controlling heat sources also are elementary steps in fire prevention. Conscientious workers can control the use of welding and cutting equipment, torches, heating equipment, spark-producing equipment, electricity, and smoking materials. The time to stop a fire is before it starts—keep heat and ignition sources away from fuel.

Table 15-1 lists common sources of ignition that cause industrial fires, gives examples in each case, and suggests preventive measures. A following section, Causes of Fire, discusses each in detail.

For many years, the principle of extinguishing fires focused on the fire triangle and the removal of any of its three sides, representative of three components. Remodeling the fire triangle into a fire tetrahedron (Figure 15-1) presents a more realistic concept of extinguishment. This is because it also takes into consideration the chemical chain reaction needed to sustain combustion. The tetrahedron has four sides or faces, one for each of the ways to extinguish a fire. Because each face is directly adjacent to and connected to the others, a tetrahedron accurately represents their interdependency. Removing one or more of the faces will put out the fire.

Consequently, to extinguish a fire, the following steps should be taken:

1. Reduce or remove the oxygen by smothering (for example, by shutting the lid over a tank of burning solvent or by covering it with

TABLE 15-1.   How to Control Sources of Ignition in Industrial Fires

Sources of Ignition	Examples	Preventive Measures
Electrical equipment	Electrical defects, generally due to poor maintenance, mostly in wiring, motors, switches, lamps, and hot elements.	Use only approved equipment. Follow *National Electrical Code.* Establish regular maintenance.
Friction	Hot bearings, misaligned or broken machine parts, choking or jamming of materials, poor adjustment.	Follow a regular schedule of inspection, maintenance, and lubrication.
Foreign substances	Tramp metal, which produces sparks when struck by rapidly revolving machinery (a common cause in textile industry).	Keep foreign material from stock. Use magnetic or other separators to remove tramp metal.
Open flames	Cutting and welding torches (chief offenders). Gas and oil burners. Misuse of gasoline torches.	Follow established welding precautions. Keep burners clean and properly adjusted. Do not use open flames near combustibles.
Smoking and matches	Dangerous near flammable liquids and in areas where combustibles are used or stored.	Smoke only in permitted areas. Use prescribed receptacles. Make sure matches are out.
Spontaneous ignition	Deposits in ducts and flues. Low-grade storage. Industrial wastes. Oily wastes and rubbish.	Clean ducts and flues frequently. Remove waste daily. Isolate stored materials likely to heat spontaneously.
Hot surfaces	Exposure of combustibles to furnaces, hot ducts or flues, electric lamps or irons, hot metal being processed.	Provide ample clearances, insulation, air circulation. Check heating apparatus before leaving it unattended.
Combustion sparks	Rubbish-burning, foundry cupolas, furnaces and fireboxes, and process equipment.	Use incinerators of approved design. Provide spark arresters on stacks. Operate equipment carefully.
Overheated materials	Abnormal process temperatures. Materials in driers. Overheating of flammable liquids.	Have careful supervision and competent operators, supplemented by well-maintained automatic temperature controls.
Static electricity	Dangerous in presence of flammable vapors. Occurs at spreading and coating rolls or where liquid flows from pipes.	Ground equipment. Use static eliminators. Humidify the atmosphere.

Adapted from *Factory Mutual Record.*

foam) or by dilution (replacing the air with an inert gas such as carbon dioxide).

2. Remove or seal off the fuel by mechanical means, or divert or shut off the flow of liquids or gases fueling the fire.

3. Cool the burning material below its ignition point with a suitable cooling agent (hose streams or water extinguishers).

4. Interrupt the chemical chain reaction of the fire (using dry chemical or halon extinguishing agents).

Once people understand the fire tetrahedron and its practical application, they will be more alert to, and aware of, fire prevention and control methods.

Determining Fire Hazards

As supervisor, you can contribute to a fire protection program in two ways: first, identify the existing fire hazards and, second, take action to resolve them. To do this, seek the best technical advice available from experts.

If you have not already done so, develop an inspection checklist that specifies as many places, materials procedures, classes of equipment, conditions, and circumstances where fire hazards are likely to exist as possible. Use these category names as headings on the list, and under each one write down the precise fire-safe practices that must be followed. Specify which personnel—for example, machine operator, electrician, maintenance crew, or porter—have specific responsibilities (see discussion of inspections in Chapter 6, Safety Inspections).

When, after a tour of inspection, you have made brief notations beside each of the described safe practices, you will have a detailed picture of how many or how few fire protection measures are actually in effect. You will be able to spot neglected precautions readily. The sample fire prevention checklist shown in Figure 15-2 can serve as a guide, but you should make your own list and include special points for conditions that are not listed in this broad example.

When conducting an inspection, you may discover that some points previously escaped your notice. Unless you have had considerable experience in fire prevention, you should ask your superior and your company's fire safety personnel, or the fire insurance company engineer, to help you conduct inspections and to make recommendations for eliminating fire hazards. If company policy permits, you can ask the local fire department for help as well.

It is important that all fire inspections be made with a critical eye. Every shortcoming should be listed. You should not hedge in listing certain hazards for fear that they might reflect poor supervision. Omission of a pertinent detail might result in a fire later; such an outcome would reflect poor supervision.

The National Fire Protection Association (NFPA) *Inspection Manual*, a pocket-sized book, is a valuable reference for the beginner as well as the experienced inspector. It covers both common and special fire hazards, their elimination or safeguarding, and human safety in all types of properties.

Informing the Work Force

Periodic inspections are an important part of any fire protection program. However, your responsibilities, under a complete program, extend further. As you become acquainted with actual or potential fire hazards, and after all physical corrections possible have been made, you should familiarize the personnel in your department with each hazard and explain how it relates to them individually.

You should inform your people of your own and management's desire for fire-safe operations. You should call their attention to all the physical safeguards that have been provided to prevent injury and destruction by fire. You should stress to individuals the precautions necessary for their jobs—the safe practices that complement mechanical protection.

If the individuals on the job understand the reason for precautions and the possible consequences if fire-safety rules are not followed, they are much more likely to comply. Patient explanation and persistent enforcement, in every case, are important fire prevention duties of the supervisor.

CAUSES OF FIRE

The supervisor should be alert for potential causes of fire, shown in Table 15-1 and discussed below. Other fire prevention concerns are also noted in the following sections.

Electrical Equipment

Electrical motors, switches, lights, and other electrical equipment exposed to flammable vapors, dusts, gases, or fibers present special fire hazards. The NFPA's code and standards pamphlets designate the standard governing a particular hazard and indicate the special protective equipment needed. *The National Electrical Code*, ANSI/NFPA 70-1990 (the *NEC*), gives the specifications for the protective equipment required. Substandard substitutions or replacements must not be made. (See Chapter 14, Electrical Safety, for more details.)

FIRE PREVENTION CHECKLIST

ELECTRICAL EQUIPMENT

- ☐ No makeshift wiring
- ☐ Extension cords serviceable
- ☐ Motors and tools free of dirt and grease
- ☐ Lights clear of combustible materials
- ☐ Safest cleaning solvents used
- ☐ Fuse and control boxes clean and closed
- ☐ Circuits properly fused or otherwise protected
- ☐ Equipment approved for use in hazardous areas (if required)
- ☐ Ground connections clean and tight and have electrical continuity

FRICTION

- ☐ Machinery properly lubricated
- ☐ Machinery properly adjusted and/or aligned

SPECIAL FIRE-HAZARD MATERIALS

- ☐ Storage of special flammables isolated
- ☐ Nonmetal stock free of tramp metal

WELDING AND CUTTING

- ☐ Area surveyed for fire safety
- ☐ Combustibles removed or covered
- ☐ Permit issued

OPEN FLAMES

- ☐ Kept away from spray rooms and booths
- ☐ Portable torches clear of flammable surfaces
- ☐ No gas leaks

PORTABLE HEATERS

- ☐ Set up with ample horizontal and overhead clearances
- ☐ Secured against tipping or upset
- ☐ Combustibles removed or covered
- ☐ Safely mounted on noncombustible surface
- ☐ Not used as rubbish burners
- ☐ Use of steel drums prohibited

HOT SURFACES

- ☐ Hot pipes clear of combustible materials
- ☐ Ample clearance around boilers and furnaces
- ☐ Soldering irons kept off combustible surfaces
- ☐ Ashes in metal containers

SMOKING AND MATCHES

- ☐ "No smoking" and "smoking" areas clearly marked
- ☐ Butt containers available and serviceable
- ☐ No discarded smoking materials in prohibited areas

SPONTANEOUS IGNITION

- ☐ Flammable waste material in closed, metal containers
- ☐ Flammable waste material containers emptied frequently
- ☐ Piled material, cool, dry, and well ventilated
- ☐ Trash receptacles emptied daily

STATIC ELECTRICITY

- ☐ Flammable liquid dispensing vessels grounded or bonded
- ☐ Moving machinery grounded
- ☐ Proper humidity maintained

HOUSEKEEPING

- ☐ No accumulations of rubbish
- ☐ Safe storage of flammables
- ☐ Passageways clear of obstacles
- ☐ Automatic sprinklers unobstructed
- ☐ Premises free of unnecessary combustible materials
- ☐ No leaks or dripping of flammables and floor free of spills
- ☐ Fire doors unblocked and operating freely with fusible links intact

EXTINGUISHING EQUIPMENT

- ☐ Proper type
- ☐ In proper location
- ☐ Access unobstructed
- ☐ Clearly marked
- ☐ In working order
- ☐ Service date current
- ☐ Personnel trained in use of equipment

Figure 15-2. This sample checklist serves as a guide for the supervisor in drawing up an inspection list. It should be reviewed regularly to keep it up-to-date.

Haphazard wiring, poor connects, and temporary repairs must be brought up to standard. Fuses should be of the proper type and size. Circuit breakers should be checked to see that they have not been locked in the closed position (which results in overloading), and to see that moving parts do not stick.

Cleaning electrical equipment with solvents can be hazardous because many solvents are flammable and toxic. It is, of course, of utmost importance to use the safest cleaning solvents available. However, keep in mind that a solvent may not present a fire hazard, but still be a health hazard. For example, carbon tetrachloride is nonflammable, but its vapors are extremely toxic. Before a solvent is used, therefore, determine both its toxic and flammable properties.

A satisfactory solvent is inhibited methyl chloroform, or a blend of Stoddard solvent and perchloroethylene. These solvents are commonly used in industry, because they are relatively nonflammable and have a fairly high Threshold Limit Value with respect to toxicity. (See Chapter 8, Industrial Hygiene, for a discussion.)

Many of the cleaning solvents encountered in

industry are not single substances but mixtures of different chemicals, usually marketed under non-descriptive trade names or code numbers. Currently, there are no absolutely safe cleaning solvents. Therefore, before any commercially available solvent is used, you must know its chemical composition. Without such knowledge, the hazards cannot be evaluated, nor the required safety controls be used (see Chapter 8).

Friction

The friction generated by overhead transmission bearings and shafting—where dust and lint accumulate in locations such as grain elevators and woodworking plants—are frequent sources of ignition. Bearings, for example, can overheat and ignite dust in the surrounding area. Keep bearings lubricated so they do not run hot, and remove accumulations of combustible dust as part of a rigid housekeeping routine. Pressure lubrication fittings should be kept in place, and oil holes of bearings should be kept covered to prevent combustible dust and grit from entering the bearings and causing overheating.

Flammable Liquids

Flammable liquids are not really a cause of fire, although they are often referred to as such. More correctly, they are contributing factors to fires. A spark or minor source of ignition, which might otherwise be harmless, can start a fire or touch off explosive forces when flammable vapors, evaporated from liquids and then mixed with air, are present. A liquid is termed flammable if the liquid emits enough vapors to burn at normal temperatures (less than 100 F). Otherwise, the liquid would be termed a combustible liquid.

Almost all industrial plants use flammable liquids. It is thus the responsibility of the supervisor to see that safe practices are followed in the storage, handling, and dispensing of such liquids. Because all flammable liquids are volatile, they are continually giving off invisible vapors. Use the following guidelines to teach workers safe handling of these materials:

- Flammable liquids should be stored in, and dispensed from, approved safety containers equipped with vapor-tight, self-closing caps or covers.
- Flammable liquids should be used only in rooms or areas having adequate and, if possible, positive ventilation. If the solvent hazard is especially high, solvents should be used only in places having local exhaust ventilation.

- When highly volatile and dangerous liquids are being used, a warning placard or sign should be placed near the operation, notifying other personnel and giving warning that all open flames are hazardous and must be kept away.
- Vapors of flammable liquids are heavier than air and will seek the floor or other lowest possible level where they may not be easily detected. This mandates adequate ventilation and the elimination of ignition sources.
- Wherever flammable liquids are used, it is essential that ignition sources—open flames and sparks—be eliminated or alternative preventive measures taken.

Depending on factors which determine how flammable liquids are stored and used, the following safeguards may be required:

- An approved flammable liquids storage vault or room
- Special explosion-proof fixtures and equipment
- Automatic suppression system
- Explosive-relief devices and panels
- Self-closing faucets and safety vents for drums
- Flammable liquids storage cabinets
- Safety cans
- Special ventilating equipment

Specific guidelines to help you determine which procedures and equipment need to be implemented at a particular site can be found in the NFPA guidelines. Fire department inspectors can also be of invaluable assistance in determining the minimal requirements of the local code.

Flammable Gases

Most flammable gases are stored in high-pressure cylinders or in bulk tanks. The gases are then piped to the user's location. Cylinder storage and gas usage areas are normally well planned for fire protection. The storage and usage areas will normally include fire-resistive separations, automatic sprinklers, special ventilation, explosion-relieving vents, the separation of gases from other materials, and the separation of incompatible gases.

Explosive Dusts

Finely divided materials not only tend to burn more readily than do bulk materials, but also can create explosive atmospheres if suspended in air. Industries that generate explosive dusts normally have planned programs to limit the amount of dust accumulations, have limits on ignition sources, and have equipment that is intended to limit the effects of the destructive forces developed in dust explosions.

The programs also include other measures, such as specific cleaning procedures. Ignition sources are limited through special electrical equipment, no-smoking policies, welding and cutting policies, and open flame limitations. Areas expected to have dust often have explosion-relief panels, and potentially, explosion-suppressive systems.

Plastics

Over the past two decades, general plastic usage has increased dramatically. Plastic containers are used for combustible, flammable, and nonflammable materials. At times, plastic containers are stored empty or within corrugated cartons. Noncontainer use of plastics has also increased.

Regardless of storage configurations, plastics tend to burn much hotter and faster, and create more smoke, than cellulosic materials such as wood or paper. Standard sprinkler systems may not control the high-challenge fires that plastics create. Supervisors of industrial processes whose product is plastic, or whose product is packaged in plastic, should be aware of the increased hazard this material represents.

Ordinary Combustibles

Paper and wood products are often referred to as ordinary combustibles. Rack storage and solid pile storage of these materials tend to provide conditions that promote fire growth. Materials that are stacked above one another provide good flue spaces while blocking sprinkler water patterns, thus acting to spread a fire. In general, sprinkler systems are designed to overcome such fire growth.

A second category of ordinary combustibles that pose rapid fire-growth hazards is wooden pallets. Most sprinkler systems will not control fires in wooden pallets stacked over 6 ft tall. If the sprinkler density has been increased, a maximum 8-ft tall stack may be possible. In either case, the supervisor in charge of areas where pallets are stored is usually faced with the problem of limiting pallet heights while meeting production and space limitations. Extra pallets are often stored outside or in separate buildings to reduce the exposure to serious fires.

Detailed information is contained in NSC Data Sheet 532, *Flammable and Combustible Liquids in Small Containers.* NFPA 30, *Flammable and Combustible Liquids*, published by the National Fire Protection Association, is also an excellent source of detailed information on this subject.

OTHER HAZARDOUS MATERIALS

There are many other types of materials that must be kept isolated to prevent fire. For example, some chemicals, such as sodium and potassium, decompose violently in the presence of water, evolve hydrogen, and ignite spontaneously. Yellow phosphorus may also ignite spontaneously on exposure to air. Other combinations, too numerous to mention here, may react with the evolution of heat and produce fire or explosion—in some cases, without air or oxygen being present. These materials must, of course, be handled in a special manner. NFPA 49, *Hazardous Chemicals Data*, lists about 100 such items and provides information on unusual shipping containers, fire hazards, life hazards, storage, fire-fighting phases, and additional data, where applicable. (Check the NFPA for other publications relating to hazardous materials.)

Some materials, principally the ethers, may become unstable during long periods of storage and eventually may explode. In such cases, using the oldest stock first is both good fire safety and sound housekeeping. Part of your job is seeing that this principle is followed.

When materials are used or stored in your department, you should find out their hazards—whether they explode when heated, react with water, heat spontaneously, yield hazardous decomposition products, or otherwise react in combination with other materials. Check the MSDS (Material Safety Data Sheet) for recommended extinguishing methods and reactivity data.

Your best course of action is to obtain detailed information on all potentially hazardous materials and to devise appropriate fire safety measures. The company's fire insurance carrier can be asked to help.

Each type of industry tends to have unique fire protection requirements. The guidelines for protecting such items as computer facilities, ovens, furnaces, boilers, and other specific equipment are covered by various NFPA design standards.

Welding and Cutting

Welding and cutting operations ideally should be conducted in a separate, well-ventilated room with a fire-resistant floor. This safety measure is not, of course, always practical.

If welding and cutting must be carried on in

other locations, these operations must not be performed until (a) the areas have been surveyed for fire safety by experts who know the hazards; (b) the necessary precautions to prevent fires have been taken; and (c) a permit has been issued. This permit must not be stretched to cover an area, item, or time not originally specified—no matter how small the job may seem, or how little time may be required to do it. (For more information on hot work permit programs, see NSC's *Accident Prevention Manual for Industrial Operations, Engineering and Technology* volume.)

If welding must be done over wood floors, they should be swept clean, wet down, and then covered with fire-retardant blankets, metal, or other noncombustible covering. Pieces of hot metal and sparks must be kept from falling through floor openings onto combustible materials.

Sheet metal or flame-resistant canvas or curtains should be used around welding operations to keep sparks from reaching combustible materials. Welding or cutting should not be permitted in or near rooms containing flammable or combustible liquids, vapors, or dusts. Nor should these operations be done in or near closed tanks that contain—or have contained—flammable liquids, until the tanks have been thoroughly drained and purged, and tested free from flammable gases or vapors.

No welding or cutting should be done on a surface until combustible coverings or deposits have been removed. It is important that combustible dusts or vapors not be created during the welding operation. Be sure to provide fire-extinguishing equipment at each welding or cutting operation. Where extra-hazardous conditions cannot be eliminated completely nor protected by isolation, consider installing the additional back-up protection of a fire hose.

Station a watcher at the welding operation who can prevent stray sparks or slag from starting fires, or who will immediately extinguish fires that do start despite all the precautions taken. The area should be under fire surveillance for at least one-half hour after welding or cutting has been completed; many fires are not detected until a few minutes after they ignite.

Open Flames

No open flames should be allowed in or near spray rooms or spray booths. Occasionally, workers may need to do indoor spray-painting or spray-cleaning outside of a standard spray room or booth. In such cases, provide adequate ventilation and eliminate all possible ignition sources, such as spark-producing devices and open flames.

Gasoline, kerosene, or alcohol torches should be placed so that their flames are at least 18 in. (46 cm) away from wood surfaces. They should not be used in the presence of dusts or vapors, near flammable or combustible liquids, paper, excelsior, or similar material. Torches should never be left unattended while they are burning.

Portable Heaters

Gasoline furnaces, portable heaters, and salamanders always present a serious fire hazard. Discourage workers from using them. Upright models of fuel oil salamanders have been involved in numerous accidents. Replace these hazards with other types of low-profile heaters.

Fuel used in portable heaters should be restricted to liquefied petroleum gas, coal, coke, fuel oil, or kerosene. The area in which they are burned must be well ventilated, since the products of combustion contain carbon monoxide.

All these heating devices require special attention with respect to clearances and mounting. A clearance of 2 ft (0.6 m) horizontally and 6 ft (1.8 m) vertically should be maintained between a heater and any combustible material. If coal or coke is used, the heater should be supported on legs 6 in. (15 cm) high or on 4 in. (10 cm) of tile blocks and set on a noncombustible surface.

Combustible material overhead should be removed or shielded by noncombustible insulating board or sheet material with an air space between it and the combustible material. A natural-draft hood and flue of noncombustible material should be installed.

As a fire precaution, each unit must be carefully watched. Heaters should be shut down and allowed to cool off before being refueled. Coal and coke salamanders should not be moved until the fire is out.

All portable heating devices must be equipped with suitable handles for safe and easy carrying. Make sure they are secured or protected against workers accidentally upsetting them.

The most serious fire hazard occurs when heaters are improvised from old steel drums or empty paint containers, with scrap wood, tar paper, or other waste used as fuel. Residue or vapors can spread the fire. Let workers know that this practice is prohibited.

Hot Surfaces

If possible, smoke pipes from heating appliances should not pass through ceilings or floors. If a smoke pipe must be run through a combustible wall, provide a galvanized double-thimble with clearance equal to the diameter of the pipe and ventilated on both sides of the wall, for protection.

Soldering irons must not be placed directly upon wood benches or other combustibles. Special insulated rests, which will prevent dangerous heat transfer, can be used instead.

Smoking and Matches

Management usually has a specific policy about cigarette/cigar/pipe smoking in the workplace. Smoking and no-smoking areas must be clearly defined and marked off with conspicuous signs. Reasons for these restrictions must be explained carefully to employees, and rigid enforcement must be maintained at all times.

Fire-safe, metal containers should be provided in places where smoking is permitted. Safety ash trays in offices, lounges, and lunchrooms should be installed. If carrying matches is prohibited, special lighter equipment should be kept in service in smoking areas. Periodically check "no-smoking" areas, especially when they include stairways and other out-of-the-way places, for evidence of discarded smoking materials.

Spontaneous Combustion

Spontaneous heating is a chemical action in which the oxidation (burning) of a fuel produces a slow generation of heat. When adjacent materials provide sufficient insulating properties, the heating process can continue until spontaneous ignition occurs. Conditions leading to spontaneous ignition exist where there is sufficient air for oxidation but not enough ventilation to carry away the heat generated by the oxidation. Any factor that accelerates the oxidation while other conditions remain constant obviously increases the likelihood of such ignition.

Materials like unslaked lime and sodium chlorate are susceptible to spontaneous ignition, especially when wet. Such chemicals should be kept cool and dry, away from combustible material. Rags and waste saturated with linseed oil, paint, or petroleum products often cause fires because no provision is made for the generated heat to escape. By keeping such refuse in air-tight metal containers with self-closing covers (Figure 15-3), you limit the oxygen supply, and a fire will quickly extinguish itself. These containers should be emptied daily.

Sound precautions against spontaneous ignition include total exclusion of air or good ventilation. The former is practical with small quantities of material through the use of air-tight containers. Ventilation can best be assured by storing material in small piles or by turning over a large pile at regular intervals.

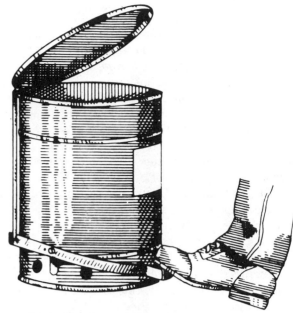

Figure 15-3. Check containers for oily waste to make sure the lid closes snugly against the top.

To determine the progress of spontaneous heating, monitor the temperature within the interior of a mass of material in various locations. Exterior temperatures are not likely to provide a good index.

Static Electricity

Sparks due to static electricity may be a hazard wherever flammable vapors, gases, or combustible dusts exist. Precautions against static electricity are required in such areas. Static charges result from friction between small particles or from the contact and separation of two substances, one or both of which are nonconductive.

Static charges can be produced in many ways—for example, by the flow of gasoline through a nonconductive hose, by passing dry and powdered materials down a nonconductive chute, or through the friction of machine parts.

Although it is impossible to prevent the generation of static electricity under the above circumstances, the hazard of static sparks can be avoided by preventing the buildup of static charges. One or more of the following preventive methods can be used:

- Grounding
- Bonding
- Maintaining a specific level of humidity
- Ionizing the atmosphere

A combination of these methods may be advisable in some instances where the accumulation of static charges presents a severe hazard. NFPA 77,

Recommended Practice on Static Electricity, gives additional information and guidance on this subject.

Grounding is accomplished by mechanically connecting a conductive machine or vessel (in which the generation of static may be a hazard) to ground by means of a low-resistance conductor. Another method is to make the entire floor and structure of the building conductive so that all equipment in contact with it will be grounded. If you use the first method, be sure to check the continuity of the ground circuit. Connections must be clean and the conductor unbroken (Figure 15-4). With the second method, make certain that the floor is free of wax, oil, or other insulating films.

As many people have learned from walking across a rug and then touching a doorknob or other conductive object, the human body can also carry an electric charge. Workers can ground themselves by wearing conductive shoes with a floor of conductive material. The conductive parts of such shoes should be made of nonferrous metal. As an added measure of safety, the conductive flooring may be made of spark-resistant metal. Ferrous contacts increase the risk of friction sparks.

When humidity is low, the hazard of static is greatest. When humidity is high, the moisture content of the air serves as a conductor to drain off static charges as they are formed. When humidity is added to prevent the accumulation of static charges, make sure that an effective relative humidity—usually 60 to 70 percent—is maintained. However, the minimum humidity required for safety may vary over a considerable range under different conditions; in fact, under some conditions the static charge cannot be controlled by humidifying the air. Consult with engineering authorities to determine the best approach.

When air is ionized, it has sufficient conductivity to prevent static accumulation. Ionization is produced by electrical discharges, radiation from radioactive substances, or gas flames. Only an electrostatic neutralizer designed for use in hazardous locations should be used; otherwise, the neutralizer may itself be a source of ignition and touch off flammable vapor or dust. Neutralizers must be kept in good condition.

EFFECTIVE HOUSEKEEPING FOR FIRE SAFETY

Good housekeeping is another important part of an effective fire protection program. As supervisor, you must maintain a positive attitude with the crew and

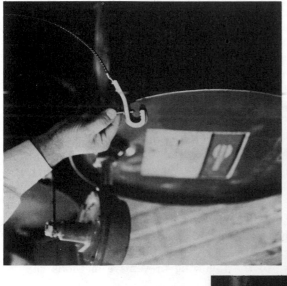

Figure 15-4. Grounds and bonds should be constructed of bare, flexible wire so that broken wires are not concealed. Wires must be attached securely.

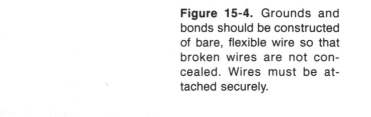

enforce housekeeping rules at all times. Each person should be held personally responsible for preventing the accumulation of unnecessary combustible materials in the work area. Workers should be held accountable for their areas at the end of their shift. Here are the precautions to take:

- Combustible materials should be present in work areas only in quantities required for the job. They should be removed to a designated, safe storage area at the end of each work day.
- Quick-burning and flammable materials should be stored only in designated locations. Such locations always should be away from ignition sources and have special fire-extinguishing provisions. Covered metal receptacles are best.
- Vessels or pipes containing flammable liquids or gases must be airtight. Any spills should be cleaned up immediately.
- Workers should be careful not to contaminate their clothing with flammable liquids. If contamination does occur, these individuals must be required to change their clothing before continuing to work.
- Passageways and fire doors should be unobstructed. Stairwell doors must never be propped open, and material should never be stored in stairwells.
- Material must never block automatic sprinklers or be piled around fire extinguisher locations or sprinkler and standpipe controls.

To obtain proper distribution of water, a minimum of 18 in. (46 cm) of clear space is required below sprinkler deflectors. However, clearance of 24 to 36 in. (60 to 90 cm) is recommended. If there are no sprinklers, clearance of 3 ft (0.9 m) between piled material and the ceiling is required to permit use of hose streams. Double these distances when stock is piled more than 15 ft (4.5 m) high.

Be sure to check applicable codes, especially *Life Safety Code*, ANSI/NFPA 101–1988.

FIRE PREVENTION INSPECTIONS

Fire prevention inspection needs will vary from facility to facility and department to department. However, all fire prevention inspections have four basic goals:

1. Minimize the size of fires by ensuring that combustible and flammable material storage is controlled.

2. Control ignition sources to reduce the possibility of fire.
3. Make sure that fire protection equipment is operational.
4. Make sure that personnel exit facilities are maintained.

You will normally have some responsibility for fire prevention inspections as part of your daily, weekly, and monthly inspection activities. Fire prevention inspections generally are conducted at the supervisory level as part of routine safety inspection activities.

The best way to ensure that the proper items are inspected is to prepare a department checklist. The checklist should be as specific as possible in defining what is, and is not, permitted. This checklist can then be used to audit the department and can be given to employees to make sure that the levels of housekeeping and other fire prevention items are defined clearly to all parties. Sample inspection checklists are included at the end of this section.

In order to prepare a checklist, the information in this chapter may be used along with information from local or corporate safety coordinators, fire department inspectors, and insurance inspectors. Once the checklist has been prepared, the inspection frequency should be defined. Deficient items should be noted and the corrective action taken as soon as possible. Follow-up is also necessary at a predetermined interval.

Fire Protection Equipment

One of the most important audit items will be to determine that the fixed fire protection equipment is on. The checklist items should include the fire protection water valves in your department which should be locked or supervised open. Other fire protection systems are normally controlled by electrical panels which can be audited to ensure that the systems are on and that no trouble signals are showing. Automatic sprinkler patterns can be audited by making sure that there are at least 18 in. between products and the sprinkler deflector.

Fire extinguishers should be tagged and marked with the date of the last inspection. Gages should show charge, and seals should be in place. The audit form should ensure that the extinguishers are hung in designated locations and that access is not blocked. The nozzle should be free and clear.

Fire Protection Systems

Fires can be extinguished by exhausting the fuel source, by manual fire fighting, or by fixed fire-fighting systems. Fire extinguishment typically occurs by lowering the temperature through dousing the fire with water, reducing the oxygen available through carbon dioxide or foam systems, or by interrupting the chemical chain reaction by using a halon or dry chemical system.

All systems and system components should be listed by recognized testing agencies according to their fire-fighting purposes. In addition, all systems should be inspected and/or tested annually or as required by applicable standards or manufacturers' instructions. Automatic sprinklers are available in wet-pipe systems, dry-pipe systems for piping subject to freezing, deluge systems for high-challenge fires, or preaction systems for locations subject to water damage.

Halon, carbon dioxide, and foam systems typically are actuated by fire detection systems. These systems generally are heat, smoke, or flame detectors. Detection systems also are used often to reduce the threat to life and property by providing early warning to initiate manual fire fighting or evacuation of the workplace.

ALARMS, EQUIPMENT, AND EVACUATION

A fully effective fire protection program, of course, includes fire extinguishment. Whenever a fire breaks out, take immediate action:

1. Sound the fire alarm right away—regardless of the size of the fire.
2. Attempt to extinguish or control the fire with appropriate fire-extinguishing equipment.

Fire Alarms

Facilities in areas where municipal fire departments are available usually have an alarm box close to the plant entrance or located in one of the buildings. Others may have auxiliary alarm boxes, connected to the municipal fire alarm system, at various points in the facility. Another system often used is a direct connection to the nearest fire station that may register water flow alarms in the sprinkler system, be activated by fire detectors, or be set off manually. In some cases, the telephone may be the only means for signaling a fire alarm.

Whatever the alarm system used, all employees should be carefully instructed in how, when, and where to report a fire. These three steps are extremely important. Many fires get out of control simply because someone did not know how, when, or where to give the alarm.

Extinguishers

As the supervisor, you need to know the four classes of fires that might break out in your area. Before employees can combat beginning fires effectively, they must be familiar with and understand these classes of fires. Briefly, they are:

ORDINARY COMBUSTIBLES

FLAMMABLE LIQUIDS

ELECTRICAL EQUIPMENT

COMBUSTIBLE METALS

Class A—Fires in ordinary combustible materials, such as wood, paper, cloth, rubber, and many plastics, where the quenching and cooling effects of water or of solutions containing large percentages of water are of prime importance.

Class B—Fires in flammable liquids, greases, oils, tars, oil-base paints, lacquers, and similar materials, where smothering or exclusion of air and interrupting the chemical chain reaction are most effective. This class also includes flammable gases.

Class C—Fires in or near live electrical equipment, where the use of a nonconductive extinguishing agent is of first importance. The material that is burning is, however, either Class A or Class B in nature.

Class D—Fires that occur in combustible metals, such as magnesium, lithium, and sodium. Special extinguishing agents and techniques are needed for fires of this type.

Each of the fire extinguishers in the department or at a job location should have on it a plate identifying the class of fire for which it is intended (Figures 15-5a and 15-5b), its operating instructions, and its servicing instructions. The data plate should have the identifying symbol/name of a recognized testing facility to indicate that the unit has been listed or approved. Equipment that does not bear an approval label should be brought to the attention of management. Only listed or approved equipment should be furnished. See "Fire Protection Equipment List" (Underwriters Laboratories Inc., Northbrook, Ill. 60062). Also, "Approved Equipment for Industrial Fire Protection" (Factory Mutual Research Corp., Norwood, Mass. 02062).

Figure 15-5a. Fire extinguishers should have a plate identifying the class of fire they can put out, the operating instructions, and servicing instructions. This one uses the triangle-square-circle-star symbols.

The information on these data plates can help you teach employees how to operate the extinguishing equipment. See to it that every worker knows the important details about each of the fire-extinguishing agents provided in the particular job area or shop (Table 15-2). Fire extinguisher manufacturers and distributors can be contacted for additional training materials, and local fire departments may be willing to provide training assistance. Information on selection, installation, use, inspection, and maintenance of extinguishers can be obtained from the National Fire Protection Association. NFPA 10, *Portable Fire Extinguishers*, is an authorative source of information and should be made available to all supervisors for reference use.

Normally, the location and installation of portable fire extinguishers, fire blankets, stretchers, and other fire safety equipment at strategic places about the plant, shop, or job site are the responsibility of higher management. However, as supervisor, you should know the location of units in your area, and recommend relocating them or obtaining additional units if such changes will afford more adequate protection.

Check to see that extinguishers are not blocked by material or equipment and that signs indicating their location are clear and conspicuous (Figure 15-6). Many companies have found that marking the area of the floor directly under a fire extinguisher is an excellent way to keep employees from placing obstructions in front of the equipment.

The location of extinguishers can be identified by painting the housing, wall area, column, or other support of the extinguisher with standard fire-protection red. Fire protection equipment itself, such as sprinkler system piping, is sometimes painted or marked in red. Make sure that each worker knows the location of the nearest unit and knows the importance of keeping areas around extinguisher units clear.

It's a good idea for you to schedule fire drills, during which workers use extinguishers appropriate for their particular work areas (Figure 15-7). The company's safety professional or fire chief, the insurance company's safety engineer, or the local fire department representative will assist in conducting such drills.

Every organization should have a specific program for periodic inspection and servicing of portable fire-extinguishing equipment. This routine work is usually outside the scope of the department supervisor and/or maintenance personnel. However, you can do some things to assist in this program. For example, during regular department inspections, you should double-check each data card to determine when each extinguisher was last inspected and serviced. Management should be notified when inspections have been missed or are outdated. If the extinguisher seal is broken, its pressure gage reads below normal, or any other unsatisfactory condition is observed, report these conditions immediately. Defective extinguishers may not work.

Fire extinguishers must meet the following requirements:

- Be kept fully charged and in their designated places.
- Be located along normal paths of travel where practical.
- Not be obstructed or obscured from view.
- Not be mounted higher than 5 ft (1.5 m) (to the top of the extinguisher) if they weigh 40 lb (18 kg) or less. If heavier than 40 lb, extinguishers must not be mounted higher than 3½ ft (1 m). There shall be a clearance of at least 4 in. (10 cm) between the bottom of the extinguisher and the floor.
- Be inspected by management or a designated employee, at least monthly, to make sure that they are in their designated places, have not

Figure 15-5b. Picture-symbol labels used by the National Association of Fire Equipment Distributors show the class of fire for which an extinguisher can be used. The symbols shown here would be on a Class A extinguisher (for extinguishing fires in trash, wood, or paper). The left symbol is blue. Because this extinguisher is not meant for use on Class B or C fires, those two picture-symbols are in black with a diagonal red line through them. A Class A/B extinguisher would have the first two symbols in blue and the third in black with a red diagonal. For use on a Class B/C, the last two would be in blue; on a Class A/B/C, all three would be in blue.

TABLE 15-2. Fire Extinguisher Selection Chart

	CLASS A						CLASS A/B	CLASS B/C			
	WATER TYPES		MULTIPURPOSE DRY CHEMICAL		AFFF FOAM	HALON 1211	AFFF FOAM	CARBON DIOXIDE	DRY CHEMICAL TYPES		HALON 1211
	STORED PRESSURE*	PUMP TANK*	STORED PRESSURE	CARTRIDGE OPERATED	STORED PRESSURE*	STORED PRESSURE	STORED PRESSURE*	SELF EXPELLING	STORED PRESSURE	CARTRDIGE OPERATED	STORED PRESSURE
SIZES AVAILABLE	2½ gal	2½ and 5-gal	2½-30 lb ALSO (Wheeled 150-350 lb)	5-30 lb ALSO (Wheeled 50-350 lb)	2½ gal	9 to 22 lb	2½ gal	5-20 lb ALSO Wheeled 50-100 lb	2½-30 lb ALSO Wheeled 150-350 lb	4-30 lb ALSO Wheeled 50-350 lb	2 to 22 lb
HORIZONTAL RANGE (APPROX.)	30 to 40 ft	30 to 40 ft	10-15 ft (Wheeled- 15-45 ft)	10-20 ft (Wheeled- 15-45 ft)	20-25 ft	14 to 16 ft	20 to 25 ft	3-8 ft (Wheeled- 10 ft)	10-15 ft (Wheeled- 15-45 ft)	10-20 ft (Wheeled- 15-45 ft)	10 to 16 ft
DISCHARGE TIME (APPROX.)	1 min.	1 to 2 min	8-25 s (Wheeled- 30-60 s)	8-25 seconds (Wheeled- 20-60 s)	50 seconds	10 to 18 seconds	50 seconds	8-15 seconds (Wheeled- 8-30 s)	8-25 seconds (Wheeled- 30-60 s)	8-25 seconds (Wheeled- 20-60 s)	8 to 18 seconds

	CLASS/A/B/C			CLASS D
	MULTIPURPOSE DRY CHEMICAL		HALON 1211	DRY POWDER
	STORED PRESSURE	CARTRIDGE OPERATED	STORED PRESSURE	CARTRIDGE OPERATED
SIZES AVAILABLE	2½-30 lb ALSO Wheeled 150-350 lb	5-30 lb ALSO Wheeled 50-350 lb	9 to 22 lb	30 lb ALSO Wheeled 150-350 lb
HORIZONTAL RANGE (APPROX.)	10-15 ft (Wheeled- 15-45 ft)	10-20 ft (Wheeled- 14-45 ft)	14-16 ft	5 ft (Wheeled- 15 ft)
DISCHARGE TIME (APPROX.)	8-25 seconds (Wheeled- 30-60 s)	8-25 seconds (Wheeled- 20-60 s)	10 to 18 seconds	20 seconds (Wheeled- 150 lb to 70 s, 350 lb 1¾ min.)

*Must be protected from freezing

Courtesy of the National Association of Fire Equipment Distributors

been tampered with or actuated, and do not have corrosion or other impairment.

- Be examined at least yearly and/or recharged or repaired to ensure operability and safety.

A tag must be attached to show the maintenance or recharge date and the signature or initials of the person performing the service.

- Be hydrostatically tested. Extinguisher serv-

Figure 15-6. Painting the floor under a fire extinguisher reminds employees not to obstruct the equipment.

icing agencies should be contacted to perform this service at appropriate intervals.

- Be selected on the basis of type of hazard, degree of hazard, and area to be protected.
- Be placed so that the maximum travel (walking) distances between extinguishers, unless there are extremely hazardous conditions, do not exceed 75 ft (23 m) for Class A extinguishers or 50 ft (15 m) where Class B extinguishers are used for hazardous-area protection. The travel distance requirement does not apply when Class B extinguishers are used for spot-hazard protection.

Fire Brigades

Many plants and construction job sites have organized fire brigades. Ordinarily, supervisors would not be required to organize such a brigade. However, they should be well enough acquainted with the form and activities of a fire brigade to supervise its actions.

In some fire brigades, you, as supervisor, or a delegated foreman is named a brigade chief or company captain. Whether or not you are chosen for this job, you should know each worker's fire-brigade assignment. On the basis of this information, you can then organize the work in your department so that brigade members may attend training and drills as designated by their brigade chief (Figure 15-8).

Regardless of your role in the fire-brigade organization, you should be familiar with the location and operation of the following items in and adjacent to your department—standpipes and valves, sprinkler system valves, electric switches for fans and lights, fire-alarm boxes and telephones, fire doors, and emergency power for equipment and/or lighting. In particular, you must make sure that enclosed stairway doors are kept closed. Such doors are provided to keep smoke, combustion gases, and heat out of the exit passageways and stairs, as well as to help prevent the rapid spread of fire. Doors should be equipped with self-closing devices and should not be propped open. In cases where the exitway is used frequently, doors may be held open with fire-actuated detectors that will automatically close the door should fire break out.

To become familiar with the items listed above, take every opportunity to accompany the fire inspector on inspection tours to learn how the items operate. Whether or not you will operate this equipment will depend upon the particular situation and specific policy and procedures in your plant. In some instances, if you know the system thoroughly, you may be the best one to use fire safety equipment. However, it can be extremely dangerous to let anyone without proper knowledge operate the equipment. Activating it at the wrong time or using it incorrectly might seriously endanger people's lives in case of fire.

Follow-up for Fire Safety

Frequent inspections of the area, correction of hazardous conditions, and instruction of the workers in fire prevention and extinguishment measures are still not enough to ensure a fire-safe work area. The supervisor must follow up on fire safety relentlessly. Make sure you instill fire-safe attitudes in your personnel so they continually observe safe work practices.

Make fire prevention and extinguishment the subject of frequent safety talks between you and each worker. Also, discuss these issues in safety meetings so that workers will be fully aware that fire prevention is a vital part of the overall departmental safety program.

Special Fire Protection Problems

Construction. When construction is going on at an existing plant site, fire protection problems are compounded. You and the construction supervisor

Figure 15-7. Employees are being trained in the most effective ways to use extinguishers.

Figure 15-8. Fire brigade members undergo training in a simulated environment requiring the use of self-contained breathing apparatus. (Courtesy Survivair Division of U.S. Divers.)

should work together in determining the hazards in each field of operations and in identifying the limits of your firefighting capabilities. On the basis of this knowledge, you should take steps to upgrade fire protection equipment and practices.

Construction activities may require measures such as using temporary wiring or portable heaters

and gasoline engines that must be refueled on the spot. As a result, manufacturing processes, normally not hazardous, may become so under these conditions. You should see that employees are aware of additional ignition sources created by the construction operations. The construction supervisor should inform workers of the additional fire hazard exposures present from the plant activity.

When the construction work involves plumbing, the water supply available for fire extinguishment may be decreased or need to be shut off. To offset this possible shortage of water, obtain additional auxiliary fire-extinguishing equipment such as temporary hose lines or water tank trucks.

Sprinkler system shutdown. Whenever a sprinkler system must be shut down, special precautions must be taken to ensure maximum fire safety. Check to see that all preliminary work is completed, set out extra firefighting equipment, and notify firefighting units which areas will be without sprinkler protection. One procedure requires that a bright red tag be attached to the sprinkler valve when it is closed (Figure 15-9) and that the insurance carrier be notified. The work is completed as quickly as possible, at which time the tag is removed and the insurance company notified that service is restored. To make certain that the valve is left wide open, it is tested and then locked in the open position. All sprinkler valves should be checked regularly to make sure they have not been tampered with, locked, or shut.

Whenever the sprinkler system must be shut down for alterations or repairs, you and your main-

Figure 15-9. Sprinkler system valves should be locked in the open position (top). When a sprinkler system must be shut down, a safe procedure is to do preliminary work first, then set out extra firefighting equipment, inform firefighting units which areas will be without sprinkler protection, notify the insurance carrier, and tag the closed sprinkler valve (bottom). When the work is completed, remove the tag, test and lock the valve in the fully open position, and notify firefighting units and insurance carrier that service is restored. All sprinkler valves should be checked regularly to ensure they have not been unlocked, tampered with, or shut.

tenance people should plan to do the work before or after normal working hours. If work must be done during the most hazardous times, special precautions may have to be taken, such as having hose lines laid and furnishing extra fire patrols.

Some plant and construction supervisors have made a practice of attending each other's safety meetings. They benefit from an exchange of their respective points of view and learn about each other's problems. This joint approach to common exposures is recommended.

Radioactive materials. The use of radioactive materials for various purposes in manufacturing processes has become more common. Their presence requires special understanding and control in the event of fire or other emergency.

In view of the potential contamination hazards to employees, supervisors in the areas or departments having radioactive materials should be thoroughly familiar with the procedures to be followed in case of fire or other emergency, and should rigidly enforce compliance with those procedures. Access to the contaminated area following the emergency

should be prohibited until authorized personnel have recovered the sources of radiation and have determined that the contamination is below the safe levels.

Valuable information on this problem can be secured from the Nuclear Regulatory Commission, 1717 H Street, N.W., Washington, D.C. 20555.

Evacuation

Prevention of fire is the primary objective of any fire protection program. Nevertheless, each program must include provisions to ensure the safety of employees in the event of a fire. Security personnel must be included in the overall plan (Figure 15-10).

Most facilities will have some form of plan to respond to emergencies. The plan may be as simple as an informal evacuation plan, or as complicated as a fully written emergency response and disaster recovery plan for various natural and man-made emergencies. Normally, your responsibilities during

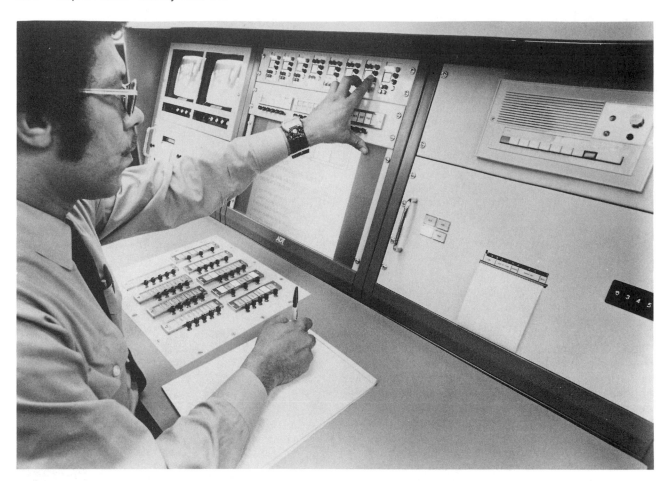

Figure 15-10. Include security personnel in your fire protection program planning. This console is part of a comprehensive proprietary system capable of detecting fire, burglary, and any unwanted fluctuations in industrial process conditions. (Courtesy ADT—American District Telegraph Company.)

an emergency are for the people and department under your authority.

You should have access to the written emergency plan for periodic review and for instruction of the workforce. You should develop a clear personal plan for complying with the facility's emergency plan and responding to evacuation, spills, fires, firefighting, natural disasters, and utility failures. Also, make sure you know the proper procedures for reporting emergencies and the meaning of the alarm signals. You should actively participate in routine emergency drills.

In some cases, you or a member of your workforce will serve on the emergency response team and will have specific duties outside of the department during an emergency. You should plan how to respond to the emergency with those members missing.

From time to time, the local fire department will tour your facility. The company can benefit by your active participation in these tours. The more the fire fighters know about the flammable and combustible materials in the facility, the quicker the fire department can respond to minimize fire losses.

You can do much to prepare for the safety of employees in the event of a serious fire. As a matter of routine, you should make sure that each man and woman knows the evacuation alarms, both primary and alternate exit and escape routes, what to do, and where to be during and after an evacuation.

You should make sure that each person knows that he or she, upon being alerted, must proceed at a fast walking pace—not a run—to an assigned exit or, if it is blocked, to the nearest clear one. Day in and day out, you should emphasize that the exit routes and any fire doors must be kept unobstructed.

Such essential steps in emergency procedures will best be instilled by periodic fire drills. If company policy assigns the responsibiity to you, plan and conduct practice evacuations at regular intervals. You should also do everything possible to integrate fire emergency training with plant-wide drills and the company's overall evacuation plan.

When it is impossible to hold drills, give oral instructions regarding evacuation procedures, and distribute printed information describing these policies to all employees.

FIRE PROTECTION EDUCATION

Fire protection education for the workforce typically involves the following:

- Emergency response
- Fire fighting
- Home fire safety

Emergency response training usually will be necessary once or twice per year. The workforce will need to know how and when to turn in alarms and how to respond to alarms that are sounded. The specific shutdown, evacuation, spill, and utility failure procedures should be rehearsed.

Workers also should practice the use of portable fire extinguishers and hoses at least yearly if they are expected to use the equipment. A yearly review of the fixed fire protection systems and manual operation of those systems is also usually necessary. If any of the employees travels, hotel fire safety is an appropriate topic to address. Fire Prevention Week in October is normally a convenient time to review these items. Safety meetings also provide a convenient forum. Brochures and written materials are available from the National Fire Protection Association in Quincy, Massachusetts.

Fire Prevention Week is also a good time to address home fire safety. The proper use of smoking materials, smoke detector usage, and home fire drills are appropriate topics. Also, the beginning of the heating season is a good time to introduce the concepts of proper use of wood-burning equipment. In December, review the hazards and protective measures regarding Christmas ornaments and natural Christmas trees.

PROTECTIVE INSURANCE REQUIREMENTS

The property (fire) insurance carrier generally conducts inspection tours of the facility on a routine basis. These tours are intended to identify items that can help to reduce loss in a fire or minimize ignition sources. The supervisors of an area will often accompany the insurance representative on the tour to understand the insurance concerns. The

representative also can be used as a resource to help solve fire protection problems.

The insurance carrier also will require specific programs that you should understand. No-smoking requirements are one program, and the cutting and welding permit system is another. The locking or supervision of water supply valves, and testing and maintenance of fire protection equipment are other typical required programs.

Another common insurance requirement involves those times when fire protection systems are removed from service. The insurance carrier often will require a call and a specific tagging system. The carrier then tracks the time when the system is out of service to ensure that the system is returned to service. The carrier also may require alternative fire protection such as standby charged hoses during those periods. The notification of the fire department is also a good procedure.

SUMMARY OF KEY POINTS

This chapter covered the following key points:

- The primary requirements for a successful fire prevention program include continuous training in fire-safe work practices, regular inspections of work areas, and close supervision of employee job performance. Ultimate responsibility for fire safety rests with the supervisor.
- Fire chemistry can be illustrated by the tetrahedron model in which elevated temperature, oxygen, fuel, and chain reaction must occur simultaneously for a fire to continue burning. Fires can be extinguished by one or more of the following: removing or reducing oxygen supply, removing fuel sources, reducing or cooling heat sources, and interrupting the chain reaction.
- The main causes of fires include faulty electrical equipment, friction, flammable liquids, flammable gases, explosive dusts, plastics, and ordinary combustibles such as paper or wood. Supervisors should see to it that electrical equipment is properly grounded, insulated, and maintained and that all flammable or combustible materials are properly stored, ventilated, and contained.
- Other causes of fires include operations such as welding and cutting, open flames, portable heaters, hot surfaces, smoking and lighted matches, spontaneous combustion, and static electricity. Stringent safe work practices and

housekeeping measures can reduce the possibility of fire from these causes.

- Supervisors must establish good housekeeping rules and guidelines and hold each employee accountable for following these procedures. These rules include preventing the accumulation of combustible materials in the workplace, changing clothing contaminated with flammable liquids, keeping passageways and fire doors unobstructed, and making sure sprinkler systems are not blocked by stacked material.

- Supervisors should make regular inspections to indentify fire hazards in their areas and take action to resolve them. Workers should be continuously informed of these hazards and of the safeguards required to prevent fires from occurring. A fire prevention inspection has four goals: (1) minimize the size of fires by controlling combustible and flammable materials, (2) controlling ignition sources, (3) making sure fire protection equipment operates, and (4) ensuring that personnel exit facilities are maintained.

- Supervisors should make sure that employees know how to operate fire alarms and firefighting equipment, and understand evacuation procedures. Employees should know how, when, and where to report a fire by activating fire alarms. They should also know Class A, B, C, and D fires and which fire-extinguishing equipment is appropriate for each class of fire. Firefighting and extinguishing equipment should have this information clearly printed on the data plate.

- Such equipment must be inspected periodi-cally and repaired or replaced if found defective. Supervisors must follow up on inspections and instruction sessions to make sure employees are complying with safety procedures.

- In some facilities, companies organize fire brigades. Supervisors must know how the fire brigade operates and the specific responsibilities of each brigade member. Supervisors should also know the location and operation of all fire safety equipment and exit facilities.

- In some cases, construction onsite, sprinkler system shutdowns, or the presence of radioactive materials present special fire safety problems. Supervisors must make sure that special fire prevention measures are taken to avoid accidents and injuries.

- Evacuation procedures should be communicated to every employee in the event of a serious fire or other emergency. Workers should know evacuation alarm signals, where to exit the facility, and how to secure their work areas before leaving. When possible, supervisors should hold drills to practice evacuation procedures.

- Fire protection education for employees involves responding to emergencies, fighting fires, and instituting home fire safety procedures. Employees should practice using fire-extinguishing equipment and understand how to fireproof their homes.

- Insurance companies generally inspect facilities on a routine basis. Supervisors can accompany these inspectors on their tours and use them as a resource to help solve fire protection problems.

Introduction to the Emergency Planning and Community Right-to-Know Act of 1986

The supervisor's role has never been easy, either from the standpoint of communicating complex issues to line employees or from the standpoint of dealing internally with management specialists whose long-term policy objectives often can conflict with those of daily work. Nothing on the horizon appears likely to make it any easier.

The Emergency Planning and Community Right-to-Know Act of 1986 emphasizes the increasing complexity of the supervisor's role in the management of hazardous chemicals. Widely seen as a legislative response to a December, 1984, chemical tragedy in which more than 2,000 citizens of Bhopal, India, died, the law for the first time provides the public and the media unprecedented access to valuable chemical information concerning their local communities.

Part of an evolving "right to know" ethic in the environmental health and safety fields, the law mandates advance contingency planning for chemical emergencies and incidents; emergency notification of accidental spills and releases of hazardous chemicals; and annual reporting of manufacturing companies' uses, inventories, and emissions of hazardous chemicals.

News reporters, if they choose, can have in their hands an unprecedented amount of plant, industry, city, county, state, regional, or federal information on about 325 toxic chemicals released by manufacturing facilities to the air, water, or land. Through newly established local committees, they can get information on inventories of various chemicals stored in their communities, and they can get information on amounts, locations, and potential effects of hazardous chemicals used or stored locally.

Because of the public attention your workplace may get from implementation of this law, it is important that your department's efforts to comply with it are thorough. Although it is unlikely you will be principally responsible for compliance, you will have an important part to play. When the local fire department arrives for an inspection, it is not just the location of doors and fire extinguishers that will be checked.

Hazardous materials response teams from municipal, town, and county agencies need to be briefed on what chemicals are stored where. If your team relocates a storage vessel, this action must be communicated to the first responders.

Should one of your employees spill 5 gallons of a certain solvent, your response can have a tremendous impact on company productivity and on the way the public and regulatory agencies see your employer. How? Suppose the vapors are strong and the plant floor must be temporarily evacuated: someone calls the fire department. It takes hours for the Hazardous Materials Response Team to arrive, investigate the situation, ventilate the room, and clean up. Operations are shut down for a half day, perhaps more. The press arrives, thinking this may be more than a "routine" response. Tomorrow, someone will be asking how this can be avoided in the future. Your suggestions will count. Training will be your responsibility.

KEY PROVISIONS OF THE LAW

The Emergency Planning and Community Right-to-Know Act, in EPA's own words, "makes citizens full partners in preparing for emergencies and managing chemical risks."

This partnership taps the supervisor in several ways: as an individual citizen, as a possible member of a local emergency planning committee, and as an employee with a responsibility to inform line workers about toxic health hazards and emergencies. To meet these challenges, supervisors must understand the law, its key provisions, and its enormous potential for providing information needed for effective emergency preparedness.

The Emergency Planning and Community Right-to-Know Act has two purposes: to encourage and support emergency planning for responding to chemical accidents, and to provide local governments and the public with timely and comprehensive information about possible chemical hazards in communities. The law operates through provisions detailed in four major sections.

Emergency Planning (Sections 301-303)

Emergency Planning requires state and local efforts to develop emergency response and preparedness capabilities based on chemical information provided by industry. These sections are designed to help communities prepare for and respond to emergencies involving hazardous substances. Every community in the United States must be part of a comprehensive emergency response plan.

The governor of each state by April 17, 1987, had to appoint a State Emergency Response Commission (SERC), which can be one or more existing state agencies, or may consist solely of individual citizens. (Some SERCs have no state agency representative and are "staffed" entirely by private citizens.) These commissions have been named in all 50 states, and the U.S. territories and possessions.

Each SERC, in turn, has divided the state into local emergency planning districts, and has appointed a Local Emergency Planning Committee (LEPC) for each district. The number of "local committees" varies widely from state to state. In some states, such as Georgia and Oregon, one committee covers the entire state; in New Jersey, on the other hand, the state is divided into 588 local committees.

There are four groups of chemicals subject to reporting under the Emergency Planning and Community Right-to-Know Act. Some chemicals appear in several groups.

Extremely Hazardous Substances (Sec- tions 301-304). This list currently contains approximately 366 chemicals. Because of their acutely toxic properties, these chemicals were chosen to provide an initial focus for chemical emergency planning. If these chemicals are released in certain amounts, they may be of *immediate* concern to the community. Releases must be reported immediately.

Hazardous Substances (Section 304). These are hazardous substances listed under previous Superfund hazardous waste cleanup regulations. This list contains about 720 substances. Releases of these chemicals above certain amounts must be reported immediately because they may represent an immediate hazard to the community. The reports are made to the state commission and to the local committee.

Hazardous Chemicals (Sections 311-312). These chemicals are not on a list at all, but are defined by Occupational Safety and Health Administration regulations as chemicals that represent a physical or health hazard. Under this definition, thousands of chemicals can be subject to reporting requirements. Inventories of these chemicals and material safety data sheets for each of them must be submitted if they are present at the facility in certain amounts.

Toxic Chemicals (Section 313). There are now about 325 chemicals or chemical categories on this list that were selected by Congress primarily because of their chronic or long-term toxicity. Estimates of releases of these chemicals into air, water, or land must be reported annually and entered into the toxic release inventory.

The LEPCs must analyze hazards and develop a plan to prepare for and respond to chemical emergencies in its district. The plan should be based on the chemical information reported to the LEPC by local industries and other facilities dealing with chemicals.

The list of 366 "extremely hazardous substances" identified by EPA as having immediate toxic health effects and hazardous properties serves as a focus for emergency planning. However, the plans are to address all hazardous materials in the community that present risks to public health and safety—including, for example, widely used fertilizers, preservatives, photographic chemicals, and insecticides.

The list of extremely hazardous substances includes a "threshold planning quantity" for each substance. If at any time this amount or more of the chemical is present at any manufacturing plant, warehouse, hospital, farm, small business, municipal installation, or any other facility, the owner or operator must notify the SERC and the LEPC. The facility's owners or operators must also name an employee as "facility coordinator." He or she participates in the district's planning process.

Emergency Release Notification (Section 304)

Emergency Release Notification requires immediate emergency notification to state and local authorities when any one of approximately 366 chemicals designated "extremely hazardous" is released to the environment or when chemicals specified under Superfund are released accidentally.

If a specified "extremely" hazardous substance is released—in an accident at a facility or on a transportation route—in an amount that exceeds the reportable quantity for the substance, facilities must *immediately* notify the LEPCs and the SERCs likely to be affected.

Chemicals covered by this section of the law include not only the 366 "extremely hazardous substances," but also about 720 other hazardous substances now subject to the emergency notification requirements of the Comprehensive Environmental Response, Compensation, and Liability Act, CERCLA or the so-called Superfund law. Some chemicals are on both lists. Superfund requires immediate notification of substance releases to the National Response Center, which alerts federal emergency response teams. Part of the significance of Section 304 is that it requires immediate notification, not just of federal agencies, but also of local and state officials.

Initial notification of a substance release can be made by telephone, radio, or in person. If the release resulted from a transportation accident, the transporter can dial 911 or the local authorities.

Hazardous Chemical Reporting (Sections 311-312)

Hazardous Chemical Reporting requires all businesses to submit information on chemicals broadly defined as "hazardous" to local and state emergency planners and local fire departments.

Accidental chemical releases are only part of what the public has a right to know about under the Emergency Planning and Community Right-to-Know Act. Under Sections 311 and 312, facilities also must report the amounts, locations, and potential effects of hazardous chemicals present in designated quantities on their property.

All companies, of any size, manufacturing or nonmanufacturing, are potentially subject to this requirement. They must report this information to the relevant LEPCs, SERCs, and local fire departments. Facilities must report on the hazardous chemicals in two different ways:

1. Material Safety Data Sheets (MSDSs). These contain information on a chemical's physical properties and health effects, and whether it presents hazards in any of these categories: immediate (acute) hazard, delayed (chronic) health hazard, fire hazard, sudden release of pressure hazard, or reactive hazard.

Under federal laws administered by OSHA, companies are required to keep MSDSs on file for all hazardous chemicals in the workplace. They also must make this information available to employees, so workers will know about the chemical hazards they are exposed to and can take necessary precautions in handling the substances.

The relevant chemicals are those defined as "hazardous chemicals" under OSHA's requirements—essentially, any chemical that poses physical or health hazards. As many as 500,000 products can be defined in this way. If they are present, they must be reported under the hazardous chemical reporting provisions.

An MSDS or list must be provided when new hazardous chemicals become present at a facility in quantities above the established threshold levels. A revised MSDS must be provided if significant new information is discovered about a chemical. Once submitted to the LEPC, SERC, and local fire department, the MSDS information is available to the public upon request.

When the Emergency Planning and Community Right-to-Know Act was passed in 1986, OSHA's regulations applied only to manufacturers—about 350,000 facilities. In the wake of court decisions applying OSHA's hazard communication standard beyond the manufacturing sector, these regulations now apply also to most facilities where workers are exposed to hazardous chemicals—about 4.5 million facilities nationwide.

2. Annual inventories. Companies must report on hazardous chemicals by submitting annual inventories to their LEPCs, SERCs, and local fire departments, under a two-tier system. Under Tier I, a facility must report the amounts and general location of chemicals in certain hazard categories. (Example: a facility stores 10,000 pounds of a substance that causes chronic health effects.) A Tier II report requires a brief description of how each chemical is stored and the specific storage location for each hazardous chemical. (Example: a facility stores 500 pounds of benzene in the northwest corner storage room of the warehouse.)

Congress gave companies the choice of filing a Tier I or Tier II form, unless the SERC, LEPC, or fire department requests Tier II. The Tier I/II forms must be submitted annually beginning March 1, 1988. Tier I forms must estimate (in ranges) the maximum amount of chemicals present at a facility at any time during the preceding calendar year, provide a range of estimates of the average daily amount of the chemicals present in each chemical category, and provide the general location of hazardous chemicals within the facility.

Tier II information must be submitted upon request from a local committee, the state commission, or a local fire department. Tier II information is to include more specific information about each substance subject to the request: chemical name or common name as indicated on the material safety data sheet, an estimate of the maximum amount of the chemical present at any time during the preceding calendar year, a brief description of how the chemical is stored, and the location of the chemical in the facility. Tier II reports also must indicate if the reporting facility has withheld location information from disclosure to the public for security reasons, such as protecting against vandalism or arson. However, the information reported under Sections 311 and 312 generally is to be available to the public through local and state governments during normal working hours. Aware of this, many companies may provide the information to reporters upon request, rather than requiring them to go through the state commission or local committee.

The civil penalty for failing to submit MSDSs or lists of MSDS chemicals is up to $10,000 a day for each violation. For noncompliance with the annual inventory requirements, the penalty is $25,000 per violation.

Toxic Chemical Release Reporting and Inventory (Section 313)

The Toxic Chemical Release Reporting and Inventory provision requires certain manufacturers to report annually, beginning July 1, 1988, on the amounts of about 325 toxic chemicals they release into the air, water, or soil. This provision applies to an estimated 25,000 to 30,000 facilities with 10 or more employees. A small business exemption frees companies with nine or fewer employees from coverage.

Toxic chemical release reports are required by facilities that use more than 10,000 pounds of a listed chemical in a calendar year. Under a phase-in schedule, the 1988 and 1989 reports were required for facilities that manufactured or processed any of these chemicals in excess of 75,000 pounds in 1987 and in excess of 50,000 pounds in 1988. Thereafter, the annual reports must be submitted by facilities that manufacture or process more than 25,000 pounds in a year.

Companies that fail to file annual toxic chemical release reports are subject to civil penalties up to $25,000 a day for each chemical they should be reporting.

Facilities must annually file a Toxic Chemical Release Inventory Form (Form R) to estimate the total amount of each chemical that they release into the environment, either by accident or as a result of routine plant operations, or that they transport as waste to another location. A complete Form R must be submitted for each chemical. Releases covered include emissions to the air from stacks, liquid waste discharged into water, wastes disposed of in landfills, and waste transported off-site to a public or private waste treatment or waste disposal facility.

Many of the chemicals covered by this section of the law, though not all, pose long-term (chronic) health and environmental hazards, such as cancer, nervous system disorders, and reproductive disorders stemming from routine exposure. Among the most commonly used substances included on the list of the approximately 325 chemicals are ammonia, chlorine, copper, lead, methanol, nickel, saccharin, silver, and zinc.

ENFORCEMENT PROVISIONS

Nongovernmental parties failing to comply with the law's key provisions—for emergency planning, emergency notification, and reporting requirements—face civil, administrative, and/or criminal penalties under the Section 325 enforcement provisions of the Emergency Planning and Community Right-to-Know Act.

Accident Case Studies

The following case studies are provided for classroom instructors and for supervisors conducting on-the-job training. The instructor should present only the case and allow the students to solve the problem through discussion and questioning. The background information, possible solutions, and summary are provided for the instructor's guidance.

CASE I

A maintenance man was working on a metal stair platform and wanted to use a ¼-inch electric drill. The drill had a three-wire power cord. An extension cord running from the source of power also was a three-wire cord, but was not long enough. In order to connect the drill to the extension cord, the man obtained another short extension cord from the tool crib.

As the man started to drill, he was electrocuted.

What caused the accident? What could have been done to prevent this accident? Similar accidents in the future?

Guide and Background Information for Case I

Explain to the group, if asked, that the drill and cord were in proper working order, no shorts. Also, the extension from the power source was okay. However, the extension cord from the tool crib had been improperly wired. The grounding lead had been connected to the "hot" terminal, so that the frame of the drill was energized. When the man grasped the grounded metal stair platform, he was electrocuted.

Once this fact is brought out, have the group pursue the question of preventing another such accident. In the state in which this accident occurred, three workmen had been electrocuted because of improperly wired electrical connections. (This report was from a State Industrial Commission.)

Following are other pertinent facts:

1. The maintenance man was not standing on damp or wet ground. The area was dry.
2. The extension cord was made up by a tool crib attendant.
3. There were no rules in effect concerning procedure for maintenance and repair of equipment.
4. The cord had not been tested.

Possible Solutions for Case I

1. All extension cords and electrical equipment should be immediately tested by a qualified electrician before allowing further use of such equipment.
2. A strict rule should be put into effect that only qualified persons be allowed to repair equipment.
3. Electric tools and equipment should be inspected on a regular schedule to make certain they are properly wired, grounded, and in proper working order. The equipment should also be marked to indicate to the user it had been inspected. The date of inspection should be shown.
4. All electric tools must be grounded regardless of use or location.
5. Make certain all workers understand they are not to make repairs to tools or equipment.
6. If tool crib attendants are to maintain equipment, they should be properly trained.

Summary

Summarize the discussion by pointing out that the unsafe condition must be eliminated—in this case, the improperly wired extension cord. The only way this can be done is to test all extensions and make certain there are no other defective ones. Once all the electrical equipment has been checked out, some procedures must be set up to make certain the equipment is kept in good condition.

Since there are also unsafe procedures involved, provisions must be made to help prevent a recurrence. In this case the tool crib attendant did not follow safe procedures and the maintenance man did not make certain the extension was okay. Safe procedures should be enacted and enforced. Stress the importance of eliminating all possible causes.

CASE II

A freight handler attempted to sharpen the point of a bale hook on a grinding wheel. The foreman of the department saw him, but assumed that "anybody can use a little bench grinder." The freight handler caught the point of his hook between the tool rest and the wheel. The wheel broke and a large piece of it struck him in the face. He was permanently disabled by the injury.

What could have been done to prevent this accident? Similar future accidents?

Guide and Background Information for Case II

Explain to the group, if asked, that this bench grinder was properly guarded and mounted. Also, the wheel was properly tested before mounting.

Following are other pertinent facts:

1. The tool rest was properly set $1/8$-inch from the wheel.
2. Eye protection was available, but obviously not used (goggles).
3. The eye shield was evidently not in place.
4. There was no regulation regarding unauthorized use of equipment.
5. The freight handler was not from this foreman's department.
6. A grinding wheel of this type is proper for sharpening bale hooks.

Possible Solutions for Case II

1. It should be obvious to the group that the primary cause of this accident was one or more unsafe procedures. The foreman made a mistake by "assuming" the freight handler knew how to use a bench grinder. The freight handler committed several unsafe procedures. First, he did not use the eye protection

that was available. Second, he must have put the hook in a position that let the point slip betwen the tool rest and the wheel.

2. While you may get such comments from the group as "Fire the foreman," the solution should focus on a rigidly enforced rule, "No Unauthorized Use of Equipment."
3. The following procedures can also be set up:
   a. A policy regarding the sharpening of tools, such as returning tools to the tool crib for sharpening or replacement.
   b. A program of training if the freight handlers are to do the sharpening.
   c. A lockout device interlocked with the starting switch to prevent the wheel from operating when the eye shield is not in the proper position.

Summary

Point out to the group that the two cases, Case I and Case II, bring out the importance of unsafe conditions and unsafe procedures as the causes of accidents. Stress the importance of **searching for all possible causes**.

Point out that in Case II there were a number of places where the accident prevention program needed tightening in order to prevent similar accidents in the future.

Emphasize the importance of checking procedures and conditions in advance to prevent accidents of this type. Mention the importance of a job safety analysis as a tool for preventing accidents.

CASE III

A tool truck driver, making his routine crib stops, picked up a crib attendant. The attendant had requested a ride to another part of the plant, since he knew the driver would be going in that direction.

The truck driver deviated from his aisle route and angled through a cleared but darkened area. This area was being prepared for new machinery installation, and at that time of night was not fully lighted.

The driver, who was sitting on the right side of the cab, suddenly noticed that the truck was headed for a steel building column. Before he could warn the driver, the left front corner of the truck struck and glanced off the column.

The impact threw the driver against the column and about 15 ft away from the truck. The truck continued for approximately 50 ft before the rider could get behind the wheel to apply the brakes and bring the truck to a stop.

The driver suffered a skull fracture, concussion, and severe injuries to the left arm and chest. He was taken immediately to the local hospital, where he died from a blood clot about three weeks later.

What could have been done to prevent this accident? Similar accidents in the future?

Guide and Background Information for Case III

Explain to the group, if asked, that these trucks are not designed to carry passengers. In order for a rider to sit on the seat, the driver must move over, which puts him in an awkward position.

Following are other pertinent facts:

1. The aisles were not marked in this particular area.
2. The area was not "roped off" and there were no signs of any type to indicate equipment was being installed.
3. The truck was not equipped with a seat belt.
4. The accident happened on the second shift (about 10 p.m.).
5. The machinery installation had been going on over an extended period of time.
6. There were no rules concerning riders.
7. The driver was experienced.

Possible Solutions for Case III

1. The driver's failure to stay within the main aisle was an important factor in the accident. Properly marked aisles might have prevented the driver from taking a short cut.
2. When there are properly marked aisles, truck drivers should be instructed and trained in proper procedures.
3. Better illumination might also have prevented the accident. Since the installation had been going on for some time, the area should have been properly illuminated.
4. The area could have been roped off or marked.
5. A rigidly enforced rule against "no riders" should have been instituted.
6. Installation of seat belts in equipment of this type is a possibility, but because of the nature of the work, not generally done.

Summary

Stress to the group that the accident was caused by a combination of factors.

The unsafe conditions were poor lighting, unmarked aisles, and lack of signs.

The unsafe procedures were the driver's "short cutting," and picking up the rider. The lack of a rule against riders did not exonerate the driver, since he had to make room for the rider and had to be aware that he was not in the best position to control the truck. The rider also must have been aware of the situation when he moved into the seat.

CASE IV

A truck driver and a millwright were in the process of moving a pump, weighing approximately 2,000 lbs., that was mounted on a skid. They were using a mobile crane, and as the operator was backing up with the crane boom elevated at about a 30-degree angle, the rear wheels of the crane rose off the ground. To compensate for this overbalanced condition, the operator, while proceeding backwards, raised the crane boom to an approximate 60-degree angle. His co-worker at the time was observing the load on the crane boom and did not notice an overhead 440-volt line.

The crane boom became entangled in the overhead 440-volt wires, located approximately 18 ft above ground. When the operator observed this, he jumped off the operating platform and was not injured. The insulation on the 440-volt line was not cut, so the crane boom did not become energized.

What could have been done to prevent this accident? Similar accident in the future?

Guide and Background Information for Case IV

Explain to the group, if the question comes up, that this man was not injured. Had the insulation been cut, this man could have been electrocuted.

Following are other pertinent facts:

1. The crane was of proper size to handle the load.
2. It was not possible to have the electric current turned off.
3. The crane was designed to operate both forward the backward.
4. Both employees were experienced and had worked together before. Both worked for the same company.
5. The normal procedure in this company was for a man to walk at the head of the mast in order to guide the movement of the crane. In this case, the millwright should have taken over the job.

Possible Solutions for Case IV

1. The millwright or another person should have walked at the head of the mast and made certain all was clear.
2. The job was poorly planned. When the rear wheels of the crane raised off the ground at the 30-degree angle, this should have been the signal to stop the operation and do some checking. The pump should have been placed on dollies and rolled under the wires, since it was quite obvious that with only 18 ft of clearance and the boom raised to 60 degrees, there was going to be contact.

Summary

Better job procedures should be instituted and definite procedures established and followed in all operations of this type. A thorough inspection should be made of the area and the route to be followed before any attempt is made to move loads of this nature.

CASE V

A warehouse employee was in the act of removing a part from the top of a storage bin. To reach it, she climbed on a wooden container instead of getting a ladder available for this purpose.

While standing on the edge of the container, it began to tip. The employee, trying to check her fall, grasped for the top of the bin with her left hand. As she did so, the ring she wore on her ring finger became caught on a bolt end, damaging the finger so badly that it had to be amputated.

What could have been done to prevent this accident? Similar accidents?

Guide and Background Information for Case V

If the question is raised, there were no rules in this company prohibiting the wearing of jewelry at work. Following are other pertinent facts:

1. The bins contained parts that were not in great demand.
2. There was only one ladder in the storage area, which was large and contained three aisles.
3. The ladder, at the time of the accident, was not in the aisle in which the employee was injured.

Possible Solutions for Case V

1. A rule prohibiting the wearing of jewelry at work could be instituted. Better training on the hazards of wearing rings would help.
2. Availability of a ladder in each aisle would make one always handy and thus more likely to be used.
3. Better housekeeping would help, since the wooden container the employee stood on actually was not in its proper place.
4. The use of posters also would help with the problem.

Summary

Point out the hazards of wearing rings at work. Stress the need for education on the subject, since it may be difficult to enforce a rule against wearing rings because of certain religious beliefs.

CASE VI

An employee asked another employee to exchange a trim knife for him at the tool crib. The man placed the knife in the side pocket of his coveralls and started for the crib. On the way, he stopped to remove some stock from a shelf 8 ft from the floor. He was unable to reach the stock from the floor, so he used the bottom shelf on the rack as a platform to stand on. As he bent his right leg to step up on the shelf, the sharp point of the knife was forced up and penetrated his body approximately one inch. A deep laceration was the only injury.

What could have been done to prevent his accident? Similar accidents in the future?

Guide and Background Information for Case VI

If asked, explain there were no holders or tool bags provided for carrying tools of this type, such as punches, awls, screw drivers, etc. Also, it was common practice for workers to exchange tools for each other.

Following are other important facts that may help:

1. The trim knife was not one of the tools used in the injured employee's own work. He did, however, use other similar tools.
2. The injured employee started out to obtain stock, but was asked to exchange the trim knife on his way.
3. There was a small stepladder nearby that the injured employee could have used.

4. There were no rules prohibiting any employee from exchanging tools, even if the attendant at the tool crib knew it was for another person.

Possible Solutions for Case VI

1. Proper belt holders should be provided to carry tools, since these employees used sharp-pointed tools frequently.
2. A rule prohibiting employees from exchanging tools for each other could be put in effect.
3. Better training in the handling of tools would help.
4. More emphasis on the hazards of using shelves as a stepladder also would help prevent accidents.

Summary

Stress the number of accidents resulting from improperly handled tools. Emphasize the importance of a sound training program on the proper use and handling of sharp and pointed tools. Stress also the need for rigid rules for carrying tools of this type.

CASE VII

A worker was assigned to clean out sludge in the large degreaser he had been operating. His supervisor advised him to use gloves and respiratory protection before entering the degreasing tank and told him to make sure he had a co-worker watch him while he worked. An hour later, the supervisor noted the co-worker eating lunch. He asked the co-worker where the degreaser operator was. The co-worker responded that the operator had told him to go on to lunch, that he would be finished in the degreaser in five minutes. The supervisor and co-worker went immediately to check on the operator, whom they saw face down in the degreasing tank. The supervisor hurriedly descended into the tank to rescue the worker, but had to leave without retrieving him when he immediately began to feel light-headed. The co-worker called the emergency squad who, equipped with self-contained breathing apparatus, retrieved the unconscious operator. The operator survived, but spent several days in the hospital recuperating and has experienced some memory loss and problems in coordination.

What could have been done to prevent this accident? Similar situations in the future?

Guide and Background Information for Case VII

Explain that this is a typical confined space accident, in which the rescuer is as likely (or more likely) to be injured as the initial victim. In over half the recorded cases, the unprepared rescuer has also been a fatality.

Other pertinent facts include:

1. The degreaser operator had not received training in proper use of respiratory protection or confined space entry procedures.
2. The operator was able to work in the pit for at least an hour before being overcome because:
   a. His organic vapor respirator did provide protection against the initial concentration of solvent vapors in the tank.
   b. The tank was not an oxygen deficient atmosphere initially (although under other circumstances, it might have been one), but became one later.
3. The worker stirred and shoveled up the sludge. Solvent vapors were released from air pockets in the solid material, so the concentration of vapors increased in the tank, and began to displace oxygen. The respirator no longer protected him because the absorbent material had become saturated with the solvent, and more important, the respirator could not supply oxygen (which had been depleted by the end of the hour). When trying to rescue the worker, the supervisor began to experience the effects of a lack of oxygen.
4. The cloth gloves the worker wore may have added to his solvent exposure, because they may have allowed the solvent to be absorbed through the skin.

Possible Solutions for Case VII

1. The job indicated poor or not planning. The operator, co-worker, and supervisor should have received training on the hazards and proper procedures for entry into confined spaces. These include:
   a. Use of a permit entry system
   b. Pre-entry sampling of the atmosphere for flammable or toxic gases and vapors
   c. Use of air-supplied or self-contained breathing apparatus where dangerous chemical exposures could occur, as well as impervious clothing
   d. Use of lifelines and belts to aid in retrieval, if necessary

e. Presence of an attendant throughout the duration of the work

2. The supervisor should have completed a confined space entry permit and communicated to the operator the equipment and precautions that were necessary, and assured that these were used.

3. The operator should have received training in the proper use of respiratory protection and protective clothing, and the distinction between air-purifying and air-supplying respirators.

4. Hazard communication training should have been provided to these workers to acquaint them with the toxic properties of the chemicals they worked with.

Summary

Better training should be provided and procedures instituted to prevent future incidents of this type. The hazards of the confined space should be assessed and understood before anyone attempts an entry.

CASE VIII

A worker in your department has taken an excessive number of sick days. You ask the company nurse what the problem might be. She says the worker has come in to ask for aspirin almost every day she has been at work for the last two months.

You go to observe the worker at her operation. Her job, which is in the rework area, involves using pliers to remove wires from a part before it is repaired. You ask the worker to demonstrate how she does her job. She shows you that she grasps the part in her left hand, and twists it as she uses the pliers to grasp and pull the wire out with her right hand.

You note that she does not use the stool provided. The worker says that it is uncomfortable; because she is short, her legs dangle. She explains that she has been absent due to pain in her back and numbness and tingling in her hands.

Guide and Background Information for Case VIII

Explain to the group, if asked, that there are at least two obvious ergonomic problems at the workstation: dangling feet and repetitive, forceful grasping combined with twisting of the wrist. The latter is probably responsible for the worker's back pain.

Other pertinent facts include:

1. Her doctor told her that her hand problem was due to hormonal changes caused by memopause and that her back pain was likely caused by stress. Her doctor had not seen her work area; however, he or she told the worker to take asperin to releive the symptoms.

2. She told you she did not want to complain about her discomfort because she was afraid of losing her job.

Possible Solutions for Case VIII

Institute an ergonomics program that could include:

- Observation of work operations with employee involvement and input
- Medical surveillance to determine if and where workers are being injured
- Management commitment to making positive ergonomic changes where necessary
- Development of an ergonomically trained team of managers, engineers, medical personnel, supervisors, workers, and others.

For this particular injured worker, the supervisor should observe which actions require forceful repetitive movement of the wrists. It is apparent that the worker had to twist both her left and right wrists while applying force to the pliers with her right hand. These repetitive motions have probably contributed to what appears to be carpal tunnel syndrome. Some solutions might be:

1. Automating the operation
2. Using a grip and clamp to hold the part, and using ergonomically designed pliers that prevent the wrist from bending
3. Reducing the work rate
4. Rotating the worker to other jobs that require use of different muscle groups

Summary

Institute an ergonomic program. Evaluate this and all similar operations to determine if other workers are experiencing similar problems. Investigate and establish specific solutions for these operations and obtain employee feedback on how well the solutions are working.

Glossary

Abrasive blasting. A process for cleaning surfaces by means of such materials as sand, alumina, or steel grit in a stream of high-pressure air.

Accident. An unplanned event, not necessarily injurious or damaging to property, interrupting the activity in process.

Accident causes. Hazards and those factors (management oversight, system limitation/failure, etc.) that, individually or in combination, directly cause accidents.

Acclimatization. The process of becoming accustomed to new conditions (i.e., heat, cold).

Accuracy (instrument). Quite often used incorrectly as precision (see Precision). Accuracy refers to the agreement of a reading or observation obtained from an instrument or a technique with the true value.

ACGIH. American Conference of Governmental Industrial Hygienists. ACGIH develops and publishes recommended occupational exposure limits for hundreds of chemical substances and physical agents. (See TLV.)

Acoustic, Acoustical. Containing, producing, arising from, actuated by, related to, or associated with sound.

Acoustic trauma. Hearing loss caused by sudden loud noise in one ear, or by sudden blow to head. In most cases, hearing loss is temporary, although there may be some permanent loss.

Acrylic. A family of synthetic resins made by polymerizing esters of acrylic acids.

Action level. Term used by OSHA and NIOSH to express the level of toxicant that requires medical surveillance, usually one half the PEL.

Acuity. The sense that pertains to the sensitivity of receptors used in hearing or vision.

Acute effect. An adverse effect on a human or animal body, with several symptoms developing rapidly and coming quickly to a crisis.

Acute toxicity. The adverse (acute) effects resulting from a single dose of or exposure to a substance.

Adhesion. The ability of one substance to stick to another. There are two types of adhesion: mechanical, which depends on the penetration of the surface, and molecular or polar adhesion, in which adhesion to a smooth surface is obtained because of polar groups such as carboxyl groups.

Administrative controls. Methods of controlling employee exposures by job rotation, work assignment, or time periods away from the hazard.

Adsorption. The condensation of gases, liquids, or dissolved substances on the surfaces of solids.

Aerosol. Liquid droplets or solid particles dispersed in air; are of fine enough particle size (0.01 to 100 micrometers) to remain so dispersed for a period of time.

Agency or agent. The principal object, such as a tool, machine, or equipment, involved in an accident; is usually the object inflicting injury or property damage.

Air. The mixture of gases that surrounds the earth; its major components are as follows: 78.08 percent nitrogen, 20.95 percent oxygen, 0.03 percent carbon dioxide, and 0.93 percent argon. Water vapor (humidity) varies. (See Standard air.)

Air cleaner. A device designed to remove atmospheric airborne impurities, such as dusts, gases, vapors, fumes, and smokes.

Air conditioning. The process of treating air so as to control simultaneously its temperature, humidity, cleanliness, and distribution to meet requirements of the conditioned space.

Air filter. An air-cleaning device to remove light particulate matter from normal atmospheric air.

Air hammer. A percussion-type pneumatic tool, fitted with a handle at one end of the shank and a tool chuck at the other, into which a variety of tools may be inserted.

Air-line respirator. A respirator that is connected to a compressed breathing air source by a hose.

Air monitoring. The sampling for and measuring of pollutants in the atmosphere.

Air-purifying respirator. Respirator that uses filters or sorbents to remove harmful substances from the air.

Air-regulating valve. An adjustable valve used to regulate airflow to the facepiece, helmet, or hood of an air-line respirator.

Air-supplied respirator. Respirator that provides a supply of breathable air from a clean source outside of the contaminated work area.

Algorithm. A precisely stated procedure or set of instructions that can be applied stepwise to solve a problem.

Aliphatic. (Derived from Greek word for oil.) Pertaining to an open-chain carbon compound. Usually applied to petroleum products derived from a paraffin base and having a straight or branched chain and saturated or unsaturated molecular structure. Substances such as methane and ethane are typical aliphatic hydrocarbons.

Alkali. Any chemical substance that forms soluble soaps with fatty acids. Alkalis are also referred to as bases. They may cause severe burns to the skin. Alkalis turn litmus paper blue and have pH values from 8 to 14.

Allergic reaction. An abnormal physiological response to chemical or physical stimuli by a sensitive person.

Allergy. An abnormal response of a hypersensitive person to chemical and physical stimuli. Allergic manifestations of major importance occur in about 10 percent of the population.

Alloy. A mixture of metals.

Alpha particle (alpha-ray, alpha-radiation). A small, electrically charged particle of very high velocity thrown off by many radioactive materials, including uranium and radium. It is made up of two neutrons and two protons. Its electric charge is positive.

Ambient noise. The all-encompassing noise associated with a given environment, being usually a composite of sounds from many sources.

Ampere. The standard unit for measuring the strength of an electrical current.

Anemometer. A device to measure air velocity.

Anesthetic. A chemical that causes a total or partial loss of sensation. Overexposure to anesthetics can cause impaired judgment, dizziness, drowsiness, headache, unconsciousness, and even death. Examples include alcohol, paint remover, and degreasers.

Anneal. To treat by heat with subsequent cooling for drawing the temper of metals, that is, to soften and render them less brittle. (See Temper.)

ANSI. The American National Standards Institute is a voluntary membership organization (run with private funding) that develops consensus standards nationally for a wide variety of devices and procedures.

Antidote. A remedy to relieve, prevent, or counteract the effects of a poison.

Antioxidant. A compound that retards deterioration by oxidation. Antioxidants for human food and animal feeds, sometimes referred to as freshness preservers, retard rancidity of fats and lessen loss of fat-soluble vitamins (A, D, E, K). Antioxidants also are added to rubber, motor lubricants, and other materials to inhibit deterioration.

Appearance. A description of a substance at normal room temperature and normal atmospheric conditions. Appearance includes the color, size, and consistency of a material.

Approved. Tested and (listed) as satisfactory meeting predetermined requirements by an authority having jurisdiction.

Arc welding. One form of electrical welding using either uncoated or coated rods.

Arc-welding electrode. A component of the welding circuit through which current is conducted between the electrode holder and the arc.

Asbestos. A hydrated magnesium silicate in fibrous form.

Asphyxia. Suffocation from lack of oxygen. Chemical asphyxia is produced by a substance such as carbon monoxide that combines with hemoglobin to reduce the blood's capacity to transport oxygen. Simple asphyxia is the result of exposure to a substance such as methane, that displaces oxygen.

Asphyxiant. A vapor or gas that can cause unconsciousness or death by suffocation (lack of oxygen).

ASTM. American Society for Testing and Materials; voluntary membership organization with members from broad spectrum of individuals, agencies, and industries concerned with materials. As the world's largest source of voluntary consensus standards for materials, products, systems, and services, ASTM is a resource for sampling and testing methods, health and safety aspects of materials, safe performance guidelines, and effects of physical and biological agents and chemicals.

Atmosphere-supply respirator. A respirator that provides breathing air from a source independent of the surrounding atmosphere. There are two types: air-line and self-contained breathing apparatus.

Atmospheric pressure. The pressure exerted in all directions by the atmosphere. At sea level, mean atmospheric pressure is 29.92 inches Hg, 14.7 psi, or 407 inches w.g.

Atmospheric tank. A storage tank that has been designed to operate at pressures from atmospheric through 0.5 psig (3.5 kPa).

Atomic energy. Energy released in nuclear reactions. Of particular interest is the energy released when a neutron splits an atom's nucleus into smaller pieces (fission) or when two nuclei are joined together under millions of degrees of heat (fusion). "Atomic energy" is really a popular misnomer. It is more correctly called "nuclear energy."

Atomic power. See Atomic energy.

Atomic waste. The radioactive ash produced by the splitting of uranium (nuclear) fuel, as in a nuclear reactor. It may include products made radioactive in such a device.

Audible range. The frequency range over which normal ears hear approximately 20 Hz through 20,000 Hz. Above the range of 20,000 Hz, the term "ultrasonic" is used. Below 20 Hz, the term "subsonic" is used.

Audiogram. A record of hearing loss of hearing level measured at several different frequencies—usually 500 to 6,000 Hz. The audiogram may be presented graphically or numerically. Hearing level is shown as a function of frequency.

Audiometer. A signal generator or instrument for measuring objectively the sensitivity of hearing in decibels.

Autoignition temperature. The temperature to which a closed, or nearly closed, container must be heated in order that a flammable liquid, when introduced into the container, will ignite spontaneously or burn.

Babbit. An alloy of tin, antimony, copper, and lead used as a bearing metal.

Background noise. Noise coming from sources other than the particular noise source being monitored. (See Background radiation.)

Background radiation. The radiation coming from sources other than the radioactive material to be measured. This "background" is primarily due to cosmic rays, which constantly bombard the earth from outer space.

Bacteria. Microscopic organisms living in soil, water, organic matter, or the bodies of plants and animals characterized by lack of a distinct nucleus and lack of ability to photosynthesize. Singular: Bacterium.

Bag house. Many different trade meanings. Term commonly used for the housing containing bag filters for recovery of fumes of arsenic, lead, sulfur, etc., from flues of smelters.

Balancing by dampers. Method of designing local exhaust system ducts using adjustable dampers to distribute airflow after installation.

Balancing by static pressure. Method of designing local exhaust system ducts by selecting the duct diameters that generate the static pressure to distribute airflow without dampers.

Ball mill. A grinding device using balls usually of steel or stone in a rotating container.

Base. A compound that reacts with an acid to form a salt. It is another term for alkali. It turns litmus paper blue.

Bauxite. Impure mixture of aluminum oxides and hydroxides; the principal source of aluminum.

Beat elbow. Bursitis of the elbow; occurs from use of heavy vibrating tools.

Beat knee. Bursitis of the knee joints due to friction or vibration; common in mining.

Beehive kiln. A kiln shaped like a large beehive; used usually for calcining ceramics.

Benign. Not malignant. A benign tumor is one that does not metastasize or invade tissue. Benign tumors may still be lethal, due to pressure on vital organs.

Beta particle (beta-radiation). A small, electrically charged particle thrown off by many radioactive materials. It is identical with the electron. Beta particles emerge from radioactive material at high speeds.

Billet. A piece of semifinished iron or steel, nearly square in section, made by rolling and cutting an ingot.

Binder. The nonvolatile portion of a coating vehicle which is the film-forming ingredient used to bind the paint pigment particles together.

Biodegradable. Capable of being broken down into innocuous products by the action of living things.

Biohazard. A combination of the words "biological hazard." Organisms or products of organisms that present a risk to humans.

Biohazard area. Any area (a complete operating complex, a single facility, a room within a facility, etc.) in which work has been or is being performed with biohazardous agents or materials.

Biohazard control. Any set of equipment and procedures utilized to prevent or minimize the exposure of humans and their environment to biohazardous agents or materials.

Biomechanics. The study of the human body as a system operating under two sets of law: the laws of Newtonian mechanics and the biological laws of life.

Black light. Ultraviolet (UV) light radiation between 3,000 and 4,000 angstroms (0.3-0.4 micrometers).

Black liquor. A liquor composed of alkaline and organic matter resulting from digestion of wood pulp and cooking acid during the manufacture of paper.

Bleaching bath. Chemical solution used to bleach colors from a garment preparatory to dyeing it, a solution of chlorine or sodium hypochlorite commonly being used.

Blind spot. Normal defect in visual field due to position at which optic nerve enters the eye.

Boiling point. The temperature at which a liquid changes to a vapor state, expressed in degrees.

BOM. Bureau of Mines of the U.S. Department of Interior. BuMines began approving air-breathing apparatus in 1918, later added all types of respirators. BOM's respirator testing/approval activities have been discontinued. BOM-approved Type 14F gas masks are still acceptable; all other BOM approvals have expired or have been replaced by NIOSH approvals.

Bonding. The interconnecting of two objects by means of a clamp and bare wire. Its purpose is to equalize the electrical potential between objects. (See Grounding.)

Brake horsepower. The horsepower required to drive a unit; it includes the energy losses in the unit and can be determined only by actual test. It does not include driver losses between motor and unit.

Brass. An alloy of copper and zinc; may also contain a small portion of lead.

Brattice. A partition constructed in underground passageways to control ventilation in mines.

Braze. To solder with any alloy that is relatively infusible.

Brazing furnace. Used for heating metals to be joined by brazing. Requires a high temperature.

Breathing tube. A tube through which air or oxygen flows to the facepiece, helmet, or hood.

Breathing zone. Imaginary globe of two-foot radius surrounding the head.

Broach. A cutting tool for cutting nonround holes.

Bubble tube. A device used to calibrate air-sampling pumps.

Buffer. Any substance in a fluid that tends to resist the change in pH when acid or alkali is added.

Bulk density. Mass of powdered or granulated solid material per unit of volume.

Bulk plant. That portion of a property where flammable or combustible liquids are received by tank vessel, pipelines, tank car, or tank vehicle and are sorted or blended in bulk for the purpose of distributing such liquids by tank vessel, pipeline, tank car, tank vehicle, or container.

Burns. Result from the application of too much heat to the skin. First-degree burns show redness of the unbroken skin; second-degree, skin blisters and some breaking of the skin; third-degree, skin blisters and destruction of the skin and underlying tissues, which can include charring and blackening.

cc. Cubic centimeter; a volume measurement in the metric system, equal in capacity to one milliliter (ml)—approximately twenty (20) drops. There are 16.4 cc in one cubic inch.

Calcination. The heat treatment of solid material to bring about thermal decomposition, to lose moisture or other volatile material, or to be oxidized or reduced.

Calender. An assembly of rollers for producing a desired finish on paper, rubber, artificial leather, plastics, or other sheet material.

Calking. The process or material used to fill seams of boats, cracks in tile, etc.

Capture velocity. Air velocity at any point in front of the hood necessary to overcome opposing air currents and to capture contaminated air by causing it to flow into the exhaust hood.

Carbon black. Essentially a pure carbon, best known as common soot. Commercial carbon black is produced by making soot under controlled conditions. It is sometimes called channel black, furnace black, acetylene black, or thermal black.

Carbon monoxide. A colorless, odorless, toxic gas produced by any process that involves the incomplete combustion of carbon-containing substances.

Carcinogen. A substance or agent that can cause a growth of abnormal tissue or tumors in humans or animals.

Carcinogenic. Cancer-producing.

Carding. The process of combing or untangling wool, cotton, etc.

Carpal tunnel. A passage in the wrist through which the median nerve and many tendons pass to the hand from the forearm.

Carpal tunnel syndrome. A common affliction caused by compression of the median nerve in the carpal tunnel. Often associated with tingling, pain, or numbness in the thumb and first three fingers.

CAS number. Identifies a particular chemical by the Chemical Abstract Service, a service of the American Chemical Society that indexes and compiles abstracts of worldwide chemical literature called "Chemical Abstracts."

Case-hardening. A process of surface-hardening metals by raising the carbon or nitrogen content of the outer surface.

Cask (or coffin). A thick-walled container (usually lead) used for tranporting radioactive materials.

Casting. The pouring of a molten material into a mold and permitting it to solidify to the desired shape.

Catalyst. A substance that changes the speed of a chemical reaction but undergoes no permanent change itself. In respirator use, a substance that converts a toxic gas (or vapor) into a less toxic gas (or vapor). Usually catalysts greatly increase the reaction rate, as in conversion of petroleum to gasoline by cracking. In paint manufacture, catalysts, which hasten the film-forming, generally become part of the final product. In most uses, however, they do not, and can often be used over again.

Cathode. The negative electrode.

Catwalk. A narrow suspended footway usually for inspection or maintenance purposes.

Caustic. Something that strongly irritates, burns, corrodes, or destroys living tissue. (See Alkali.)

Ceiling limit (C). An airborne concentration of a toxic substance in the work environment, that should never be exceeded. ACGIH terminology. (See TLV.)

Cellulose ($C_6H_{10}OH_5$). A carbohydrate that makes up the structural material of vegetable tissues and fibers.

Celsius. The Celsius temperature scale is a designation of the scale previously known as the centigrade scale.

Cement, portland. Portland cement commonly consists of hydraulic calcium silicates to which the addition of certain materials in limited amounts is permitted. Ordinarily, the mixture consists of calcareous materials such as limestone, chalk, shells, marl, clay, shale, blast furnace slag, etc. In some specifications, iron ore and limestone are added. The mixture is fused by calcining at temperatures usually up to 1,500 C.

Central nervous system. The brain and spinal cord. These organs supervise and coordinate the activity of the entire nervous system. Sensory impulses are transmitted into the central nervous system, and motor impulses are transmitted out.

Centrifuge. An apparatus that uses centrifugal force to separate or remove particulate matter suspended in a liquid.

Ceramic. A term applied to pottery, brick, and the tile products molded from clay and subsequently calcined.

CFR. Code of Federal Regulations. A collection of the regulations that have been promulgated under U.S. law.

Chemical burns. Generally similar to those caused by heat. After emergency first aid, their treatment is the same as that for thermal burns.

Chemical cartridge. A chemical cartridge is used with a respirator for removal of low concentrations of specific vapors and gases.

Chemical cartridge respirator. A respirator that uses various chemical substances to purify inhaled air of certain gases and vapors.

Chemical engineering. That branch of engineering concerned with the development and application of manufacturing processes in which chemical or certain physical changes of materials are involved.

Chemical reaction. A change in the arrangement of atoms or molecules to yield substances of different composition and properties. Common types of reaction are combination, decomposition, double decomposition, replacement, and double replacement.

CHEMTREC. Chemical Transportation Emergency Center.

Chronic. Persistent, prolonged, repeated.

Chronic toxicity. Adverse (chronic) effects resulting from repeated doses of or exposures to a substance over a relatively prolonged period of time.

Circuit. A complete path over which electrical current may flow.

Circuit breaker. A device that automatically interrupts the flow of an electrical current when the current becomes excessive.

Clays. A great variety of aluminum-silicate-bearing rocks, plastic when wet, hard when dry. Used in pottery, stoneware, tile, bricks, cements, fillers, and abrasives. Kaolin is one type of clay.

Some clay deposits may include appreciable quartz. Commercial grades of clays may contain up to 20 percent quartz.

Clean Air Act. Federal law enacted to regulate/reduce air pollution. Administered by EPA.

Clean Water Act. Federal law enacted to regulate/reduce water pollution. Administered by EPA.

Coagulation. Formation of a clot or gelatinous mass.

Coated welding rods. The coatings of welding rods vary. For the welding of iron and most steel, the rods contain manganese, titanium, and a silicate.

Code of Federal Regulations (CFR). The rules promulgated under U.S. law and published in the *Federal Register* and actually in force at the end of a calendar year are incorporated in this code.

Coffin. A thick-walled container (usually lead) used for transporting radioactive materials; also called cask.

Cohesion. Molecular forces of attraction between particles of like composition.

Colloid mill. A machine that grinds materials into very fine state of suspension, often simultaneously placing this in suspension in a liquid.

Combustible. Able to catch on fire and burn.

Combustible liquids. Combustible liquids are those having a flash point at or above 37.8 C (100 F).

Comfort ventilation. Airflow intended to maintain comfort of room occupants (heat, humidity, and odor).

Comfort zone. The average range of effective temperatures over which the majority (50 percent or more) of adults feel comfortable.

Common name. Any designation or identification such as code name, code number, trade name, brand name, or generic name used to identify a chemical other than by its chemical name.

Communicable. Refers to a disease whose causative agent is readily transferred from one person to another.

Compaction. The consolidation of solid particles between rolls, or by tamp, piston, screw, or other means of applying mechanical pressure.

Compound. A substance composed of two or more elements joined according to the laws of chemical combination. Each compound has its own characteristic properties different from those of its constituent elements.

Compressed gas. (1) Any gas or mixture of gases having, in a container, an absolute pressure exceeding 40 psi at 70 degrees F (21.2 degrees C); or (2) a gas or mixture of gases having, in a container, an absolute pressure exceeding 104 psi at 130 degrees F (54.4 degrees C) regardless of the pressure at 70 degrees F (21.2 degrees C); or (3) a liquid having a vapor pressure exceeding 40 psi at 100 degrees F (37.8 degrees C) as determined by ASTM D-323-72.

Concentration. The amount of a given substance in a stated unit of measure. Common methods of stating concentration are percent by weight or by volume, weight per unit vlume, normality, etc.

Condensate. The liquid resulting from the process of condensation.

Condensation. Act or process of reducing from one form to another denser form, such as steam to water.

Conditions to avoid. Conditions encountered during handling or storage that could cause a substance to become unstable.

Conductive hearing loss. Type of hearing loss; not caused by noise exposure, but due to any disorder in the middle or external ear that prevents sound from reaching the inner ear.

Confined space. Any area that has limited openings for entry and exit that would make escape difficult in an emergency, has a lack of ventilation, contains known and potential hazards, and is not intended nor designated for continuous human occupancy. (See "Permit Required Confined Space," 29 *CFR* 1910.146(b)(23) proposed.)

Contact dermatitis. Dermatitis caused by contact with a substance—gaseous, liquid, or solid. May be due to primary irritation or an allergy.

Controlled area. A specified area in which exposure of personnel to radiation or radioactive material is controlled and which is under the supervision of a person who has knowledge of the appropriate radiation protection practices, including pertinent regulations, and who has responsibility for applying them.

Convection. The motions in fluids resulting from differences in density and the action of gravity.

Coolants. Coolants are transfer agents used in a flow system to convey heat from its source.

Corrective lens. A lens ground to the wearer's individual prescription.

Corrosion. Physical change, usually deterioration or destruction, brought about through chemical or electrochemical action as contrasted with erosion caused by mechanical action.

Corrosive (COR). As defined by DOT, a corrosive material is a liquid or solid that causes visible destruction or irreversible alterations in human skin tissue at the site of contact or—in the case of leakage from its packaging—a liquid that has a severe corrosion rate on steel. Two common examples are caustic soda and sulfuric acid.

Cosmic rays. High-energy rays of great penetrating power that bombard the earth from outer space.

Cottrell precipitator. A device for dust collection using high-voltage electrodes.

Counter. A device for counting. (See Geiger counter and Scintillation counter.)

Covered electrode. A composite filler metal electrode consisting of a core of bare electrode or metal-cored electrode to which a covering (sufficient to provide a slag layer on the weld metal) has been applied; the covering may contain materials providing such functions as shielding from the atmosphere, deoxidation, and arc stabilization and can serve as a source of metallic additions to the weld.

CPR. Cardiopulmonary resuscitation.

CPSC. Consumer Products Safety Commission; federal agency with responsibility for regulating hazardous materials when they appear in consumer goods. For CPSC purposes, hazards are defined in the Hazardous Substances Act and the Poison Prevention Packaging Act of 1970.

Cristobalite. A crystalline form of free silica, extremely hard and inert chemically; very resistant to heat. Quartz in refractory bricks and amorphous silica in diatomaceous earth are altered to cristobalite when exposed to high temperatures (calcined).

Critical pressure. The pressure under which a substance may exist as a gas in equilibrium with a liquid at the critical temperature.

Critical temperature. The temperature above which a gas cannot be liquefied by pressure alone.

Crucible. A heat-resistant, barrel-shaped pot used to hold metal during melting in a furnace.

Crude petroleum. Hydrocarbon mixtures that have a flash point below 65.6 C (150 F), and that have not been processed in a refinery.

Cry-, cryo- (prefix). Very cold.

Cryogenics. The field of science dealing with the behavior of matter at very low temperatures.

Cubic centimeter (cc). A volumetric measurement that is also equal to one milliliter (mL.)

Cubic meter (m³). A measure of volume in the metric system.

Cutting fluids (oils). The cutting fluids used in industry today are usually an oil or an oil-water emulsion used to cool and lubricate a cutting tool. Cutting oils are usually light or heavy petroleum fractions.

CW laser. Continuous wave laser. (See Laser.)

Cyanide (as CN). Cyanides inhibit tissue oxidation upon inhalation or ingestion and cause death.

Cyanosis. Blue appearance of the skin, especially on the face and extremities, indicating a lack of sufficient oxygen in the arterial blood.

Cyclone separator. A dust-collecting device that has the ability to separate particles by size. Typically used to collect respirable dust samples.

Damage risk criterion. The suggested base line of noise tolerance which, if not exceeded, should result in no hearing loss due to noise. A damage risk criterion may include in its statement a specification of such factors as time of exposure, noise level and frequency, amount of hearing loss that is considered significant, percentage of the population to be protected, and method of measuring the noise.

Damp. A harmful gas or mixture of gases occurring in coal mining.

Dampers. Adjustable sources of airflow resistance used to distribute airflow in a ventilation system.

Dangerous to life or health, immediately (IDLH). Used to describe very hazardous atmospheres where employee exposure can cause serious injury or death within a short time or serious delayed effects.

dBA. Sound level in decibels read on the A-scale of a sound level meter. The A scale discriminates against very low frequencies (as does the human ear) and is, therefore, better for measuring general sound levels. (See also Decibel.)

Decibel (dB). A unit used to express sound power level (L_w). Sound power is the total acoustic output of a sound source in watts (W). By definition, sound power level, in decibels, is: $L_w = 10 \log W$ divided by W_o, where W is the sound power of the source and W_o is the reference sound power.

Decomposition. Breakdown of a material or substance (by heat, chemical reaction, electrolysis, decay, or other process) into parts or elements or simpler compounds.

Decontaminate. To make safe by eliminating poisonous or otherwise harmful substances, such as noxious chemicals or radioactive material.

Density. The mass (weight) per unit volume of a substance. For example, lead is much more dense than aluminum.

Depressant. A substance that reduces a functional activity or an instinctive desire of the body, such as appetite.

Dermal toxicity. Adverse effects resulting from skin exposure to a substance. Ordinarily used to denote effects in experimental animals.

Dermatitis. Inflammation of the skin.

Dermatosis. A broader term than dermatitis; it includes any cutaneous abnormality, encompassing folliculitis, acne, pigmentary changes, and nodules and tumors.

Diaphragm. (1) The musculomembranous partition separating the abdominal and thoracic cavities; (2) any separating membrane or structure; (3) a disk with one or more openings in it, or with an adjustable opening, mounted in relation to a lens, by which part of a light may be excluded from an area.

Diatomaceous earth. A soft, gritty, amorphous silica composed of minute siliceous skeletons of small aquatic plants. Used in filtration and decolorization of liquids, insulation, filler in dynamite, wax, textiles, plastics, paint, and rubber. Calcined and flux-calcined diatomaceous earth contains appreciable amounts of cristobalite, and dust levels should be controlled the same as for cristobalite.

Die. A hard metal or plastic form used to shape material to a particular contour or section.

Differential pressure. The difference in static pressure between two locations.

Diffusion rate. A measure of the tendency of one gas or vapor to disperse into or mix with another gas or vapor.

Dike. A barrier constructed to control or confine substances and prevent their movement.

Diluent. A liquid that is blended with a mixture to reduce concentration of the active agents.

Dilution. The process of increasing the proportion of solvent or diluent (liquid) to solute or particulate matter (solid).

Dilution ventilation. Airflow designed to dilute contaminants to acceptable levels. Also referred to as general ventilation or exhaust. (See General exhaust.)

Direct-reading instrumentation. Those instruments that give an immediate indication of the concentration of matter.

Disinfectant. An agent that frees from infection by killing the vegetative cells of microorganisms.

Dispersion. The general term describing systems consisting of particulate matter suspended in air or other fluid; also, the mixing and dilution of contaminant in the ambient environment.

Distal. Away from the central axis of the body.

Distillery. A plant or that portion of a plant where flammable or combustible liquids produced by fermentation are concentrated, and where the concentrated products may also be mixed, stored, or packaged.

DOL. U.S. Department of Labor; includes the Occupational Safety and Health Administration (OSHA) and Mine Safety and Health Administration (MSHA).

Dose. A term used (1) to express the amount of a chemical or of ionizing radiation energy absorbed in a unit volume or an organ or individual. Dose rate is the dose delivered per unit of time. (See also Roentgen, Rad, Rem.) (2) Used to express amount of exposure to a chemical substance.

Dose equivalent, maximum permissible (MPD). The largest equivalent received within a specified period which is permitted by a regulatory agency or other authoritative group on the assumption that receipt of such dose equivalent creates no appreciable somatic or genetic injury. Different levels of MPD may be set for different groups within a population. (By popular usage, "dose, maximum permissible," is an accepted synonym.)

Dosimeter (dose meter). An instrument used to determine the full-shift exposure a person has received to a physical hazard.

DOT hazard class. DOT requires that hazardous materials offered for shipment be labeled with the proper DOT hazard class. These classes include corrosive, flammable liquid, organic peroxide, ORM-E, poison B, etc. The DOT hazard class may not adequately describe all the hazard properties of the material.

Drier. Any catalytic material that, when added to a drying oil, accelerates drying or hardening of the film.

Drop forge. To forge between dies by a drop hammer or drop press.

Droplet. A liquid particle suspended in a gas. The liquid particle is generally of such size and density that it settles rapidly and remains airborne for an appreciable length of time only in a turbulent atmosphere.

Dross. The scum that forms on the surface of molten metals, largely oxides and impurities.

Dry chemical. A powdered fire-extinguishing agent usually composed of sodium bicarbonate, monoammonium phosphate, potassium bicarbonate, etc.

Duct. A conduit used for conveying air at low pressures.

Dust collector. An air-cleaning device to remove heavy particulate loadings from exhaust systems before discharge to outdoors; usual range is loadings of 0.003 gr/cu ft (0.007 mg/m^3) and higher.

Dusts. Solid particles generated by handling, crushing, grinding, rapid impact, detonation, and decrepitation of organic or inorganic materials, such as rock, ore, metal, coal, wood, and grain. Dusts do not tend to flocculate except under electrostatic forces; they do not diffuse in air but settle under the influence of gravity.

Dynometer. Apparatus for measuring force or work output external to a subject. Often used to compare external output with associated physiological phenomena to assess physiological work efficiency.

EAP. Employee Assistance Program.

Ear. The entire hearing apparatus, consisting of

three parts: external ear, the middle ear or tympanic cavity, and the inner ear or labyrinth. Sometimes the pinna is called the ear.

Effective temperature. An arbitrary index that combines into a single value the effect of temperature, humidity, and air movement on the sensation of warmth and cold on the human body.

Effective temperature index. An empirically determined index of the degree of warmth perceived on exposure to different combinations of temperature, humidity, and air movement. The determination of effective temperature requires simultaneous determinations of dry bulb and wet bulb temperatures.

Effluent. Generally something that flows out or forth, like an outflow of a sewer, storage tank, canal, or other channel.

Electrical current. The flow of electricity measured in amperes or milliamperes.

Electrical precipitator. A device that removes particles from an airstream by charging the particles and collecting the charged particles on a suitable surface.

Electrolysis. The process of conduction of an electric current by means of a chemical solution.

Electromagnetic radiation. The propagation of varying electric and magnetic fields through space at the speed of light, exhibiting the characteristics of wave motion.

Electroplate. To cover with a metal coating (plate) by means of electrolysis.

Element. Solid, liquid, or gaseous matter that cannot be further decomposed into simpler substances by chemical means.

Elutriator. A device used to separate particles according to mass and aerodynamic size by maintaining a laminar flow system at a rate which permits the particles of greatest mass to settle rapidly while the smaller particles are kept airborne by the resistance force of the flowing air for longer times and distances. The various times and distances of deposit may be used to determine representative fractions of particle mass and size.

Emery. Aluminum oxide, natural and synthetic abrasive.

Emission factor. Statistical average of the amount of a specific pollutant emitted from each type of polluting source in relation to a unit quantity of material handled, processed, or burned.

Emission inventory. A list of primary air pollutants emitted into a given community's atmosphere in amounts per day, by type of source.

Emission standards. The maximum amount of pollutant permitted to be discharged from a single polluting source.

Emulsifier or emulsifying agent. A chemical that holds one insoluble liquid in suspension in another. Casein, for example, is a natural emulsifier in milk, keeping butterfat droplets dispersed.

Emulsion. A suspension, each in the other, of two or more unlike liquids that usually will not dissolve in each other.

Enamel. A paintlike oily substance that produces a glossy finish to a surface to which it is applied. Often contains various synthetic resins. It is lead free. In contrast is the ceramic enamel, that is, porcelain enamel, which contains lead.

Engineering controls. Methods of controlling employee exposures by modifying the source or reducing the quantity of hazards.

Entrance loss. The loss in static pressure of a fluid that flows from an area into and through a hood or duct opening. The loss in static pressure is due to friction and turbulence resulting from the increased gas velocity and configuration of the entrance area.

Entry loss. Loss in pressure caused by air flowing into a duct or hood.

Environmental toxicity. Information obtained as a result of conducting environmental testing designed to study the effects of various substances on aquatic and plant life.

Enzymes. Delicate chemical substances, mostly proteins, that enter into and bring about chemical reactions in living organisms.

EPA. U. S. Environmental Protection Agency.

EPA number. The number assigned to chemicals regulated by the Environmental Protection Agency.

Ergonomics. A multidisciplinary activity dealing with interactions between people and their total working environment plus stresses related to such environmental elements as atmosphere, light, and sound as well as all tools and equipment of the workplace.

Esters. Organic compounds that may be made by interaction between an alcohol and an acid, and by other means. Esters are nonionic compounds, including solvents and natural fats.

Etch. To cut or eat away material with acid or other corrosive substance.

Evaporation. The process by which a liquid is changed into the vapor state.

Evaporation rate. The ratio of the time required to evaporate a measured volume of a liquid to the time required to evaporate the same volume of a reference liquid (ethyl ether) under ideal test conditions. The higher the ratio, the slower the evaporation rate.
   FAST evaporation rate greater than 3.0. Examples: Methyl Ethyl Ketone (MEK) = 3.8, Acetone = 5.6, Hexane = 8.3.
   MEDIUM evaporation rate 0.8 to 3.0. Examples: 190 proof (95%) Ethyl Alcohol = 1.4, VM&P Naphtha = 1.4, MIBK = 1.6.
   SLOW evaporation rate less than 0.8. Examples: Xylene = 0.6, Isobutyl Alcohol = 0.6, Normal Butyl Alcohol = 0.4, Water = 0.3, Mineral Spirits = 0.1.

Exhalation valve. A device that allows exhaled air to leave a respirator and prevents outside air from entering through the valve.

Exhaust ventilation. The removal of air (usually by mechanical means) from any space. The flow of air between two points is due to the occurrence of a pressure difference between the two points. This pressure difference will cause air to flow from the high pressure to the low pressure zone.

Explosive limit. See Flammable limit.

Explosive. A reaction that causes a sudden, almost instantaneous release of pressure, gas, and heat.

Exposure. Contact with a chemical, biological, or physical hazard.

Extinguishing medium. The fire-fighting substance to be used to control a material in the event of a fire. It is usually named by its generic name, such as fog, foam, water, etc.

Extrusion. The forcing of raw material through a die or a form in either a heated or cold state, and in a solid state or in partial solution.

Eyepiece. Gastight, transparent window(s) in a full facepiece through which the wearer may see.

Eye protection. Recommended safety glasses, chemical splash goggles, face shields, etc. to be utilized when handling a hazardous material.

Face velocity. Average air velocity into the exhaust system measured at the opening into the hood or booth.

Facepiece. That portion of a respirator that covers the wearer's nose and mouth in a half-mask facepiece, or nose, mouth, and eyes in a full facepiece.

Facing. In foundry work, the final touch-up work of the mold surface to come in contact with metal is called the facing operation and the fine powdered material used is called the facing.

Fainting. Technically called syncope, a temporary loss of consciousness as a result of a diminished supply of blood to the brain.

Fallout. Dust particles that contain radioactive fission products resulting from a nuclear explosion. The wind can carry fallout particles many miles.

Fan static pressure. The pressure added to a system by a fan. It equals the sum of pressure losses in the system minus the velocity pressure in the air at the fan inlet.

Farmer's lung. Fungus infection and ensuing hypersensitivity from grain dust.

Fatal accident. An accident that results in one or more deaths within one year.

FDA. The U.S. Food and Drug Administration; under the provisions of the federal Food, Drug and Cosmetic Act, the FDS establishes requirements for the labeling of foods and drugs to protect consumers from misbranded, unwholesome, ineffective, and hazardous products. FDA also regulates materials used in contact with food and the conditions under which such materials are approved.

Federal Register. Publication of U.S. government documents officially promulgated under the law, documents whose validity depends upon such publication. (See *Code of Federal Regulations.*)

Fertilizer. Plant food usually sold in mixed formula containing basic plant nutrients: compounds of nitrogen, potassium, phosphorus, sulfur, and sometimes other minerals.

Fever. A condition in which the body temperature is above its regular or normal level.

FIFRA. Federal Insecticide, Fungicide, and Rodenticide Act; regulations administered by EPA under this Act require that certain useful poisons, such as chemical pesticides, sold to the public contain labels that carry health hazard warnings to protect users.

Film badge. A piece of masked photographic film worn by nuclear workers. Because it is darkened by nuclear radiation, radiation exposure can be checked by inspection of the film.

Filter. (1) A device for separating components of a signal on the basis of its frequency. It allows components in one or more frequency bands to pass relatively unattenuated, and it attenuates greatly components in other frequency bands. (2) A fibrous medium used in respirators to remove solid or liquid particles from the airstream entering the respirator. (3) A sheet of material that is interposed between patient and the source of x-rays to absorb a selective part of the x-rays. (4) A fibrous or membrane medium used to collect air samples of dust, fume, or mist.

Filter, HEPA. High-efficiency particulate air filter that is at least 99.97 percent efficient in removing thermally generated monodisperse dioctylphthalate smoke particles with a diameter of 0.3 μ.

Firebrick. A special clay that is capable of resisting high temperatures without melting or crumbling.

Fire damp. In mining, the accumulation of an explosive gas, chiefly methane gas. Miners refer to all dangerous underground gases as "damps."

Fire point. The lowest temperature at which a material can evolve vapors to support continuous combustion.

First aid. Emergency measures to be taken before regular medical help can be obtained.

Fission. The splitting of an atomic nucleus into two parts accompanied by the release of a large amount of radioactivity and heat.

Flame propagation. See Propagation of flame.

Flammable limits. Flammables have a minimum concentration below which propagation of flame does not occur on contact with a source of ignition. This is known as the lower flammable explosive limit (LEL). There is also a maximum concentration of vapor or gas in air above which propagation of flame does not occur. This is known as the upper flammable explosive limit (UEL). These units are expressed in percent of gas or vapor in air by volume.

Flammable liquid. Any liquid having a flash point below 37.8 C (100 F).

Flammable range. The difference between the lower and upper flammable limits, expressed in terms of percentage of vapor or gas in air by volume; is also often referred to as the "explosive range."

Flange. In an exhaust system, a rim or edge added to a hood to reduce the quantity of air entering the hood from behind the hood.

Flashback. Occurs when flame from a torch burns back into the tip, the torch, or the hose.

Flash blindness. Temporary visual disturbance resulting from viewing an intense light source.

Flash ignition. See Flash point.

Flash point. The lowest temperature at which a liquid gives off enough vapor to form an ignitable mixture with air and produce a flame when a source of ignition is present.

Flask. In foundry work, the assembly of the cope and the drag constitutes the flask. It is the wooden or iron frame containing sand into which molten metal is poured. Some flasks may have three or four parts.

Flocculation. The process of forming a very fluffy mass of material held together by weak forces of adhesion.

Flotation. A method of ore concentration in which the mineral is caused to float due to chemical frothing agents while the impurities sink.

Flow coefficient. A correction factor used for figuring volume flow rate of a fluid through an orifice. This factor includes the effects of contraction and turbulence loss (covered by the coefficient of discharge), plus the compressibility effect, and the effect of an upstream velocity other than zero. Since the latter two effects are negligible in many instances, the flow coefficient is often equal to the coefficient of discharge.

Flow meter. An instrument for measuring the rate of flow of a fluid.

Fluid. A substance tending to flow or conform to the outline of its container. It may be liquid, vapor, gas, or solid (like raw rubber).

Fluorescence. Emission of light from a crystal, after the absorption of energy.

Fluorescent screen. A screen coated with a fluorescent substance so that it emits light when irradiated with x-rays.

Flux. Usually refers to a substance used to clean

surfaces and promote fusion in soldering. However, fluxes of various chemical nature are used in the smelting of ores, in the ceramic industry, in assaying silver and gold ores, and in other endeavors. The most common fluxes are silica, various silicates, lime, sodium, and potassium carbonate, and litharge and red lead in the ceramic industry. (See Galvanizing.)

Fly ash. Finely divided particles of ash entrained in flue gases arising from the combustion of fuel.

Foot-candle. A unit of illumination. The illumination at a point on a surface which is one foot from, and perpendicular to, a uniform point source of one candle.

Foot-pound. A unit of work equal to the amount of energy needed to raise one pound one foot upward.

Foot-pounds of torque. A measurement of the physiological stress exerted upon any joint during the performance of a task. The product of the force exerted and the distance from the point of application to the point of stress. Physiologically, torque that does not produce motion nonetheless causes work stress, the severity of which depends on the duration and magnitude of the torque. In lifting an object or holding it elevated, torque is exerted and applied to the lumbar vertebrae.

Force. That which changes the state of rest or motion in matter.

Frequency (in hertz or Hz). Rate at which pressure oscillations are produced. One hertz is equivalent to one cycle per second.

Friable. Particles or compounds in a crumbled or pulverized state.

Friction factor. A factor used in calculating loss of pressure due to friction of a liquid, solid, or gas flowing through a pipe or duct.

Friction loss. The pressure loss due to friction.

Fume. Airborne particulate formed by the evaporation of solid materials, e.g., metal fume emitted during welding. Usually less than one micron in diameter.

Fuse. A wire or strip of easily melted metal, usually set in a plug, placed in an electrical cuicuit as a safeguard. If the current becomes too strong, the metal melts, thus breaking the circuit.

Fusion. The joining of atomic nuclei to form a heavier nucleus, accomplished under conditions of extreme heat (millions of degrees). If two nuclei of light atoms fuse, the fusion is accompanied by the release of a great deal of energy. The energy of the sun is believed to be derived from the fusion of hydrogen atoms to form helium. In welding, the melting together of filler metal and base metal (substrate), or of base metal only, which results in coalescense.

Gage pressure. Pressure measured with respect to atmospheric pressure.

Galvanizing. An old but still used method of providing a protective coating for metals by dipping them in a bath of molten zinc.

Gamma-rays (gamma radiation). The most penetrating of all radiation. Gamma-rays are very high-energy x-rays.

Gangue. In mining or quarrying, useless chipped rock.

Gas. A state of matter in which the material has very low density and viscosity; can expand and contract greatly in response to changes in temperature and pressure; easily diffuses; neither a solid nor liquid.

Gas metal arc-welding (GMAW). An arc-welding process that produces coalescense of metals by heating them with an arc between a continuous filler metal (consumable) electrode and the work; shielding is obtained entirely from an externally supplied gas or gas mixture; some methods of this process are called MIG or CO_2 welding.

Gas tungsten arc-welding (GTAW). An arc-welding process that produces coalescence of metals by heating them with an arc between a tungsten (nonconsumable) electrode and the work; shielding is obtained from a gas or gas mixture. Pressure may or may not be used and filler metal may or may not be used. (This process has sometimes been called TIG welding.)

Gate. A groove in a mold to act as a passage for molten metal.

Geiger counter. A gas-filled electrical device that counts the presence of atomic particles or rays by detecting the ions produced. (Sometimes called a "Geiger-Mueller" counter.)

General exhaust. A system for exhausting air containing contaminants from a general work area; usually accomplished via dilution.

General ventilation. System of ventilation consisting of either natural or mechanically induced fresh air movements to mix with the dilute contaminants in the workroom air.

Generic name. A nonproprietary name for a material.

Genetic effects. Mutations or other changes that are produced by irradiation of the germ plasm.

Genetically significant dose (GSD). The dose which, if received by every member of the population, would be expected to produce the same total genetic injury to the population as do the actual doses received by various individuals.

Glove box. A sealed enclosure in which all handling of items inside the box is carried out through long impervious gloves sealed to ports in the walls of the enclosure.

Gob. "Gob pile" is waste mineral material such as from coal mines sometimes containing sufficient coal that gob fires may arise from spontaneous combustion.

Grab sample. A sample that is taken within a very short time period.

Gram (g). A metric unit of weight; one ounce equals 28.4 grams.

Gravity, specific. The ratio of the mass of a unit volume of a substance to the mass of the same volume of a standard substance at a standard temperature. Water at 4 C (39.2 F) is the standard substance usually referred to. For gases, dry air, at the same temperature and pressure as the gas, is often taken as the standard substance.

Gravity, standard. A gravitational force that will produce an acceleration equal to 9.8 m/sec^2 or 32.17 ft/sec^2. The actual force of gravity varies slightly with altitude and latitude. The standard was arbitrarily established as that at sea level and 45-degree latitude.

Gray iron. The same as cast iron and, in general, any iron containing high carbon.

Ground-fault circuit interrupter (GFCI). A device that measures the amount of current flowing to and from an electrical source. When a difference is sensed, indicating a leakage of current that could cause an injury, the device very quickly breaks the circuit.

Grounding. The procedure used to carry an electrical charge to ground through a conductive path. (See Bonding.)

Grooving. Designing a tool with grooves on the handle to accommodate the fingers of the user. A bad practice because of the great variation in the size of workers' hands. Grooving interferes with sensory feedback. Intense pain may be caused by the grooves to an arthritic hand.

Half-life, radioactive. For a single radioactive decay process, the time required for the activity to decrease to half its value by that process.

Halogenated hydrocarbon. A chemical material that has carbon plus one or more of the halogen elements: chlorine, fluorine, bromine, or iodine.

Hammer mill. A machine for reducing the size of stone or other bulk material by means of hammers usually placed on a rotating axle inside a steel cylinder.

Hand protection. Specific type of gloves or other hand protection required to prevent harmful exposure to hazardous materials.

Hardness. A relative term to describe the penetrating quality of radiation. The higher the energy of the radiation, the more penetrating (harder) is the radiation.

Hazard. An unsafe condition which, if left uncontrolled, may contribute to an accident.

Hazardous decomposition products. Any hazardous materials that may be produced in dangerous amounts if the material reacts with other agents, burns, or is exposed to other processes, such as welding. Examples of hazardous decomposition products formed when certain materials are heated include carbon monoxide and carbon dioxide.

Hazardous material. Any substance or compound that has the capability of producing adverse effects on the health and safety of humans.

Heading. In mining, a horizontal passage or drift of a tunnel, also the end of a drift or gallery. In tanning, a layer of ground bark over the tanning liquor.

Health hazard. A chemical for which there is statistically significant evidence based on at least one study conducted in accordance with established scientific principles that acute or chronic health effects may occur in exposed employees.

Hearing conservation. The prevention or minimizing of noise-induced deafness through the use of hearing protection devices; also, the control of noise through engineering methods, annual audiometric tests, and employee training.

Hearing level. The deviation in decibels of an individual's hearing threshold from the zero reference of the audiometer.

Heat stress. Relative amount of thermal strain from the environment.

Heat stress index. Index which combines the environmental heat and metabolic heat of the body into an expression of stress in terms of requirement for evaporation of sweat.

Heatstroke. A serious disorder resulting from exposure to excess heat. It results from sweat suppression and increased storage of body heat. Symptoms include hot, dry skin, high temperature, mental confusion, convulsions, and coma. Heatstroke is fatal if not treated promptly.

Heat treatment. Any of several processes of metal modification such as annealing.

Helmet. A device that shields the eyes, face, neck, and other parts of the head.

HEPA filter. See Filter, HEPA.

Hertz. The frequency measured in cycles per second. 1 cps = 1 Hz.

High-frequency loss. Refers to a hearing deficit starting with 2,000 Hz and higher.

Homogenizer. A machine that forces liquids under high pressure through a perforated shield against a hard surface to blend or emulsify the mixture.

Hood. (1) Enclosure, part of a local exhaust system; (2) a device that completely covers the head, neck, and portions of the shoulders.

Horsepower. A unit of power, equivalent to 33,000 foot-pounds per minute (746 W). (See Brake horsepower.)

Hot. In addition to meaning "having a relatively high temperature," this is a colloquial term meaning "highly radioactive."

Human-equipment interface. Area of physical or perceptual contact between man and equipment. The design characteristics of the human-equipment interface determine the quality of information. Poorly designed interfaces may lead to excessive fatigue or localized trauma, e.g., calluses.

Humidify. To add water vapor to the atmosphere; to add water vapor or moisture to any material.

Humidity. (1) Absolute humidity is the weight of water vapor per unit volume, pounds per cubic foot or grams per cubic centimeter. (2) Relative humidity is the ratio of the actual partial vapor pressure of the water vapor in a space to the saturation pressure of pure water at the same temperature.

Hydration. The process of converting raw material into pulp by prolonged beating in water; to combine with water or the elements of water.

Hydrocarbons. Organic compounds composed solely of carbon and hydrogen.

ICC. Interstate Commerce Commission.

IDLH. Immediately dangerous to life or health.

Ignitable. Capable of being set afire.

Immiscible. Not miscible. Any liquid that will not mix with another liquid, in which case it forms two separate layers or exhibits cloudiness or turbidity.

Impaction. The forcible contact of particles of matter; a term often used synonymously with impingement, but generally reserved for the case where particles are contacting a dry surface.

Impervious. A material that does not allow another substance to pass through or penetrate it.

Impingement. As used in air sampling, impingement refers to a process for the collection of particulate matter in which particles containing gas are directed against a wetted glass plate and the particles are retained by the liquid.

Inches of mercury column. A unit used in measuring pressure. One inch of mercury column equals a pressure of 1.66 kPa (0.491 lb per sq in.).

Inches of water column. A unit used in measuring pressure. One inch of water column equals a pressure of 0.25 kPa (0.036 lb per sq in.).

Incidence rate (as defined by OSHA). The number of injuries and/or illnesses or lost workdays per 100 full-time employees per year or 200,000 hours of exposure.

Incompatible. Materials that could cause dangerous reactions from direct contact with one another.

Incubation. Holding cultures of microorganisms under conditions favorable to their growth.

Induration. Heat hardening that may involve little more than thermal dehydration.

Industrial hygiene. The science (or art) devoted

to the recognition, evaluation, and control of those environmental factors or stresses (i.e., chemical, physical, biological, and ergonomic) that may cause sickness, impaired health, or significant discomfort to employees or residents of the community.

Inert gas. A gas that does not normally combine chemically with another substance.

Inert gas, welding. An electric welding operation utilizing an inert gas such as helium to flush away the air to prevent oxidation of the metal being welded.

Infrared radiation. Electromagnetic energy with wavelengths from 770 nm to 12,000 nm.

Ingestion. (1) The process of taking substances into the stomach, as food, drink, medicine, etc. (2) With regard to certain cells, the act of engulfing or taking up bacteria and other foreign matter.

Ingot. A block of iron or steel cast in a mold for ease in handling before processing.

Inhalation. The breathing in of a substance in the form of a gas, vapor, fume, mist, or dust.

Inhalation valve. A device that allows respirable air to enter the facepiece and prevents exhaled air from leaving the facepiece through the intake opening.

Injury. Damage or harm to the body as the result of violence, infection, or other source.

Inorganic. Term used to designate compounds that generally do not contain carbon. Source: matter, other than vegetable or animal. Examples: sulfuric acid and salt. Exceptions are carbon monoxide, carbon dioxide.

Insoluble. Incapable of being dissolved.

Intermediate. A chemical formed as a "middle-step" in a series of chemical reactions, especially in the manufacture of organic dyes and pigments. In many cases, it may be isolated and used to form a variety of desired products. In other cases, the intermediate may be unstable or used up at once.

Intoxication. Means either drunkenness or poisoning.

Inverse square law. The propagation of energy through space is inversely proportional to the square of the distance it must travel. An object 3 m (9.8 ft) away from an energy source receives $\frac{1}{9}$ as much energy as an object 1 m (3.3 ft) away.

Ionizing radiation. Refers to (1) electrically charged or neutral particles, or (2) electromagnetic radiation that will interact with gases, liquids, or solids to produce ions. There are five major types: alpha, beta, x (or x-ray), gamma, and neutrons.

Irradiation. The exposure of something to radiation.

Irritant. A substance that produces an irritating effect when it contacts skin, eyes, nose, or respiratory system.

Isotope. One of two or more atomic species of an element differing in atomic weight but having the same atomic number. Each contains the same number of protons but a different number of neutrons.

Jigs and fixtures. Often used interchangeably; precisely, a jig holds work in position and guides the tools acting on the work, while a fixture holds but does not guide.

Kaolin. A type of clay composed of mixed silicates and used for refractories, ceramics, tile, and stoneware. In some deposits, free silica may be present as an impurity.

kg. Kilogram; a metric unit of weight, about 2.2 U.S. pounds. (Also see mg.)

Kilogram (kg). A unit of weight in the metric system equal to 2.2 lb.

Kinetic energy. Energy due to motion. (See Work.)

L. Liter; a metric unit of capacity. A liter is about the same as one quart (0.946 L).

Lacquer. A collodial dispersion or solution of nitrocellulose, or similar film-forming compounds, resins, and plasticizers in solvents and diluents used as a protective and decorative coating for various surfaces.

Lapping. The operation of polishing or sanding surfaces such as metal or glass to a precise dimension.

Laser. The acronym for Light Amplification by Stimulated Emission of Radiation. The gas laser is a type in which laser action takes place in a gas medium, usually a continous wave (CW) laser, which has a continuous output—with an off time of less than 1 percent of the pulse duration time.

Laser light region. A portion of the electromagnetic spectrum that includes ultraviolet, visible, and infrared light.

Lathe. A machine tool used to perform cutting operations on wood or metal by the rotation of the workpiece.

Latex. Original meaning: milky extract from rubber tree, containing about 35 percent rubber hydrocarbon; the remainder is water, proteins, and sugars. This word also is applied to water emulsions of synthetic rubbers or resins. In emulsion paints, the film-forming resin is in the form of latex.

LC. Lethal concentration; a concentration of a substance being tested that will kill a test animal.

LD. Lethal dose; a concentration of a substance being tested that will kill a test animal.

Lead poisoning. Poisoning produced when lead compounds are swallowed or inhaled. Inorganic lead compounds commonly cause symptoms of lead colic and lead anemia. Organic lead compounds can attack the nervous system.

Leakage radiation. Radiation emanating from diagnostic equipment in addition to the useful beam; also, radiation produced when the exposure switch or timer is not activated.

LEL. See Lower explosive limit.

Lethal. Capable of causing death.

LFL. Lower flammable limit. See LEL.

Liquefied petroleum gas. A compressed or liquefied gas usually composed of propane, some butane, and lesser quantities of other light hydrocarbons and impurities; obtained as a by-product in petroleum refining. Used chiefly as a fuel and in chemical synthesis.

Liquid. A state of matter in which the substance is a formless fluid that flows in accord with a law of gravity.

Liter (L). A metric measure of capacity—one quart equals 0.9 L.

Local exhaust. A system for capturing and exhausting contaminants from the air at the point where the contaminants are produced.

Local exhaust ventilation. A ventilation system that captures and removes contaminants at the point they are being produced before they escape into the workroom air.

Loudness. The intensive attribute of an auditory sensation, in terms of which sounds may be ordered on a scale extending from soft to loud. Loudness depends primarily upon the sound pressure of the stimulus, but it also depends upon the frequency and wave form of the stimulus.

Low-pressure tank. A storage tank designed to operate at pressures at more than 0.5 psig but not more than 15 psig (3.5 to 103 kPa).

Lower explosive limit (LEL). The lower limit of flammability of a gas or vapor at ordinary ambient temperatures expressed in percent of the gas or vapor in air by volume.

LP-gas. See Liquefied petroleum gas.

Lumen. The flux on one square foot of a sphere, one foot in radius, with a light source of one candle at the center that radiates uniformly in all directions.

M. Meter; a unit of length in the metric system. One meter is about 39 inches.

M^3. Cubic meter; a metric measure of volume, about 35.3 cubic feet or 1.3 cubic yards.

Makeup air. Clean, tempered outdoor air supplied to a workspace to replace air removed by exhaust ventilation or some industrial process.

Manometer. Instrument for measuring pressure; essentially a U-tube partially filled with a liquid (usually water, mercury, or a light oil), so constructed that the amount of displacement of the liquid indicates the pressure being exerted on the instrument.

Maser. Microwave Amplification by Stimulated Emission of Radiation. When used in the term "optical maser," it is often interpreted as molecular amplification by stimulated emission of radiation.

Maximum permissible concentration (MPC). The concentrations set by the National Committee on Radiation Protection (NCRP). They are recommended maximum average concentrations of radionuclides to which a worker may be exposed, assuming that he or she works 8 hours a day, 5 days a week, 50 weeks a year.

Maximum permissible dose (MPD). Currently, a permissible dose is defined as the dose of ionizing radiation that, in the light of present knowledge, is not expected to cause appreciable bodily injury to a person at any time during their lifetime.

Maximum use concentration (MUC). The product of the protection factor of the respiratory protection equipment and the permissible exposure limit (PEL).

Mechanical filter respirator. A respirator used

factory organ in the nasal cavity is the sensing element that detects odors and transmits information to the brain through the olfactory nerves.

Ophthalmologist. A physician who specializes in the structure, function, and diseases of the eye.

Oral. Used in or taken into the body through the mouth.

Oral toxicity. Adverse effects resulting from taking a substance into the body via the mouth. Ordinarily used to denote effects in experimental animals.

Organic. Term used to designate chemicals that contain carbon. To date nearly one million organic compounds have been synthesized or isolated. Many occur in nature; others are produced by chemical synthesis. (See also Inorganic.)

Orifice. The opening that serves as an entrance and/or outlet of a body cavity or organ, especially the opening of a canal or a passage.

Orifice meter. A flowmeter, employing as the measure of flow rate the difference between the pressures measured on the upstream and downstream sides of a restriction within a pipe or duct.

Oscillation. The variation, usually with time, of the magnitude of a quantity with respect to a specified reference when the magnitude is alternately greater and smaller than the reference.

OSHA. Occupational Safety and Health Administration; federal agency in the U.S. Department of Labor with safety and health regulatory and enforcement authority for most U.S. industry and business.

Osmosis. The passage of fluid through a semipermeable membrane as a result of osmotic pressure.

Overexposure. Exposure to a hazardous material beyond the allowable exposure levels.

Oxidation. Process of combining oxygen with some other substance; technically, a chemical change in which an atom loses one or more electrons whether or not oxygen is involved. Opposite of reduction.

Oxygen deficiency. An atmosphere having less than the percentage of oxygen found in normal air.

Paraffins, Paraffin series. Those straight- or branched-chain hydrocarbon components of crude oil and natural gas whose molecules are saturated (i.e., carbon atoms attached to each other by single bonds) and therefore very stable. Examples: methane and ethane. Generalized formula: C_nH_{2n+2}.

Particle. A small discrete mass of solid or liquid matter.

Particulate. A particle of solid or liquid matter.

Particulate matter. A suspension of fine solid or liquid particles in air, such as dust, fog, fume, mist, smoke, or sprays. Particulate matter suspended in air is commonly known as an aerosol.

PEL. Permissible exposure limit; the legally enforced exposure limit for a substance established by OSHA regulatory authority. The PEL indicates the permissible concentration of air contaminants to which nearly all workers may be repeatedly exposed 8 hours a day, 40 hours a week, over a working lifetime (30 years) without adverse health effects.

Pelleting. In various industries, a material in powder form that may be made into pellets or briquettes for convenience. The pellet is a distinctly small briquette. (See Pelletizing.)

Pelletizing. Refers primarily to extrusion by pellet mills. The word "pellet," however, carries its lay meaning in that it is also applied to other small extrusions and to some balled products. Pellets are generally regarded as being larger than grains but smaller than briquettes.

Percent impairment of hearing. Percent hearing loss; an estimate of a person's ability to hear correctly. It is usually based, by means of an arbitrary rule, on the pure tone audiogram. The specific rule for calculating this quantity from the audiogram varies from state to state according to rule or law.

Percent volatile. The percentage of a liquid or solid (by volume) that will evaporate at an ambient temperature of 70 F (unless some other temperature is stated). Examples: butane, gasoline, and paint thinner (mineral spirits) are 100% volatile; their individual evaporation rates vary, but over a period of time each will evaporate completely.

Permanent disability (or permanent impairment). Any degree of permanent, nonfatal injury. It includes any injury that results in the partial loss, to complete loss of use, of any part of the body, or any permanent impairment of functions of the body or a part thereof.

Permissible exposure limit (PEL). An exposure limit that is published and enforced by U.S. OSHA as a legal standard.

Personal protective equipment. Devices worn by workers to protect against hazards in the environment.

Pesticides. General term for that group of chemicals used to control or kill such pests as rats, insects, fungi, bacteria, weeds, etc., that prey on man or agricultural products. Among these are insecticides, herbicides, fungicides, rodenticides, miticides, fumigants, and repellents.

Petrochemical. A term applied to chemical substances produced from petroleum products and natural gas.

pH. Means used to express the degree of acidity or alkalinity of a solution with neutrality indicated at seven.

Phosphors. Materials capable of absorbing energy from suitable sources, such as visible light, cathode rays, or ultraviolet radiation, and then emitting a portion of the energy in the ultraviolet, visible, or infrared region of the electromagnetic spectrum. In short, they are fluorescent or luminescent.

Pig. (1) A container (usually lead) used to ship or store radioactive materials. The thick walls protect the person handling the container from radiation. (2) In metal refining, a small ingot from the casting of blast furnace metal.

Pigment. A finely divided, insoluble substance that imparts color to the material to which it is added.

Pilot plant. Small-scale operation preliminary to a major enterprise. Common in the chemical industry.

Pink noise. Noise that has been weighted, especially at the low end of the spectrum, so that the energy per band (usually octave band) is about constant over the spectrum.

Pitot tube. A device consisting of two concentric tubes, one serving to measure the total or impact pressure existing in an airstream, the other to measure the static presure only. When the annular space between the tubes and the interior of the center tube are connected across a pressure-measuring device, the pressure difference automatically nullifies the static pressure, and the velocity pressure alone is registered.

Plasma arc welding (PAW). An arc welding process that produces coalescence of metals by heating them with a constricted arc between an electrode and the workpiece (transferred arc) or the electrode and the constricting nozzle (nontransferred arc). Shielding is obtained by the hot, ionized gas issuing from the orifice, which may be supplemented by an auxiliary source of shielding gas. Shielding gas can be an inert gas or a mixture of gases. Pressure may or may not be used, and filler metal may or may not be supplied.

Plastic. Officially defined as any one of a large group of materials that contains as an essential ingredient an organic substance of large molecular weight. Two basic types: thermosetting (irreversibly rigid) and thermoplastic (reversibly rigid). Before compounding and processing, plastics often are referred to as (synthetic) resins. Final form may be as film, sheet, solid, or foam; flexible or rigid.

Plenum chamber. An air compartment connected to one or more ducts or connected to a slot in a hood used for air distribution.

Plutonium. A heavy element that undergoes fission under the impact of neutrons. It is a useful fuel in nuclear reactors. Plutonium cannot be found in nature, but can be produced and "burned" in reactors.

Point source. A source of radiation whose dimensions are small enough compared with distance between source and receptor for those dimensions to be neglected in calculations.

Poison. A material introduced into a reactor core to absorb neutrons. Any substance that, when taken into the body, is injurious to health.

Poison, Class A. A DOT hazard class for extremely dangerous poisons, that is, poisonous gases or liquids of such nature that a very small amount of the gas, or vapor of the liquid, mixed with air is dangerous to life. Some examples: phosgene, cyanogen, hydrocyanic acid, nitrogen peroxide.

Poison, Class B. A DOT hazard class for liquid, solid, paste, or semisolid substances—other than Class A poisons or irritating materials— that are known (or presumed on the basis of animal tests) to be so toxic to man as to afford a hazard to health during transportation. Some examples: arsenic, beryllium chloride, cyanide, mercuric oxide.

Pollution. Contamination of soil, water, or atmosphere beyond that which is natural.

Polymer. A high molecular weight material formed by the joining together of many simple molecules (monomers). There may be hundreds or even thousands of the original molecules linked end to end and often crosslinked. Rubber and cellulose are naturally occurring polymers. Most resins are chemically produced polymers.

Portal. Place of entrance.

Portland cement. See Cement, portland.

Positive displacement pump. Any type of air mover pump in which leakage is negligible, so that the pump delivers a constant volume of fluid, building up to any pressure necessary to deliver that volume (unless, of course, the motor stalls or the pump breaks).

Potential energy. Energy due to position of one body with respect to another or to the relative parts of the same body.

Power. Time rate at which work is done; units are the watt (one joule per second) and the horsepower (33,000 foot-pounds per minute). One horsepower equals 746 watts.

PPE, personal protective equipment. Includes items such as gloves, goggles, respirators, and protective clothing.

ppm. Parts per million; part of air by volume of vapor or gas or other contaminant.

Precision. The degree of agreement of repeated measurements of the same property, expressed in terms of dispersion of test results about the mean result obtained by repetitive testing of a homogeneous sample under specified conditions.

Pressure. Force applied to, or distributed over a surface; measured as force per unit area. (See Atmospheric pressure, Gage pressure, Standard temperature and pressure, Static pressure, and Velocity pressure.)

Pressure vessel. A storage tank or vessel designed to operate at pressures greater the 15 psig (103 kPa).

PRF laser. A pulsed recurrence frequency laser, which is a pulsed-typed laser with properties similar to a CW laser if the frequency is very high.

Probe. A tube used for sampling or for measuring pressures at a distance from the actual collection or measuring apparatus. It is commonly used for reaching inside stacks or ducts.

Propagation of flame. The spread of flame through the entire volume of a flammable vapor-air mixture from a single source of ignition.

Protection factor (PF). With respiratory protective equipment, the ratio of the ambient airborne concentration of the contaminant to the concentration inside the facepiece.

Protective atmosphere. A gas envelope surrounding the part to be brazed, welded, or thermal sprayed, with the gas composition controlled with respect to chemical composition, dew point, pressure, flow rate, etc.

Protective coating. A thin layer of metal or organic material, such as paint, applied to a surface primarily to protect it from oxidation, weathering, and corrosion.

psi. Pounds per square inch. For technical accuracy, pressure must be expressed as psig (pounds per square inch gage) or psia (pounds per square inch absolute); that is, gage pressure plus sea level atmospheric pressure, or psig plus about 14.7 pounds per square inch. (See also mmHg.)

Pulsed laser. A class of laser characterized by operation in a pulsed mode, i.e., emission occurs in one or more flashes of short duration (pulse length).

Pumice. A natural silicate, such as volcanic ash or lava. Used as an abrasive.

Push-pull hood. A hood consisting of an air supply system on one side of the contaminant source blowing across the source and into an exhaust hood on the other side.

Pyrethrum. A pesticide obtained from the dried, powdered flowers of the plant of the same name; mixed with petroleum distillates, it is used as an insecticide.

Pyrophoric. A chemical that will ignite spontaneously in air at a temperature of 130 F (54.4 C) or below.

Q fever. Disease caused by rickettsial organism that infects meat and livestock handlers; similar but not identical to tick fever.

Q-switched laser (also known as Q-spoiled). A pulsed laser capable of extremely high peak powers for very short durations (pulse length of several nanoseconds).

Quality. A term used to describe the penetrating power of x-rays or gamma-rays.

Quartz. Vitreous, hard, chemically resistant, free silica, the most common form in nature. The main constituent in sandstone, igneous rocks, and common sands.

Quenching. A heat-treating operation in which metal raised to the desired temperature is quickly cooled by immersion in an oil bath.

rad. Radiation absorbed dose.

Radiation (nuclear). The emission of atomic particles or electromagnetic radiation from the nucleus of an atom.

Radiation (thermal). The transmission of energy by means of electromagnetic waves longer than visible light. Radiant energy of any wavelength may, when absorbed, become thermal energy and result in the increase in the temperature of the absorbing body.

Radiation protection guide (RPG). The radiation dose that should not be exceeded without careful consideration of the reasons for doing so; every effort should be made to encourage the maintenance of radiation doses as far below this guide as practical.

Radiation source. An apparatus or a material emitting or capable of emitting ionizing radiation.

Radiator. That which is capable of emitting energy in wave form.

Radioactive. The property of an isotope or element that is characterized by spontaneous decay to emit radiation.

Random noise. A sound or electrical wave whose instantaneous amplitudes occur as a function of time, according to a normal (Gaussian) distribution curve. Random noise is an oscillation whose instantaneous magnitude is not specified for any given instant of time.

Rated line voltage. The range of potentials in volts of an electrical supply line.

Reaction. A chemical transformation or change; the interaction of two or more substances to form new substances.

Reactive. See Unstable.

Reactivity. A description of the tendency of a substance to undergo chemical reaction with the release of energy. Undesirable effects—such as pressure buildup, temperature increase, formation of noxious, toxic, or corrosive by-products—may occur because of the reactivity of a substance to heating, burning, direct contact with other materials, or other conditions in use or in storage.

Reagent. Any substance used in a chemical reaction to produce, measure, examine, or detect another substance.

Recoil energy. The emitted energy that is shared by the reaction products when a nucleus undergoes a nuclear reaction, such as fission or radioactive decay.

Reducing agent. In a reduction reaction (which always occurs simultaneously with an oxidation reaction), the reducing agent is the chemical or substance that (1) combines with oxygen or (2) loses electrons to the reaction.

Refractory. A material that is especially resistant to the action of heat, such as fireclay, magnetite, graphite, and silica; used for lining furnaces, etc.

Regimen. A regulation of the mode of living, diet, sleep, exercise, etc., for a hygienic or therapeutic purpose; sometimes mistakenly called regime.

Relative humidity. See Humidity.

Reliability. The degree to which an instrument, component, or system retains its performance characteristics over a period of time.

Rem. Roentgen equivalent man, a dose unit that equals the dose in rads multiplied by the appropriate value of RBE for the particular radiation.

Resistance. (1) In electricity, any condition that retards current or the flow of electrons; it is measured in ohms. (2) Opposition to the flow of air, as through a canister, cartridge, particulate filter, or orifice. (3) A property of conductors, depending on their dimensions, material, and temperature, that determines the current produced by a given difference in electrical potential.

Respirable size particulates. Particles in the size range that permits them to penetrate deep into the lungs upon inhalation.

Respirator. A device to protect the wearer from inhalation of harmful contaminants.

Respiratory protection. Devices that will protect the wearer's respiratory system from overexposure by inhalation of airborne contaminants.

Respiratory system. Consists of (in descending order)—the nose, mouth, nasal passages, nasal pharynx, pharynx, larynx, trachea, bronchi, bronchioles, air sacs (alveoli) of the lungs, and muscles of respiration.

Reverbatory furnace. A furnace in which heat is supplied by burning of fuel in a space between the charge and a low roof.

Riser. In metal casting, a channel in a mold to permit escape of gases.

Roentgen (R). A unit of radioactive dose, or exposure was called a roentgen (pronounced rentgen). (See rad.)

Route of entry. The path by which chemicals can enter the body; primarily inhalation, ingestion, and skin absorption.

Rosin. Specifically applies to the resin of the pine tree and chiefly derives from the manufacture of turpentine. Widely used in the manufacture of soap, flux.

Rotameter. A flow meter, consisting of a precision-bored, tapered, transparent tube with a solid float inside.

Rotary kiln. Any of several types of kilns used to heat material, such as in the portland cement industry.

Rouge. A finely powdered form of iron oxide used as a polishing agent.

Safety. The control of recognized hazards to attain an acceptable level of risk.

Safety can. An approved container, of not more than 19 L (5 gal) capacity, having a spring-closing lid and spout cover, and so designed that it will safely relieve internal pressure when subjected to fire exposure.

Safety program. Activities designed to assist employees in the recognition, understanding, and control of hazards in the workplace.

Salamander. A small furnace, usually cylindrical in shape, without grates, used for heating.

Salt. A product of the reaction between an acid and a base.

Sampling. The withdrawal or isolation of a fractional part of a whole.

Sandblasting. A process for cleaning metal castings and other surfaces with sand by a high-pressure airstream.

Sandhog. Any worker doing tunneling work requiring atmospheric pressure control.

Sanitize. To reduce the microbial flora in or on articles, such as eating utensils, to levels judged safe by public health authorities.

SCBA. See Self-contained breathing apparatus.

Sealed source. A radioactive source sealed in a container or having a bonded cover, where the container or cover has sufficient mechanical strength to prevent contact with and dispersion of the radioactive material under the conditions of use and wear for which it was designed.

Self-contained breathing apparatus (SCBA). A respiratory protection device that consists of a supply or a means of obtaining respirable air, oxygen, or oxygen-generating material carried by the wearer.

Self-ignition. See Auto-ignition temperature.

Sensible. Capable of being perceived by the sense organs.

Sensitivity. The minimum amount of contaminant that can repeatedly be detected by an instrument.

Sensitization. The process of rendering an individual sensitive to the action of a chemical.

Sensitizer. A substance which, on first exposure, causes little or no reaction in man or test animals but which, on repeated exposure, may cause a marked response not necessarily limited to the contact site.

Sensory feedback. Use of external signals perceived by sense organs (e.g., eye, ear) to indicate quality or level of performance of an event triggered by voluntary action.

Shakeout. In the foundry industry, the separation of the solid, but still not cold, casting from its molding sand.

Shale. Many meanings in industry, but in geology a common fossil rock formed from clay, mud, or silt somewhat stratified but without characteristic cleavage.

Shale oil. Some shale is bituminous and on distillation yields a tarry oil.

Shield, shielding. Interposed material (like a wall) that protects workers from harmful radiations released by radioactive materials.

Shielded metal arc welding (SMAW). An arc-welding process that produces coalescence of metals by heating them with an arc between a covered metal electrode and the work. Shielding is obtained from decomposition of the electrode covering. Pressure is not used and filler metal is obtained from the electrode.

Shock. Primarily the rapid fall in blood pressure following an injury, operation, or the administration of anesthesia.

Shootblasting. A process for cleaning of metal castings or other surfaces by small steel shot in a high-pressure airstream. This process is a substitute for sandblasting to avoid silicosis.

Short-term exposure limit (STEL). ACGIH-recommended exposure limit. Maximum concentration to which workers can be exposed for a short period of time (15 minutes) for only four times throughout the day with at least one hour between exposures.

Silica gel. A regenerative absorbent consisting of the amorphous silica manufactured by the action of HCl on sodium silicate. Hard, glossy, quartzlike in appearance. Used in dehydrating and in drying and as a catalyst carrier.

Silicates. Compounds of silicon, oxygen, and one or more metals with or without hydrogen. Silicate dusts cause nonspecific dust reactions, but generally do not interfere with pulmonary function or result in disability.

Silicon. A nonmetallic element being, next to oxygen, the chief elementary constituent of the earth's crust.

Silicones. Unique group of compounds made by molecular combination of the element silicon or certain of its compounds with organic chemicals. Produced in a variety of forms, including silicone fluids, resins, and rubber. Silicones have special properties, such as water repellency, wide temperature resistance, high durability, and great dielectric strength.

Silver solder. A solder of varying components but usually containing an appreciable amount of cadmium.

Sintering. Process of making coherent powder of earthy substances by heating, but without melting.

Skin absorption. Ability of some hazardous chemicals to pass directly through the skin and enter the bloodstream.

Skin dose. A special instance of tissue dose, referring to the dose immediately on the surface of the skin.

Skin sensitizer. See Sensitizer.

Skin toxicity. See Dermal toxicity.

Slag. The dross of flux and impurities that rise to the surface of molten metal during melting and refining.

Sludge. In general, any muddy or slushy mass.

Slurry. A thick, creamy liquid resulting from the mixing and grinding of limestone, clay, and other raw materials with water.

Smelting. One step in the procurement of metals from ore—hence, to reduce, to refine, to flux, or to scorify.

Smog. Irritating hazard resulting from the sun's effect on certain pollutants in the air.

Smoke. An air suspension (aerosol) of particles, originating from combustion or sublimation.

Synthesis. The reaction or series of reactions by which a complex compound is obtained from simpler compounds or elements.

Soaking pit. A device in steel manufacturing in which ingots with still molten interiors stand in a heated, upright chamber until solidification is complete.

Soap. Ordinarily a metal salt of a fatty acid, usually sodium stearate, sodium oleate, sodium palmitate, or some combination of these.

Soapstone. Complex silicate of varied composition similar to some talcs with wide industrial application such as in rubber manufacture.

Solder. A material used for joining metal surfaces together by filling a joint or covering a junction.

Solid-state laser. A type of laser that utilizes a solid crystal such as ruby or glass. This type is most commonly used in impulse lasers.

Solubility in water. A term expressing the percentage of a material (by weight) that will dissolve in water at ambient temperature. Solubility information can be useful in determining spill cleanup methods and fire-extinguishing agents and methods for diluting a material. Terms used to express solubility are:

Negligible	Less than 0.1 percent
Slight	0.1 to 1.0 percent
Moderate	1 to 10 percent
Appreciable	More than 10 percent
Complete	Soluble in all proportions

Solution. Mixture in which the components lose their identities and are uniformly dispersed. All solutions consist of a solvent (water or other fluid) and the substance dissolved, called the "solute." A true solution is homogeneous such as salt in water.

Solvent. A substance that dissolves another substance.

Soot. Agglomerations of particles of carbon impregnated with "tar," formed in the incomplete combustion of carbonaceous material.

Sorbent(s). (1) A material that removes toxic gases and vapors from air inhaled through a canister or cartridge. (2) Material used to collect gases and vapors during air-sampling. (3) Nonreactive materials used to clean up chemical spills. Examples: clay and vermiculite.

Sound. An oscillation in pressure, stress, particle displacement, particle velocity, etc., which is propagated in an elastic material, in a medium with internal forces (e.g., elastic, viscous), or the superposition of such propagated oscillations.

Sound absorption. The change of sound energy into some other form, usually heat, as it passes through a medium or strikes a surface.

Sound analyzer. A device for measuring the band pressure level or pressure-spectrum level of a sound as a function of frequency.

Sound level. A weighted sound pressure level, obtained by the use of metering characteristics and the weighting A, B, or C specified in ANSI S1.4.

Sound-level meter and octave-band analyzer. Instruments for measuring sound pressure levels in decibels referenced to 0.0002 microbar.

Sound pressure level (SPL). The level, in decibels, of a sound is 20 times the logarithm to the base 10 of the ratio of the pressure of the sound to the reference pressure. The reference pressure must be explicitly stated.

Sour gas. Slang for either natural gas or a gasoline contaminated with odor-causing sulfur compounds. In natural gas, the contaminant is usually hydrogen sulfide; in gasolines, usually mercaptans.

Source. Any substance that emits radiation. Usually refers to a piece of radioactive material conveniently packaged for scientific or industrial use.

Spasm. Tightening or contraction of any set(s) of muscles.

Special fire-fighting procedures. Special procedures and/or personal protective equipment that is necessary when a particular substance is involved in a fire.

Specific gravity. The weight of a material compared to the weight or an equal volume of water; an expression of the density (or heaviness) of the material. (See Gravity, specific.)

Specific weight. The weight per unit volume of a substance, same as density.

Spectrophotometer. An instrument used for comparing the relative intensities of the corresponding colors produced by chemical reaction.

Spectrum. The distribution in frequency of the magnitudes (and sometimes phases) of the components of a wave. Spectrum also is used to signify a continuous range of frequencies, usually wide in extent, within which waves have some specified common characteristics. Also, the pattern of red-to-violet light observed when a beam of sunlight passes through a prism and then projects upon a surface.

Speech interference level (SIL). The speech interference level of a noise is the average, in decibels, of the sound pressure levels of the noise in the three octave bands of frequency 600-1,200, 1,200-2,400, and 2,400-4,800 Hz.

Speech perception test. A measurement of hearing acuity by the administration of a carefully controlled list of words. The identification of correct responses is evaluated in terms of norms established by the average performance of normal listeners.

Spontaneously combustible. A material that ignites as a result of retained heat from processing, or which will oxidize to generate heat and ignite, or which absorbs moisture to generate heat and ignite.

Spot welding. One form of electrical-resistance welding in which the current and pressure are restricted to the spots of metal surfaces directly in contact.

Spray coating painting. The result of the application of a spray in painting as a substitute for brush painting or dipping.

Stability. An expression of the ability of a material to remain unchanged. For MSDS purposes, a material is stable if it remains in the same form under expected and reasonable conditions of storage or use. Conditions which may cause instability (dangerous change) are stated—i.e., temperatures above 150 F, shock from dropping, etc.

Stain. A dye used to color microorganisms as an aid to visual inspection.

Stamping. Many different usages in industry, but a common one is the crushing of ores by pulverizing.

Standard air. Air at standard temperature and pressure. The most common values are 21.1 C (70 F) and 101.3 kPa (29.92 in. Hg).

Standard conditions. In industrial ventilation, 21.1 C (70 F), 50 percent relative humidity, and 101.3 kPa (29.92 in. of mercury) atmosphere pressure.

Standard Industrial Classification (SIC) Code. A classification system for places of employment according to major type of activity by U.S. Government.

Standard man. A theoretical physically fit man of standard (average) height, weight dimensions, and other parameters (blood composition, percentage of water, mass of salivery glands, to name a few), used in studies of how heat or ionizing radiation affects humans.

Standard temperature and pressure. See Standard air.

Static pressure. The potential pressure exerted in all directions by a fluid at rest.

STEL. Short-term exposure limit; ACGIH terminology. See TLV.

Sterile. Free of living microorganisms.

Sterility. Inability to reproduce.

Sterilization. The process of making sterile; the killing of all forms of life.

Stink damp. In mining, hydrogen sulfide.

Stress. A physical, chemical, or emotional factor that causes bodily or mental tension and may be a factor in disease causation or fatigue.

Stressor. Any agent or thing causing a condition of stress.

Strip mine. A mine in which coal or ore is extracted from the earth's surface after removal of overlayers of soil, clay, and rock.

Supplied-air respirators. Air-line respirators or self-contained breathing apparatus.

Supplied-air suit. A one- or two-piece suit that is impermeable to most particulate and gaseous contaminants and is provided with an adequate supply of respirable air.

Surface-active agent or **surfactant.** Any of a group of compounds added to a liquid to modify surface of interfacial tension. In synthetic detergents, which is the best known use of surface-active agents, reduction of interfacial tension provides cleansing action.

Surface coating. Term used to include paint, lacquer, varnish, and other chemical compositions used for protecting and/or decorating surfaces. See Protective coating.

Suspect carcinogen. A material that is believed to be capable of causing cancer but for whichs there is limited scientific evidence.

Sweating. (1) Visible perspiration; (2) the process of uniting metal parts by heating solder so that it runs between the parts.

Sweetening. The process by which petroleum products are improved in odor by chemically changing certain sulfur compounds of objectionable odor into compounds having little or no odor.

Swing grinder. A large power-driven grinding wheel mounted on a counterbalanced swivel-supported arm guided by two handles.

Synthesis. The reaction or series of reactions by which a complex compound is obtained from simpler compounds or elements.

Synthetic. (From Greek work *synthetikos*—that which is put together.) "Man-made 'synthetic' should not be thought of as a substitute for the natural," states *Encyclopedia of the Chemical Process Industries;* it adds, "Synthetic chemicals are frequently more pure and uniform than those obtained naturally." Example: synthetic indigo.

Synthetic detergents. Chemically tailored cleaning agents soluble in water or other solvents. Originally developed as soap substitutes. Because they do not form insoluble precipitates, they are especially valuable in hard water. They may be composed of surface-active agents alone, but generally are combinations of surface-active agents and other substances, such as complex phosphates, to enhance detergency.

Synthetic rubber. Artificially made polymer with rubberlike properties. Various types have varying composition and properties. Major types designated as S-type, butyl, neoprene (chloroprense polymers), and N-type. Several synthetics duplicate the chemical structure of natural rubber.

Systemic toxicity. Adverse effects caused by a substance that affects the body in a general rather than local manner.

Tailings. In mining or metal recovery processes, the gangue rock residue from which all or most of the metal has been extracted.

Talc. A hydrous magnesium silicate used in ceramics, cosmetics, paint, and pharmaceuticals, and as filler in soap, putty, and plaster.

Tall oil. (Name derived from Swedish word tallolja; material first investigated in Sweden—not synonymous with U.S. pine oil). Natural mixture of rosin acids, fatty acids, sterols, high-molecular weight alcohols, and other materials derived primarily from waste liquors of sulfate wood pulp manufacture. Dark brown, viscous, oily liquid often called liquid rosin.

Tar. A loose term embracing wood, coal, or petroleum exudations. In general, represents complex mixture of chemicals of top fractional distillation systems.

Tar crude. Organic raw material derived from distillation of coal tar and used for chemicals.

Tare. A deduction of weight, made in allowance for the weight of a container or medium. The initial weight of a filter, for example.

Temper. To relieve the internal stresses in metal or glass and to increase ductility by heating the material to a point below its critical temperature and cooling slowly.

Tempering. The process of heating or cooling makeup air to the proper temperature.

Temporary threshold shift (TTS). The hearing loss suffered as the result of noise exposure, all or part of which is recovered during an arbitrary period of time when one is removed from the noise.

Temporary total disability. An injury that does not result in death or permanent disability, but which renders the injured person unable to perform regular duties or activities on one or more calendar days after the day of injury. (This is a definition established by OSHA.)

Tendon. Fibrous component of a "muscle." It frequently attaches at the area of application of tensile force. When its cross section is small, stresses in the tendon are high, particularly because the total force of many muscle fibers is applied at the single terminal tendon. (See Tenosynovitis.)

Tenosynovitis. Inflammation of the connective tissue sheath of a tendon.

Teratogen. A substance or agent to which exposure of a pregnant female can result in malformations in the fetus. An example is the drug thalidomide.

Terminal velocity. The terminal rate of fall of a particle through a fluid as induced by gravity or other external force; the rate at which frictional drag balances the accelerating force (or the external force).

Thermal pollution. Discharge of heat into bodies of water to the point that increased warmth activates all sewage, depletes the oxygen the water needs to cleanse itself, and eventually destroys some of the fish and other organisms in the water.

Thermonuclear reaction. A fusion reaction, that is, a reaction in which two light nuclei combine to form a heavier atom, releasing a large amount of energy. This is believed to be the sun's source of energy. It is called thermonuclear because it occurs only at a very high termperature.

Thermoplastic. Capable of being repeatedly softened by heat.

Thermosetting. Capable of undergoing a chemical change from a soft to a hardened substance when heated.

Thermosetting plastics. Those that are heat-set in their final processing to a permanently hard state. Examples: phenolics, ureas, and melamines.

Thinner. A liquid used to increase the fluidity of paints, varnishes, and shellac.

Threshold. The level where the first effects occur; also the point at which a person just begins to notice a tone is becoming audible.

Time-weighted average concentration (TWA). Refers to concentrations of airborne toxic materials that have been weighted for a certain time duration, usually 8 hours.

Tinning. Any work with tin, such as tin roofing; but in particular in soldering, the primary coating with solder of the two surfaces to be united.

TLV. *Threshold Limit Value*; a term used by ACGIH to express the airborne concentration of a material to which *nearly* all persons can be exposed day after day, without adverse effects. ACGIH expresses TLVs in three ways:

TLV-C. The *Ceiling* limit—the concentration that should not be exceeded even instantaneously.

TLV-STEL. The *Short-Term Exposure Limit*, or maximum concentration for a continuous 15-minute exposure period (maximum of four such periods per day, with at least 60 minutes between exposure periods, and provided that the daily TLV-TWA is not exceeded).

TLV-TWA. The allowable *Time Weighted Average* concentration for a normal 8-hour work day or 40-hour work week.

Tolerance. (1) The ability of a living organism to resist the usually anticipated stress. (2) Also, the limits of permissible inaccuracy in the fabrication of an article above or below its design specifications.

Tolerance dose. See Maximum permissible concentration and MPL.

Toxic substance. Any substance that can cause acute or chronic injury to the human body, or that is suspected of being able to cause diseases or injury under some conditions.

Toxicity. The sum of adverse effects resulting from exposure to a material, generally by the mouth, skin, or respiratory tract.

Toxin. A poisonous substance that is derived from an organism.

Trade name. The commercial name or trademark by which something is known.

Trade secret. Any confidential formula pattern, process, device, information or compilation of information (including chemical name or other unique chemical identifier) that is used in an employer's business, and that gives the employer an opportunity to obtain an advantage over competitors who do not know or use it.

Trauma. An injury or wound brought about by an outside force.

TSCA. Toxic Substances Control Act; federal environmental legislation, administered by EPA, for regulating the manufacture, handling, and use of materials classified as "toxic substances."

Tumbling. An industrial process, such as in founding, in which small castings are cleaned by friction in a rotating drum (tumbling mill, tumbling barrel), which may contain sand, sawdust, stone, etc.

Turbid. Cloudy.

Turning vanes. Curved pieces added to elbows or fan inlet boxes to direct air and so reduce turbulence losses.

TWA. See TLV-TWA.

UEL. See Upper explosive limit. (Also see Lower explosive limit.)

Ultraviolet. Those wavelengths of the electromagnetic spectrum that are shorter than those of visible light and longer than x-rays, 10 cm to 100 cm wavelength.

UN number. A registry number assigned to dangerous commonly carried goods by the United Nations Committee of Experts on the Transport of Dangerous Goods. The UN number is required in shipping documentation and on packaging as part of the DOT regulations for shipping hazardous materials.

Unsafe condition. That part of the work environment that contributed or could have contributed to an accident.

Unstable. A chemical that, in the pure state, or as produced or transported, will vigorously polymerize, decompose, condense, or become self-reactive under conditions of shock, pressure, or temperature. Such chemicals are also referred to as reactive.

Upper explosive limit (UEL). The highest concentration (expressed in percent vapor or gas in the air by volume) of a substance that will burn or explode when an ignition source is present.

USC. *United States Code,* the official compilation of federal statutes.

USDA. U.S. Department of Agriculture.

Valve. A device that controls the direction of air or fluid flow or the rate and pressure at which air or fluid is delivered, or both.

Vapor. The gaseous form of substances that are normal by in the solid or liquid state (at room temperature and pressure).

Vat dyes. Water insoluble, complex coal tar dyes that can be chemically reduced in a heated solution to a soluble form that will impregnate fibers. Subsequent oxidation then produces insoluble color dyestuffs that are remarkably fast to washing, light, and chemicals.

Vector. (1) Term applied to an insect or any living carrier that transports a pathogenic microorganism from the sick to the well, inoculating the latter; the organism may or may not pass through any developmental cycle. (2) Anything (e.g., velocity, mechanical force, electromotive force) having magnitude, direction, and sense that can be represented by a straight line of appropriate length and direction.

Velocity. A vector that specifies the time rate of change of displacement with respect to a reference.

Velocity pressure. The kinetic pressure in the direction of flow necessary to cause a fluid at rest to flow at a given velocity. When added to static pressure, it gives total pressure.

Velometer. A device for measuring air velocity.

Ventilation. Circulating fresh air to replace contaminated air.

Ventilation, dilution. Airflow designed to dilute contaminants to acceptable levels.

Ventilation, mechanical. Air movement caused by a fan or other air-moving device.

Ventilation, natural. Air movement caused by wind, temperature difference, or other non-mechanical factors.

Vibration. An oscillation motion about an equilibrium position produced by a disturbing force.

Vinyl. A general term applied to a class of resins such as polyvinyl chloride, acetate, butyral, etc.

Viscose rayon. The type of rayon produced from the reaction of carbon disulfide with cellulose and the hardening of the resulting viscous fluid by passing it through dilute sulfuric acid, causing the evolution of hydrogen sulfide gas.

Viscosity. The tendency of a fluid to resist internal flow without regard to its density.

Visual acuity. Ability of the eye to sharply perceive the shape of objects in the direct line of vision.

Volatile. The percentage of a liquid or solid (by volume) that will evaporate at an ambient temperature of 70 F (unless some other temperature is stated). Examples: butane, gasoline, and paint thinner (mineral spirits) are 100% volatile; their individual evaporation rates vary, but over a period of time each will evaporate completely.

Volt. The practical unit of electromotive force or difference in potential between two points in an electrical field. Just as pressure in a water pipe causes water to flow, electrical pressure—or voltage—pushes the current of electrons through wires to receptacles, such as light fixtures and outlets.

Volume flow rate. The quantity (measured in units of volume) of a fluid flowing per unit of time, as cubic feet per minute, gallons per hour, or cubic meters per second.

Vulcanization. Process of combining rubber (natural, synthetic, or latex) with sulfur and accelerators in presence of zinc oxide under heat and usually pressure in order to change the material permanently from a thermoplastic to a thermosetting composition, or from a plastic to an elastic condition. Strength, elasticity, and abrasion resistance also are improved.

Water column. A unit used in measuring pressure. (See also Inches of water column.)

Water curtain or waterfall booth. In spray painting, the water running down a wall into which excess paint spray is drawn or blown by fans; the water carries the paint downward to a collecting point.

Waterproofing agents. These usually are formulations of three distinct materials: (1) a coating material, (2) a solvent, and (3) a plasticizer. Among the materials used in waterproofing are cellulose esters and ether, polyvinyl chloride resins or acetates, and variations of vinyl chloride-vinylidine chloride polymers.

Water-reactive. A chemical that reacts with water to release a gas that is either flammable or presents a health hazard.

Watt (w). A unit of electrical power, equal to one joule per second.

Weight. The force with which a body is attracted toward the earth.

Weld. A localized coalescense of metals or nonmetals produced either by heating the materials to suitable temperatures, with or without the application of pressure, or by the application of pressure alone, and with or without the use of filler material.

Welding. The several types of welding are electric arc welding, oxyacetylene welding, spot welding, and inert or shielded gas welding utilizing helium or argon. The hazards involved in welding stem from (1) the fumes from the weld metal such as lead or cadmium metal, or (2) the gases created by the process, or (3) the fumes or gases arising from the flux.

Welding rod. A rod or heavy wire that is melted and fused into metals in arc welding.

White damp. In mining, carbon monoxide.

White noise. A noise whose spectrum density (or spectrum level) is substantially independent of frequency over a specified range.

Work. When a force acts against a resistance to produce motion in a body, the force is said to do work. Work is measured by the product of the force acting and the distance moved through against a resistance. The units of measurement are the erg (the joule is 1×10^7 ergs) and the footpound.

Workers' Compensation. Insurance plan that compensates injured workers or their survivors.

Work-hardening. The property of a metal to become harder and more brittle on being "worked," that is, bent repeatedly or drawn.

Work hours. The total number of hours worked by all employees.

Work injuries (including occupational illnesses). Those that arise out of and in the course of gainful employment regardless of where the accident occurs. Excluded are work injuries to private household workers and injuries occurring in connection with farm chores, which are classified as home injuries.

Work strain. The natural physiological response reaction of the body to the application of work stress. The locus of the reaction may often be remote from the point of application of work stress. Work strain is not necessarily traumatic, but may appear as trauma when excessive, either directly or cumulatively, and must be considered by the industrial engineer in equipment and task design. Thus, a moderate increase of heart rate is nontraumatic work strain resulting from physical work strain if caused by undue work stress on the wrists.

Work stress. Biomechanically, any external force acting on the body during the performance of a task. It always produces work strain. Application of work stress to the human body is the inevitable consequence of performance of any task, and is, therefore, only synonymous with "stressful work conditions" when excessive. Work stress analysis is an integral part of task design.

Workers. All persons gainfully employed, including owners, managers, other paid employees, the self-employed, and unpaid family workers, but excluding private household workers.

Conversion of Units

ALL PHYSICAL UNITS OF MEASUREMENT can be reduced to three basic dimensions—mass, length, and time. Not only does reducing units to these basic dimensions simplify the solution of problems, but standardization of units makes comparison between operations (and between operations and standards) easier.

For example, air flows are usually measured in liters per minute, cubic meters per second, or cubic feet per minute. The total volume of air sampled can be easily converted to cubic meters or cubic feet. In another situation, the results of atmospheric pollution studies and stack sampling surveys are often reported as grains per cubic foot, grams per cubic foot, or pounds per cubic foot. The degree of contamination is usually reported as parts of contaminant per million parts of air.

If physical measurements are made or reported in different units, they must be converted to the standard units if any comparisons are to be meaningful.

To save time and space in reporting data, many units have standard abbreviations. Because the metric system (SI) is becoming more frequently used, conversion factors are given for the standard units of measurement.

FUNDAMENTAL UNITS

Conversion factors for various measurement units are listed in the tables in this section. To use a table to find the numerical value of the quantity desired, locate the unit to be converted in the first column. Then multiply this value by the number appearing at the intersection of the row and the column containing the desired unit. The answer will be the numerical value in the desired unit.

Various English systems and metric system units are given for the reader's convenience. The new system of measurement, however, is the International System of Units (SI). The official conversion factors and an explanation of the system are given to 6- or 7-place accuracy in ASTM Standard E 380-76 (ANSI Z210.1-1976).

Briefly, the SI System being adopted throughout the world is a modern version of the MKSA (meter, kilogram, second, ampere) system. Its details are published and controlled by an international treaty organization, the International Bureau of Weights and Measures (BIPM), set up by the Metre Convention signed in Paris, France, on May 20, 1875. The United States and Canada are member states of this Convention, as implemented by the Metric Conversion Act of 1975 (Public Law 94-168).

HELPFUL ORGANIZATIONS

The following four groups in the U.S. and Canada are deeply involved in planning and implementing metric conversion:

American National Metric Council
5410 Grosvenor Lane
Bethesday, MD. 20814

Metric Commission Canada
240 Sparks Street
Ottawa K1A 0H5, Canada

U.S. Metric Association, Inc.
Boulder, Colo. 80302

Office of Metric Programs
U.S. Dept. of Commerce
Washington, D.C. 20230

Conversion of Units

FAHRENHEIT-CELSIUS CONVERSION TABLE

Fahrenheit-Celsius Conversion.—A simple way to convert a Fahrenheit temperature reading into a Celsius temperature reading or vice versa is to enter the accompanying table in the center or boldface column of figures. These figures refer to the temperature in either Fahrenheit or Celsius degrees. If it is desired to convert from Fahrenheit to Celsius degrees, consider the center column as a table of Fahrenheit temperatures and read the corresponding Celsius temperature in the column at the left. If it is desired to convert from Celsius to Fahrenheit degrees, consider the center column as a table of Celsius values, and read the corresponding Fahrenheit temperature on the right.

To convert from "degrees Fahrenheit" to "degrees Celsius" (formerly called "degrees centigrade"), use the formula:

$$t_c = \frac{(t_f - 32)}{1.8} \text{ or } \frac{5}{9}(t_f - 32)$$

Conversely,
$$t_f = 1.8\, t_c + 32 \text{ or } \frac{9}{5}\, t_c + 32$$

Example, convert the boiling point of water in F to C:

$$212 \text{ F} - 32 = 180$$

$$\frac{5}{9}(180) = 100 \text{ C}$$

Fahrenheit — Celsius Conversion Table

Deg C		Deg F	Deg C		Deg F	Deg C		Deg F	Deg C		Deg F
−273	−459.4	...	−129	−200	−328	−13.9	7	44.6	1.1	34	93.2
−268	−450	...	−123	−190	−310	−13.3	8	46.4	1.7	35	95.0
−262	−440	...	−118	−180	−292	−12.8	9	48.2	2.2	36	96.8
−257	−430	...	−112	−170	−274	−12.2	10	50.0	2.7	37	98.6
−251	−420	...	−107	−160	−256	−11.7	11	51.8	3.3	38	100.4
−246	−410	...	−101	−150	−238	−11.1	12	53.6	3.9	39	102.2
−240	−400	...	− 96	−140	−220	−10.6	13	55.4	4.4	40	104.0
−234	−390	...	− 90	−130	−202	−10.0	14	57.2	5.0	41	105.8
−229	−380	...	− 84	−120	−184	− 9.4	15	59.0	5.6	42	107.6
−223	−370	...	− 79	−110	−166	− 8.9	16	60.8	6.1	43	109.4
−218	−360	...	− 73	−100	−148	− 8.3	17	62.6	6.7	44	111.2
−212	−350	...	− 68	− 90	−130	− 7.8	18	64.4	7.2	45	113.0
−207	−340	...	− 62	− 80	−112	− 7.2	19	66.2	7.8	46	114.8
−201	−330	...	− 57	− 70	− 94	− 6.7	20	68.0	8.3	47	116.6
−196	−320	...	− 51	− 60	− 76	− 6.1	21	69.8	8.9	48	118.4
−190	−310	...	− 46	− 50	− 58	− 5.6	22	71.6	9.4	49	120.2
−184	−300	...	− 40	− 40	− 40	− 5.0	23	73.4	10.0	50	122.0
−179	−290	...	− 34	− 30	− 22	− 4.4	24	75.2	10.6	51	123.8
−173	−280	...	− 29	− 20	− 4	− 3.9	25	77.0	11.1	52	125.6
−169	−273	−459.4	− 23	− 10	14	− 3.3	26	78.8	11.7	53	127.4
−168	−270	−454	−17.8	0	32−	− 2.8	27	80.6	12.2	54	129.2
−162	−260	−436	−17.2	1	33.8	− 2.2	28	82.4	12.8	55	131.0
−157	−250	−418	−16.7	2	35.6	− 1.7	29	84.2	13.3	56	132.8
−151	−240	−400	−16.1	3	37.4	− 1.1	30	86.0	13.9	57	134.6
−146	−230	−382	−15.6	4	39.2	− 0.6	31	87.8	14.4	58	136.4
−140	−220	−364	−15.0	5	41.0	0−	32	89.6	15.0	59	138.2
−134	−210	−346	−14.4	6	42.8	0.6	33	91.4	15.6	60	140.0

Deg C		Deg F	Deg C		Deg F	Deg C		Deg F	Deg C		Deg F
16.1	**61**	141.8	50.0	**122**	251.6	83.9	**183**	361.4	276.7	**530**	986
16.7	**62**	143.6	50.6	**123**	253.4	84.4	**184**	363.2	282.2	**540**	1004
17.2	**63**	145.4	51.1	**124**	255.2	85.0	**185**	365.0	287.8	**550**	1022
17.8	**64**	147.2	51.7	**125**	257.0	85.6	**186**	366.8	293.3	**560**	1040
18.3	**65**	149.0	52.2	**126**	258.8	86.1	**187**	368.6	298.9	**570**	1058
18.9	**66**	150.8	52.8	**127**	260.6	86.7	**188**	370.4	304.4	**580**	1076
19.4	**67**	152.6	53.3	**128**	262.4	87.2	**189**	372.2	310.0	**590**	1094
20.0	**68**	154.4	53.9	**129**	264.2	87.8	**190**	374.0	315.6	**600**	1112
20.6	**69**	156.2	54.4	**130**	266.0	88.3	**191**	375.8	321.1	**610**	1130
21.1	**70**	158.0	55.0	**131**	267.8	88.9	**192**	377.6	326.7	**620**	1148
21.7	**71**	159.8	55.6	**132**	269.6	89.4	**193**	379.4	332.2	**630**	1166
22.2	**72**	161.6	56.1	**133**	271.4	90.0	**194**	381.2	337.8	**640**	1184
22.8	**73**	163.4	56.7	**134**	273.2	90.6	**195**	383.0	343.3	**650**	1202
23.3	**74**	165.2	57.2	**135**	275.0	91.1	**196**	384.8	348.9	**660**	1220
23.9	**75**	167.0	57.8	**136**	276.8	91.7	**197**	386.6	354.4	**670**	1238
24.4	**76**	168.8	58.3	**137**	278.6	92.2	**198**	388.4	360.0	**680**	1256
25.0	**77**	170.6	58.9	**138**	280.4	92.8	**199**	390.2	365.6	**690**	1274
25.6	**78**	172.4	59.4	**139**	282.2	93.3	**200**	392.0	371.1	**700**	1292
26.1	**79**	174.2	60.0	**140**	284.0	93.9	**201**	393.8	376.7	**710**	1310
26.7	**80**	176.0	60.6	**141**	285.8	94.4	**202**	395.6	382.2	**720**	1328
27.2	**81**	177.8	61.1	**142**	287.6	95.0	**203**	397.4	387.8	**730**	1346
27.8	**82**	179.6	61.7	**143**	289.4	95.6	**204**	399.2	393.3	**740**	1364
28.3	**83**	181.4	62.2	**144**	291.2	96.1	**205**	401.0	398.9	**750**	1382
28.9	**84**	183.2	62.8	**145**	293.0	96.7	**206**	402.8	404.4	**760**	1400
29.4	**85**	185.0	63.3	**146**	294.8	97.2	**207**	404.6	410.0	**770**	1418
30.0	**86**	186.8	63.9	**147**	296.6	97.8	**208**	406.4	415.6	**780**	1436
30.6	**87**	188.6	64.4	**148**	298.4	98.3	**209**	408.2	421.1	**790**	1454
31.1	**88**	190.4	65.0	**149**	300.2	98.9	**210**	410.0	426.7	**800**	1472
31.7	**89**	192.2	65.6	**150**	302.0	99.4	**211**	411.8	432.2	**810**	1490
32.2	**90**	194.0	66.1	**151**	303.8	100.0	**212**	413.6	437.8	**820**	1508
32.8	**91**	195.8	66.7	**152**	305.6	104.4	**220**	428.0	443.3	**830**	1526
33.3	**92**	197.6	67.2	**153**	307.4	110.0	**230**	446.0	448.9	**840**	1544
33.9	**93**	199.4	67.8	**154**	309.2	115.6	**240**	464.0	454.4	**850**	1562
34.4	**94**	201.2	68.3	**155**	311.0	121.1	**250**	482.0	460.0	**860**	1580
35.0	**95**	203.0	68.9	**156**	312.8	126.7	**260**	500.0	465.6	**870**	1598
35.6	**96**	204.8	69.4	**157**	314.6	132.2	**270**	518.0	471.1	**880**	1616
36.1	**97**	206.6	70.0	**158**	316.4	137.8	**280**	536.0	476.7	**890**	1634
36.7	**98**	208.4	70.6	**159**	318.2	143.3	**290**	554.0	482.2	**900**	1652
37.2	**99**	210.2	71.1	**160**	320.0	148.9	**300**	572.0	487.8	**910**	1670
37.8	**100**	212.0	71.7	**161**	321.8	154.4	**310**	590.0	493.3	**920**	1688
38.3	**101**	213.8	72.2	**162**	323.6	160.0	**320**	608.0	498.9	**930**	1706
38.9	**102**	215.6	72.8	**163**	325.4	165.6	**330**	626.0	504.4	**940**	1724
39.4	**103**	217.4	73.3	**164**	327.2	171.1	**340**	644.0	510.0	**950**	1742
40.0	**104**	219.2	73.9	**165**	329.0	176.7	**350**	662.0	515.6	**960**	1760
40.6	**105**	221.0	74.4	**166**	330.8	182.2	**360**	680.0	521.1	**970**	1778
41.1	**106**	222.8	75.0	**167**	332.6	187.8	**370**	698.0	526.7	**980**	1796
41.7	**107**	224.6	75.6	**168**	334.4	193.3	**380**	716.0	532.2	**990**	1814
42.2	**108**	226.4	76.1	**169**	336.2	198.9	**390**	734.0	537.8	**1000**	1832
42.8	**109**	228.2	76.7	**170**	338.0	204.4	**400**	752.0	565.6	**1050**	1922
43.3	**110**	230.0	77.2	**171**	339.8	210	**410**	770.0	593.3	**1100**	2012
43.9	**111**	231.8	71.8	**172**	341.6	215.6	**420**	788	621.1	**1150**	2102
44.4	**112**	233.6	78.3	**173**	343.4	221.1	**430**	806	648.9	**1200**	2192
45.0	**113**	235.4	78.9	**174**	345.2	226.7	**440**	824	676.7	**1250**	2282
45.6	**114**	237.2	79.4	**175**	347.0	232.2	**450**	842	704.4	**1300**	2372
46.1	**115**	239.0	80.0	**176**	348.8	237.8	**460**	860	732.2	**1350**	2462
46.7	**116**	240.8	80.6	**177**	350.6	243.3	**470**	878	760.0	**1400**	2552
47.2	**117**	242.6	81.1	**178**	352.4	248.9	**480**	896	787.8	**1450**	2642
47.8	**118**	244.4	81.7	**179**	354.2	254.4	**490**	914	815.6	**1500**	2732
48.3	**119**	246.2	82.2	**180**	356.0	260.0	**500**	932	1093.9	**2000**	3632
48.9	**120**	248.0	82.8	**181**	357.8	265.6	**510**	950	1648.9	**3000**	5432
49.4	**121**	249.8	83.3	**182**	359.6	271.1	**520**	968	2760.0	**5000**	9032

From Machinery's Handbook, *18th ed.* (*The Industrial Press*)

Above 1000 in the center column, the table increases in increments of 50. To convert 1462 degrees F to Celsius, for instance, add to the Celsius equivalent of 1400 degrees F ⅝ths of 62 or 34 degrees, which equals 794 C.

LENGTH

To Obtain → Multiply Number of by ↘	meter (m)	centimeter (cm)	millimeter (mm)	micron (μ) or micrometer (μm)	angstrom unit, (A)	inch (in.)	foot (ft)
meter	1	100	1000	10^6	10^{10}	39.37	3.28
centimeter	0.01	1	10	10^4	10^8	0.394	0.0328
millimeter	0.001	0.1	1	10^3	10^7	0.0394	0.00328
micron	10^{-6}	10^{-4}	10^{-3}	1	10^4	3.94×10^{-5}	3.28×10^{-6}
angstrom	10^{-10}	10^{-8}	10^{-7}	10^{-4}	1	3.94×10^{-9}	3.28×10^{-10}
inch	0.0254	2.540	25.40	2.54×10^4	2.54×10^8	1	0.0833
foot	0.305	30.48	304.8	304,800	3.048×10^9	12	1

AREA

To Obtain → Multiply Number of By ↘	square meter (m²)	square inch (sq in.)	square foot (sq ft)	square centimeter (cm²)	square millimeter (mm²)
square meter	1	1,550	10.76	10,000	10^6
square inch	6.452×10^{-3}	1	6.94×10^{-3}	6.452	645.2
square foot	0.0929	144	1	929.0	92,903
square centimeter	0.0001	0.155	0.001	1	100
square millimeter	10^{-6}	0.00155	0.00001	0.01	1

DENSITY

To Obtain → Multiply Number of By ↘	gm/cm³	lb/cu ft	lb/gal
gram/cubic centimeter	1	62.43	8.345
pound/cubic foot	0.01602	1	0.1337
pound/gallon (U.S.)	0.1198	7.481	1

1 grain/cu ft = 2.28 mg/m³

FORCE

To Obtain → Multiply Number of By →	dyne	newton (N)	kilogram-force	pound-force (lbf)
dyne	1	1.0×10^{-5}	1.02×10^{4}	2.248×10^{4}
newton	1.0×10^{5}	1	0.1020	0.2248
kilogram-force	9.807×10^{-5}	9.807	1	2.205
pound-force	4.448×10^{-5}	4.448	0.4536	1

MASS

To Obtain → Multiply Number of By →	gram (gm)	kilogram (kg)	grains (gr)	ounce (avoir) (oz)	pound (avoir) (lb)
gram	1	0.001	15.432	0.03527	0.00220
kilogram	1,000	1	15,432	35.27	2.205
grain	0.0648	6.480×10^{-5}	1	2.286×10^{-3}	1.429×10^{-4}
ounce	28.35	0.02835	437.5	1	0.0625
pound	453.59	0.4536	7,000	16	1

VOLUME

To Obtain → Multiply Number of By →	cu ft	gallon (U.S. liquid)	liters	cm³	m³
cubic foot	1	7.481	28.32	28,320	0.0283
gallon (U.S. liquid)	0.1337	1	3.785	3,785	3.79×10^{-3}
liter	0.03531	0.2642	1	1,000	1×10^{-3}
cubic centimeters	3.531×10^{-5}	2.64×10^{-4}	0.001	1	10^{-6}
cubic meters	35.31	264.2	1,000	10^{6}	1

VELOCITY

To Obtain → Multiply Number of By	cm/s	m/s	km/hr	ft/s	ft/min	mph
centimeter/second	1	0.01	0.036	0.0328	1.968	0.02237
meter/second	100	1	3.6	3.281	196.85	2.237
kilometer/hour	27.78	0.2778	1	0.9113	54.68	0.6214
foot/second	30.48	0.3048	18.29	1	60	0.6818
foot/minute	0.5080	0.00508	0.0183	0.0166	1	0.01136
mile per hour	44.70	0.4470	1.609	1.467	88	1

PRESSURE

To Obtain → Multiply Number of By	lb/sq in. (psi)	atm	in. (Hg) 32 F 0 C	mm (Hg) 32 F 0 C	kPa (kN/m²)	ft (H₂O) 60 F 15 C	in. (H₂O)	lb/sq ft
pound/square inch	1	0.068	2.036	51.71	6.895	2.309	27.71	144
atmospheres	14.696	1	29.92	760.0	101.32	33.93	407.2	2,116
inch (Hg)	0.4912	0.033	1	25.40	3.386	1.134	13.61	70.73
millimeter (Hg)	0.01934	0.0013	0.039	1	0.1333	0.04464	0.5357	2.785
kilopascals	0.1450	9.87×10^{-3}	0.2953	7.502	1	0.3460*	4.019	20.89
foot (H₂0)(15 C)	0.4332	0.0294	0.8819	22.40	2.989*	1	12.00	62.37
inch (H₂0)	0.03609	0.0024	0.073	1.867	0.2488	0.0833	1	5.197
pound/square foot	0.0069	4.72×10^{-4}	0.014	0.359	0.04788	0.016	0.193	1

*at 4C

HEAT, ENERGY, OR WORK

To Obtain → Multiply Number of By	joule	ft-lb	kwh	hp-hour	kcal	cal	Btu
joules	1	0.737	2.773×10^{-7}	3.725×10^{-7}	2.39×10^{-4}	0.2390	9.478×10^{-4}
foot-pound	1.356	1	3.766×10^{-7}	5.05×10^{-7}	3.24×10^{-4}	0.3241	1.285×10^{-3}
kilowatt-hour	3.6×10^6	2.66×10^6	1	1.341	860.57	860,565	3,412
hp-hour	2.68×10^6	1.98×10^6	0.7455	1	641.62	641,615	2,545
kilocalorie	4,184	3,086	1.162×10^{-3}	1.558×10^{-3}	1	1,000	3.9657
calorie	4.184	3.086	1.162×10^{-6}	1.558×10^{-6}	0.001	1	.00397
British thermal unit	1,055	778.16	2.930×10^{-4}	3.93×10^{-4}	0.252	252	1

Bibliography

American Conference of Governmental Industrial Hygienists, 6500 Glenway Avenue, Building D-5, Cincinnati, OH 45211. "Threshold Limit Values for Chemical Substances and Physical Agents in the Workroom Environment." (Issued annually.)

———. Committee on Industrial Ventilation, P.O. Box 16153, Lansing, MI 48901. *Industrial Ventilation—A Manual of Recommended Practice.* (Latest edition.)

American Industrial Hygiene Association, 345 White Pond Drive, Akron, OH 44320.

American National Red Cross, 17th and D Streets N.W., Washington, DC, 20006. *First Aid Textbook.* (Latest edition.)

American National Standards Institute, 1430 Broadway, New York, NY, 10018. "Catalog of American National Standards." (Issued annually.)

American Society for Training and Development (ASTD), 1630 Duke Street, P.O. Box 1443, Alexandria, VA 22313. *Literature Index.* (Quarterly.)

American Welding Society (AWS), 550 LeJeune Road N.W., P.O. Box 351040, Miami, FL 33135. *Welding Handbook.*

Bird, Frank E., Jr. *Management Guide to Loss Control.* Loganville, GA: Institute Publications ILCI, 1974.

Bird, Frank E., Jr., and Robert G. Loftus. *Loss Control Management.* Loganville, GA: Institute Publications ILCI, 1976.

Blanchard, Kenneth, et al. *The One Minute Manager Gets Fit.* New York: Nightingale-Conant, 1989.

"Care and Operating Instructions for Various Electric Tools." Milwaukee Electric Tool Corp., 13171 West Lisbon Road, Brookfield, WI 53005.

DeCristofores, RJ. *Complete Book of Stationary Power Tool Techniques.* New York: Sterling, 1988.

Factory Mutual Engineering Corporation, 1151 Boston-Providence Turnpike, Norwood, MA 02062.
Factory Mutual System Approval Guide. (Annual.)
Handbook of Property Conservation.
Loss Prevention Data Books. (Periodic.)

Fallon, William K., ed. *AMA Management Handbook.* 2nd ed. New York: AMACOM, 1983.

Ferry, Ted S. *Modern Accident Investigation and Analysis.* New York: John Wiley & Sons, 1988.

Firenze, Robert J. *The Process of Hazard Control.* Dubuque, IA: Kendall/Hunt Publishing Company, 1978.

Grandjean, Etienne. *Fitting the Task to the Man: A Textbook of Occupational Ergonomics*, 4th ed. London: Taylor and Francis, 1988.

Hammer, Willie. *Occupational Safety Management and Engineering*, 4th ed. Englewood Cliffs, NJ: Prentice-Hall, Inc., 1988.

Heinrich, Herbert W., Dan Petersen, and Nestor Roos. *Industrial Accident Prevention: A Safety Management Approach.* 5th rev. ed. New York: McGraw-Hill Book Company, 1980.

Human Factors Section, Eastman Kodak Laboratory. *Ergonomics Design for People at Work*, vol. 1. New York: VanNostrand Reinhold, 1983.

Illuminating Engineering Society, 345 East 47th Street, New York: 10017.
IES Lighting Handbook (The Standard Lighting Guide). Practice for Industrial Lighting (ANSI/IES RP7).

International Labor Office, 1750 New York Avenue, Suite 330, Washington, DC 20006. *Encyclopedia of Occupational Health and Safety.*

Johnson, William G. *MORT Safety Assurance Systems.* New York: Marcel Dekker, 1980.

Kroemer, KHE. Coupling the hand with the handle. *Human Factors* 28 (1986): 337-39.

———. Engineering Anthropometry, in Salvendy, G., ed. *Handbook of Human Factors/Ergonomics.* New York: John Wiley & Sons, 1987, pp. 154-68.

———. Office Ergonomics: Work Station Dimensions. In Alexander, DC, and Pulat, BM, eds. *Industrial Ergonomics.* Norcross, GA: Institute of Industrial Engineers, 1985, pp. 187-201.

———. Testing individual capability to lift material: Repeatability of a dynamic test compared with static testing. *Journal of Safety Research* 16(1985):1-7.

———. VDT Workstation Design. In Helander, MG, ed. *Handbook of Human Computer Interaction.* New York: Elsevier Publications, 1987.

National Association of Suggestion Systems (NASS), 230 North Michigan Avenue, Suite 1200, Chicago, IL 60601. *NASS Horizons.* (Quarterly.)

National Fire Protection Association. Batterymarch

Park, Quincy, MA 02269.
Fire Protection Handbook.
Fire Protection Guide on Hazardous Materials.
Flammable and Combustible Liquids Code Handbook.
Industrial Fire Hazards Handbook.
Inspection Manual.
Life Safety Code Handbook.
National Electrical Code Handbook.
"Standards and Recommended Practices." (Catalog available.)

National Instutite of Occupational Safety and Health (NIOSH), Public Health Service, Centers for Disease Control, Department of Health and Human Service, 1600 Clifton Road, Atlanta, GA 30333.
Certified Personal Protective Equipment.
"Criteria Documents."
The Industrial Environment: Its Evaluation and Control.
Machine Guarding—Assessment of Need.
"Publications Catalog"
Safety Program Practices in High Versus Low Accident Rate Companies.
Toxic Substances List.
"Welding Safely."

National Safety Council, 444 North Michigan Avenue, Chicago, IL 60611.
Accident Facts (Annually.)
Accident Investigation: A New Approach.
Accident Prevention Manual for Industrial Operations (2 volumes).
Electrical Inspection Illustrated.
Family Safety and Health. (Quarterly.)
"5 Minute Safety Talks." (Series.)
Forging Safety Manual.
Fundamentals of Industrial Hygiene.
Guards Illustrated.
Industrial Data Sheets. (Index available.)
Out in Front: Effective Supervision in the Workplace.
Power Press Safety Manual.
Safety and Health. (Monthly.)
You Are the Safety and Health Committee.

National Society to Prevent Blindness, 500 East Remington Road, Schaumburg, IL 60173. Eyesight in Industry.

Ottoboni, M. Alice. *The Dose Makes the Poison: A Plain Language Guide to Toxicology.* Berkeley, CA: Vincente Books, 1984.

Petersen, Dan C. *Techniques of Safety Management*, 3rd ed. Goshen, NY: Aloray, Inc., 1989.

Sax, NI. *Dangerous Properties of Industrial Materials*, 7th ed., 3 vols. New York: VanNostrand Reinhold, 1989.

Simonds, Rollin H., and John V. Grimaldi. *Safety Management*, 5th ed. Homewood, IL: Richard D. Irwin, Inc., 1989.

Steil, Lyman, et al. *Listening: It Can Change Your Life.* New York: John Wiley & Sons, 1984.

Underwriters Laboratories Inc., 333 Pfingsten Road, Northbrook, IL 60062. "Product Directories."

U.S. Department of Labor, Occupational Safety and Health Administration, 200 Constitution Avenue, Washington, DC 20210.
An Illustrated Guide to Electrical Safety, Pub. 3073.
A Brief Guide to Recordkeeping Requirements for Occupational Injuries and Illnesses. O.M.B. No. 1220-0029. June 1986.

U.S. General Services Administration, National Archives and Records Administration, Office of the Federal Register, Washington, DC 20408.
Code of Federal Regulations:
Title 10—"Energy."
Title 29—"Labor."
Title 40—"Protection of the Environment."
Title 49—"Transportation."
Note: Government publications are available from the Superintendent of Documents, U.S. Government Printing Office, Washington, DC 20402.

Wolvin, Andrew D., and Carolyn Coakley. *Listening*, 3rd ed. Dubuque, IA: William C. Brown Company, Publishers, 1988.

Index